Time-Lapse Microscopy in In-Vitro Fertilization

Time-Lapse Microscopy in In-Vitro Fertilization

Edited by

Marcos Meseguer

*Scientific Supervisor and Senior Embryologist, IVI Clinic Valencia, and Assistant Professor in
Biotechnology, Valencia University, Valencia, Spain*

CAMBRIDGE
UNIVERSITY PRESS

CAMBRIDGE
UNIVERSITY PRESS

University Printing House, Cambridge CB2 8BS, United Kingdom

Cambridge University Press is part of the University of Cambridge.

It furthers the University's mission by disseminating knowledge in the pursuit of education, learning and research at the highest international levels of excellence.

www.cambridge.org
Information on this title: www.cambridge.org/9781107593268

© Cambridge University Press 2016

First published 2016

Printed in the United Kingdom by Clays, St Ives plc

A catalog record for this publication is available from the British Library

ISBN 978-1-107-59326-8 Mixed Media

ISBN 978-1-107-13206-1 Hardback

ISBN 978-1-316-41549-8 Cambridge Books Online

This book is dedicated to those professional pioneers as mechanical engineers and biologists who made it possible to bring this technology into the clinical field, especially to Dr. Niels Ramsing and Dr. Jens Gundersen for their vast contribution in this task.

Contents

Contributors

Inge E. Agerholm
The Fertility Clinic, Hospital Horsens, Horsens, Denmark

Belén Aparicio-Ruiz
Embryologist, IVI Clinic Valencia, Valencia, Spain

Natalia Basile
Embryologist, IVI Clinic Madrid, Madrid, Spain

Maura Caiazzo
Embryologist, IVI Clinic Madrid, Madrid, Spain

Mariabeatrice Dal Canto
Biogenesi Reproductive Medicine Centre, Istituti Clinici Zucchi, Monza, Italy

Alice A. Chen
Auxogyn, Inc., Menlo Park, CA, USA

Giovanni Coticchio
Biogenesi Reproductive Medicine Centre, Istituti Clinici Zucchi, Monza, Italy

Thomas Ebner
Landes-Frauen- und Kinderklinik, Kinderwunsch Zentrum Linz, and Faculty of Medicine, Johannes Kepler University, Linz, Austria

Rubens Fadini
Biogenesi Reproductive Medicine Centre, Istituti Clinici Zucchi, Monza, Italy

Thomas Freour
Reproductive Medicine Unit, Faculty of Medicine, University Hospital of Nantes, and INSERM UMR1064, Nantes, France

Tine Q. Kajhøj
Vitrolife A/S Denmark, Aarhus, Denmark

Marcos Meseguer
Scientific Supervisor and Senior Embryologist, IVI Clinic Valencia, and Assistant Professor in Biotechnology, Valencia University, Valencia, Spain

Markus Montag
ilabcomm GmbH, St. Augustin, Germany

Dean E. Morbeck
Division of Reproductive Endocrinology and Infertility, Division of Laboratory Genetics, Mayo Clinic, Rochester, MN, USA

Csaba Pribenszky
Senior Researcher, Vitrolife Kft., Budapest, and Assistant Professor, University of Veterinary Science, Budapest, Hungary

Mario Mignini Renzini
Biogenesi Reproductive Medicine Centre, Istituti Clinici Zucchi, Monza, Italy

Zev Rosenwaks
Director, Claudia Cohen Center for Reproductive Medicine, and Professor of Reproductive Medicine and Obstetrics and Gynecology, Weill Cornell Medical College, New York, NY, USA

Maria Jose de los Santos
IVF Laboratory Director, IVI Clinic Valencia, Valencia, Spain

Shehua Shen
Auxogyn, Inc., Menlo Park, CA, USA

Lei Tan
Auxogyn, Inc., Menlo Park, CA, USA

Alberto Tejera
Embryologist, IVI Clinic Valencia, Valencia, Spain

Nikica Zaninovic
Director, Embryology Laboratory, and Associate
Professor of Embryology in Reproductive Medicine,
Claudia Cohen Center for Reproductive Medicine,
Weill Cornell Medical College, New York, NY, USA

Qiansheng Zhan
Assistant Professor of Embryology in Reproductive
Medicine, Claudia Cohen Center for Reproductive
Medicine, Weill Cornell Medical College, New York,
NY, USA

Preface

Assisted reproduction is a field constantly seeking for improvement. For those who work in it, fascination is inevitable during the first microscopic observations of the cellular material. Male and female gametes can enthrall the viewer, giving rise to a mysterious structure that develops according to a pre-defined plan: the human embryo. The understanding of this complex process definitely exceeds its morphological evaluation and it welcomes new technologies on a cellular and molecular level.

The study of the embryo's metabolism was crucial for the formulation of suitable culture media. In the early 1990s a groundbreaking technique, ICSI, revolutionized the treatment of male infertility. Later on, and as we entered the world of genomics, we began to understand the extraordinarily detailed genetic assemblage that determines the fate of an embryo, a process governed by molecular mechanisms and signaling pathways. Reconciliation of the social evolution of modern women and their maternity came along with vitrification, and the wave of the "omics" initiated an era of non-invasiveness to study embryo development in the IVF laboratory. Assisted reproduction has evolved indeed; however, there will always be room for improvement.

In the past 30 years, cell biology has benefited from the achievements in the image analysis technology. During the 1980s, the use of analog videos greatly expanded the use of the microscope as an analytical tool and, most recently, analog systems have been replaced by digital ones. The culmination of this initial approach is presented to us in the form of time-lapse systems, a technology leading us to evolve from single static observations to the continuous surveillance of human embryos in the IVF laboratory.

The idea of this time-lapse atlas was forged by Marcos Meseguer, undoubtedly a pioneer in the field, in combination with the University of Cambridge. The objective was to create a dynamic and highly visual atlas of human embryo development. For this purpose a group of different leaders in the clinical and scientific fields reunite to share their experience on this new technology. The atlas gives the reader the opportunity to review known aspects of human embryo development from a different approach, as well as to learn and visualize new and useful concepts related to human embryo development. It is an absolute "must-see-read" for all the clinicians and scientists involved in the field of assisted reproduction.

Natalia Basile
Embryologist
IVI Clinic Madrid
Spain

Introduction: IVF concepts, embryo development, and embryo selection

Thomas Ebner

1.1. Introduction

The birth of Louise Brown [1] represents a milestone in the history of assisted reproductive technologies (ART). While at the beginning of the ART era, treatment was only possible for patients showing normozoospermia or mild forms of male subfertility, severe male factor infertility could not be treated at all. In other words, conventional in vitro fertilization (IVF) worked well for female factor infertility, e.g., bilateral tubal blockage, but those patients with bad-quality semen samples or failed fertilization after IVF still had to face childlessness. This unwanted scenario led to intensive research in order to solve the problem of suboptimal sperm count and quality. What all the upcoming techniques had in common is the effort to bring the male gamete closer to the oocyte as routinely done in conventional IVF.

The first step was to breach the zona pellucida (ZP) and inseminate the collected cumulus-oocyte-complexes (COC). This approach, known as partial zona dissection or PZD [2], worked to a certain degree; however, the rate of polyspermic fertilization was rather high since the artificial opening in the ZP allowed bypass of the cortical reaction usually preventing the egg from polyspermy. The consequent next step was to place less sperm in the perivitelline space by subzonal injection (SUZI) [3]. Although this technique led to a reduction of polyspermy associated with a slight increase of regular fertilization, characterized by the presence of two pronuclei (2Pn), it was immediately replaced by intracellular sperm injection (ICSI) once Palermo and co-workers [4] placed a single spermatozoon into the ooplasm (Video 1.1), realizing that by doing so fertilization rates up to 70–80% could be observed while all drawbacks were excluded.

All of a sudden, all those male factor patients starting from oligo-asthenoteratozoospermia through

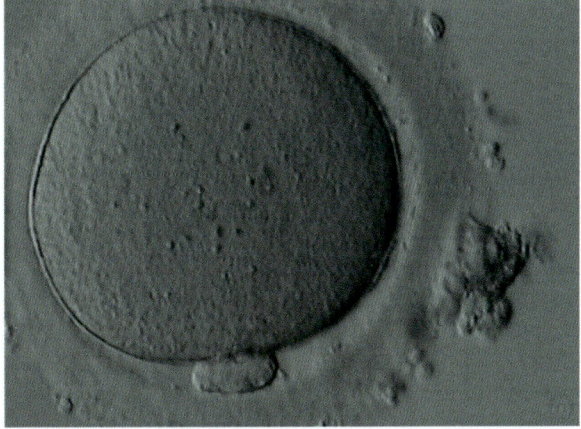

Video 1.1 Intracytoplasmic sperm injection (ICSI). Inner diameter of ICSI pipette (MicroTech, Gynemed, Lensahn, Germany) is 5 μm.

to azoospermia were treatable. In the latter scenario several sites in the genitourinary tract could serve as a sperm source. Either spermatozoa could be aspirated from the epididymis [5], a technique called microsurgical epididymal sperm aspiration (MESA), or via TESE, which refers to a testicular biopsy [6].

1.2. Controlled ovarian hyperstimulation and ovarian puncture

Of course spermatozoa are only one side of the coin. The female counterpart can also be a limiting factor. In the early years of IVF exclusively natural cycles were used. Oocyte collection was routinely performed via laparoscopy. The combination of both approaches increased the risk of ending up without oocytes for further fertilization.

The development of transvaginal ovarian puncture [7, 8] and, particularly, of controlled ovarian hyperstimulation (COH) steadily increased the number of mature oocytes available for IVF or ICSI [9, 10].

In principle, to increase the number of follicles (and consequently of eggs) per ovary two main stimulation protocols are used. The first regimen is called "long protocol" since down-regulation of the pituitary gland using GnRH-agonists is started in the preceding cycle whereas ovarian stimulation applying human menopausal or recombinant gonadotrophins (LH and/or FSH) is done in the treatment cycle. The shorter antagonist protocol, on the other hand, represents a combination of gonadotrophin stimulation and immediate suppression of a potential LH surge with GnRH-antagonists, the latter being done if at least one follicle reaches the diameter of 12 mm and serum estradiol level appears adequate. Irrespective of the stimulation regimen, final maturation of the oocyte has to be performed by utilizing either human choriogonadotrophin (hCG) or a GnRH-agonist [11].

Based on hormonal parameters and follicular size oocyte retrieval is routinely carried out 36 hours after hCG application under ultrasound control. Where there is an indication of ovarian hyperstimulation syndrome ovulation induction can alternatively be done with a gonadotrophin agonist. However, the negative pressure exerted by the aspiration pump guarantees that the COCs located in the cumulus oophorus gather in the follicular fluid which is collected in special tubes. These vials then are handed over to the IVF laboratory where the embryologists separate the COCs from the sometimes hemorrhagic follicular fluid in order to transfer them in an optimized culture medium.

1.3. IVF

After a resting period of 2 to 3 hours, harvested COCs are evaluated by their appearance, particularly by the expansion of the corona radiata and the outer cumulus cells. Based on such criteria eggs within the cumulus matrix are roughly classified as either mature (metaphase II) or immature (pro- and metaphase I). In more detail, an expanded and/or luteinized complex and a radiant corona radiata indicate completion of nuclear maturation, while the absence of an expanded cumulus is associated with immaturity [12]. Recent publications, however, showed that nuclear maturation

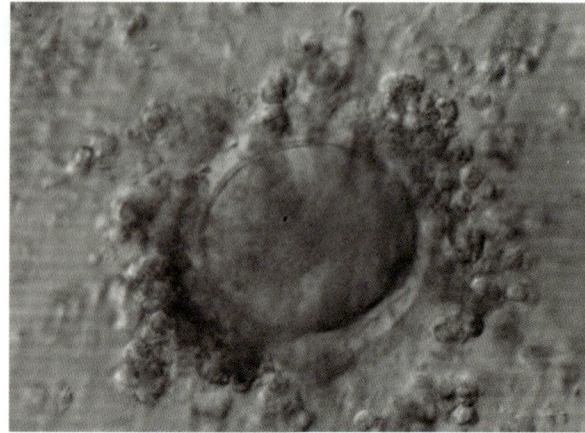

Video 1.2 In vitro fertilization (IVF).

and oocyte quality cannot be predicted adequately by scoring the COC appearance [13, 14].

Frequently, blood clots or other amorphous clumps [14, 15] are present in the COC. One tends to cut off these contaminated areas using needles but apart from the mechanical stress to the oocyte, this procedure is of no need since there is evidence that COCs showing blood clots have already been harmed during folliculogenesis and, thus, their developmental capacity may not be retained [14]. Thus, in the presence of a (borderline) normozoospermic sperm sample in conventional IVF (Video 1.2), COCs in their original state are inseminated with a sufficient number of spermatozoa, which turned out to be 25,000 to 50,000 per complex depending on the quality of the original ejaculate.

It is important to note that insemination is not performed with raw semen (Video 1.3), rather the sample is processed with different preparation techniques such as swim-up, mini swim-up or density gradient centrifugation, to name but a few. Since these processing methods include one or two centrifugation steps which could potentially harm the spermatozoa sperm selection chambers have been developed which avoid the above mentioned stressor. In such chambers spermatozoa accumulate by motility (Video 1.4) and as a collateral benefit contact between the gametes and seminal plasma (e.g., containing erythrocytes or inflammatory cells etc.) is kept at a minimum.

Video 1.3 Native ejaculate.

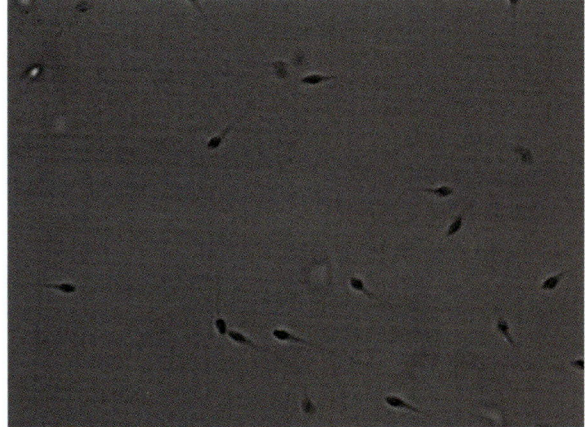

Video 1.4 Same semen as in Video 1.3 processed with sperm selection chamber (swim-over time 30 minutes).

1.4. ICSI

For successful ICSI it is critical that the majority of cumulus cells are removed from the female gamete. Apart from technical problems, such as incorporation of foreign somatic DNA [16], control of oocyte maturity and quality can only be checked after proper denudation [17]. Stripping of the oocytes is traditionally achieved using enzymatic digestion of the outer matrix followed by mechanical denudation through pipetting. Since it has been shown that a dislocation between the first polar body and the meiotic spindle can occur if the mechanical part is performed inadequately, e.g., using pipettes of an inappropriate inner diameter, in routine work, hyaluronidase, which is the enzyme degrading hyaluronic acid (the major component of the extracellular matrix of oocytes), is used at the beginning of the denudation process. Most of the commercially available products have a concentration of 80 IU/l, which is only a tenth of the critical threshold above which parthenogenetic activation does occur [18]. Alternatively, plant or recombinant human products can be utilized [19, 20].

Once the ejaculate is processed in order to separate motile from non-viable sperms, an a priori motile spermatozoon is selected, which has to be immobilized prior to injection [21]. Immobilization of the sperm in ICSI has two beneficial effects: on the one hand, any theoretical damage to the cytoskeleton caused by motile sperms is negligible, and on the

Video 1.5 Mechanical immobilization as in a single sperm.

other, permeabilization of the sperm membrane will ensure that phospholipase C zeta, a sperm-derived oocyte-activating factor, immediately enters the ooplasm, thus initiating oocyte activation [22]. Sperm immobilization is usually performed towards the end of the tail (Video 1.5); however, permeabilizing other sites is an alternative [23]. It can be performed using either mechanical [21, 24] or piezoelectrical manipulation [25, 26]. Recently, a laser-assisted permeabilization technique (Video 1.6) was introduced into the field of ICSI [27–29]. In detail, male gametes were immobilized with a non-contact diode laser applying two successive laser irradiations per spermatozoon, the first aimed near the middle of the tail and the

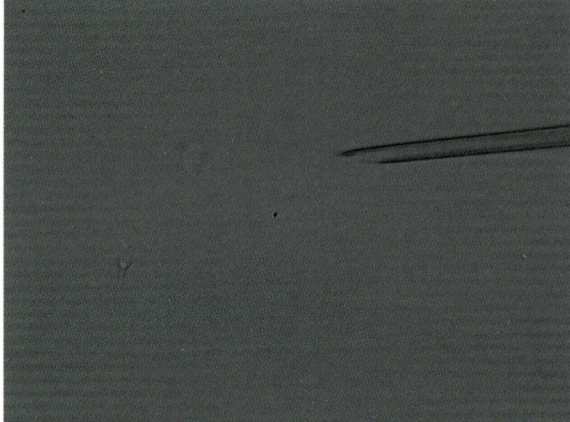

Video 1.6 Laser-assisted immobilization of a single sperm. Laser shot can be divined in the middle of the image section.

Video 1.7 Fresh testicular biopsy showing one non-progressive motile sperm with residual body at the end of the video.

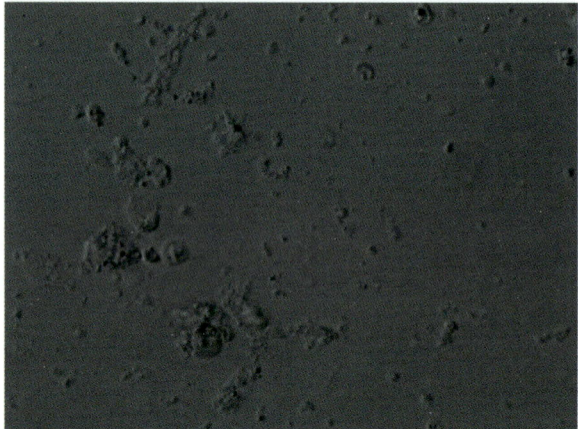

Video 1.8 Same biopsy after cryopreservation. Restoration of progressive motility has been done with a ready-to-use theophylline (GM501 SpermMobil, Gynemed, Lensahn, Germany).

second directly at the end of the tail. This strategy minimized the total energy dose male gametes were exposed to.

Regardless of the source of the sperm, ICSI is done according to a standardized procedure. The oocyte is held in place with a holding pipette at the 9 o'clock position (provided that the first polar body is located at the 6 or 12 o'clock position). Once the equatorial plane of the oocyte is in focus, the ICSI-pipette has to be pressed against the zona pellucida (3 o'clock) creating a characteristic funnel. After penetrating both the zona and the oolemma, a small volume of

cytoplasm should be aspirated into the glass tool to activate the egg and to ensure rupture [30]. A single permeabilized spermatozoa is then gently placed near the horizontal axis. Withdrawal has to be done carefully to prevent leakage from the oocyte.

Sometimes, e.g., in case of cryopreserved or testicular sperm, embryologists have to deal with exclusive presence of immotile sperm (Video 1.7). This scenario bears the risk of selecting non-viable spermatozoa if viability is not estimated with either an elasticity test [31], a hypo-osmotic swelling test [32], or by the usage of laser pulses [33]. Most elegantly this viability check is done by the help of dimethylxanthines, which allow for partial restoration of original motility [34] (Video 1.8).

The most important prerequisite for a best prognosis ICSI performance is a well-calculated time schedule. With respect to this, the period between oocyte collection and subsequent ICSI should never exceed 6 hours [35, 36] in order to avoid in vitro aging of the oocytes. However, it is rather irrelevant whether the COCs are denuded immediately after collection (e.g., leaving denuded eggs in culture for later ICSI) or if the manipulation is performed directly prior to ICSI after a resting period of several hours [35].

1.5. In vitro culture

Immediately after IVF or ICSI, the inseminated COCs or injected eggs, respectively, are transferred to special culture dishes for prolonged in vitro culture (up to blastocyst stage if possible). Primarily, two

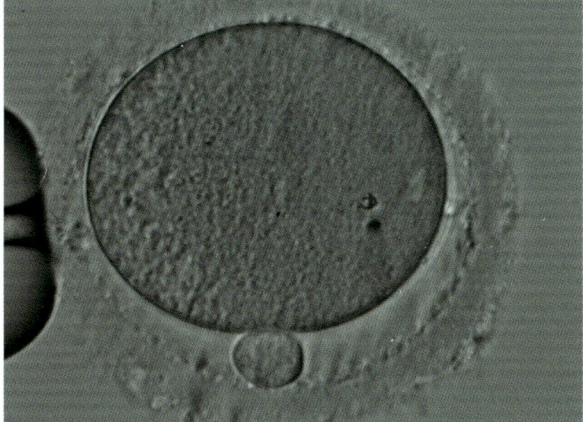

Video 1.9 Biopsy of first polar body. Zona opening is done with a diode laser at the 5 o'clock position. Inner diameter of biopsy pipette (MicroTech, Gynemed, Lensahn, Germany) is 15 µm.

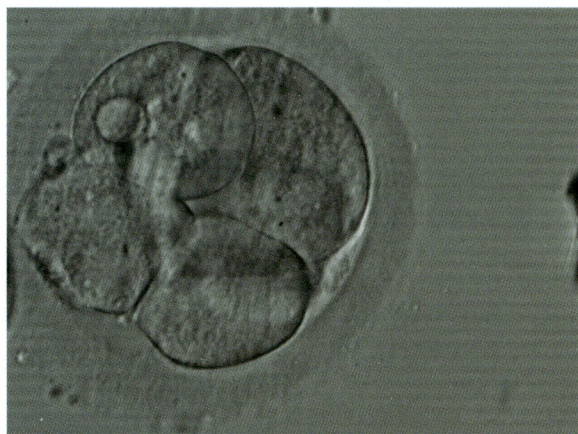

Video 1.10 Blastomere biopsy in a donated embryo showing a large blastomere with binucleation (out of focus). It should be noted that the embryo has already started to compact. Inner diameter of biopsy pipette (MicroTech, Gynemed, Lensahn, Germany) is 30 µm.

parameters influence the yield and quality of blastocysts out of a pool of fertilized oocytes.

Firstly, embryologists have to decide which type of culture medium they want to use. Advocates of the "back to nature" principle would rely on sequential culture media whereas followers of the "let the embryo choose" philosophy would rather use a global medium. The main difference between these types is that sequential media are composed of two different media in order to mimic the milieu in the oviduct as well as in the uterus. Global media are more universal and can be used from the very beginning up to blastocyst stage. Meanwhile it turned out that both concepts will result in a comparable number of viable blastocysts [37].

Embryo density is the second parameter that might positively affect blastocyst outcome. In other words, reduction of the incubation volume as well as grouping of the embryos can increase blastocyst development in humans [38]. This is most likely the result of the stimulating capacity of embryotrophic ligands such as platelet-activating factor. On the other hand, there is the potential risk of embryo-toxic ammonium accumulation if embryo density is too high. It has been stated that group culture of embryos creates both deleterious and beneficial effects for the embryos, with a net beneficial outcome [39].

1.6. Embryo selection techniques

Irrespective of whether IVF or ICSI was used to generate embryos in culture it is of utmost

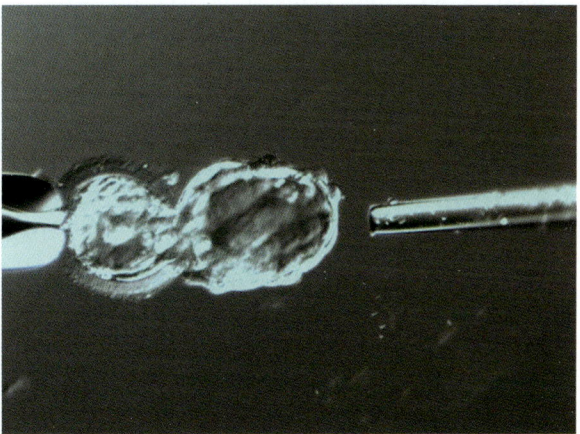

Video 1.11 Trophectoderm biopsy (courtesy of Dr Amparo Mercader, IVI Valencia).

importance to effectively select the embryo with the highest implantation potential. This approach would help to limit the number of embryos transferred and as a consequence multiple pregnancy rate, the most severe complication in assisted reproductive technologies, would be significantly reduced.

Naturally, invasive techniques, such as polar body (Video 1.9), embryo (Video 1.10), or trophectoderm biopsy (Video 1.11), most accurately reflect the actual state of health of gametes and embryos, with the latter approach being the most reliable one, since it maximizes insight while minimizing mechanical harm due to the biopsy process.

Alternatively, the health state of the somatic granulosa cells surrounding the oocyte may be used non-invasively, addressing their mutual dependence [40]. Since both cell types grow in the same surrounding and experience identical conditions in terms of nutrition during folliculogenesis, somatic cells can serve as an indirect marker of the health of the female gamete. With respect to this the rate of apoptosis [41, 42] and the relative telomere length in cumulus cells [43] as well as their transcriptomic profile [44] was successfully used for accurately predicting treatment outcome.

However, selection of the best embryo for transfer in routine IVF/ICSI programs is much rather based on techniques such as morphological analysis. Prediction of implantation can be enhanced if morphological information on gametes, zygotes, and embryos at different stages of pre-implantation development is pooled [45]. Nevertheless, this non-invasive method is far from being a perfect predictor of implantation. Recently, adding time course of mitotic divisions to static morphological observations started the era of morphokinetics, a field that may provide additional help in specifying implantation [46].

1.7. Non-invasive embryo selection based on morphology

Both approaches, morphological evaluation and morphokinetics, focus on proper embryonic development within the first days of in vitro culture. Since not all prospective studies do support prolonged embryo culture up to blastocyst stage it should only be considered in patients with sufficient embryos with good prognosis of blastulation.

Independently of whether transfer is planned at cleavage or blastocyst stage, proper identification of viable concepti is a prerequisite for high rates of implantation, pregnancy, and live birth. Most importantly, repeated screening for morphological features attributed to chromosomal aberrations is strongly recommended from oocyte stage onwards because spontaneous post-zygotic errors do occur and show in vitro persistence throughout pre-implantation development [47]. As a consequence of accurate deselection of oocytes and embryos having a high risk for aneuploidy, euploid concepti will accumulate in culture [45].

It is generally acknowledged that giant oocytes, apart from prophase I eggs showing a germinal vesicle, are the only type of female gamete that are 100% diploid which would cause digynic triploidy in case of ICSI [48]. Although not related to aneuploidy, oocytes with clusters of the smooth endoplasmic reticulum should also be used with caution since cases of stillbirth, malformation, and imprinting disorders have been published [49–51].

At zygote stage, concepti with uneven pronuclear size have to be discarded since the vast majority of them showed aneuploidy [52]. In addition, zygotes with pronuclei located in the periphery of the ooplasm or without abuttal have a bad prognosis in terms of further development [53, 54]. Gianaroli *et al.* [54] reported that the type of pronuclei (e.g., not abutted, unequal size, fragmented), the distribution and size of nucleoli (e.g., small and scattered), and the orientation of polar bodies with respect to pronuclei were highly predictive for the presence of complex chromosomal abnormalities in the developing embryos.

At cleavage stage embryos with bi- or multinucleation and/or uneven blastomeres should be eliminated in order to prevent accidental transfer in case of blastocyst formation [55]. One additional dysmorphism closely related to aneuploidy is the presence of a tetrafoliate-clover shaped arrangement of blastomeres on day 2 [56].

On day 5 of pre-implantation development blastocyst formation is checked and blastocysts are scored according to the quality (based on cell number and cohesion) of inner cell mass and trophectoderm [57].

Both cell lineages are not distinguishable at early blastocyst stage but from full blastocyst stage onwards. Although the majority of embryologists tend to choose blastocysts with optimal inner cell mass rather than an optimal trophectoderm, recent papers stress that in terms of live birth actually the latter is of utmost importance [58].

1.8. Embryo transfer

The final step of ART is to transfer one or two thoroughly selected embryos/blastocysts into the uterus approximately 1 cm from the fundus [59]. Embryologists have to carefully load a transfer catheter by aspirating a definite sequence of liquid and/or gaseous phases according to the internal guidelines of their laboratories [60]. It is of importance that in the catheter the culture medium containing the embryos considered for transfer is of minimal volume (e.g.,

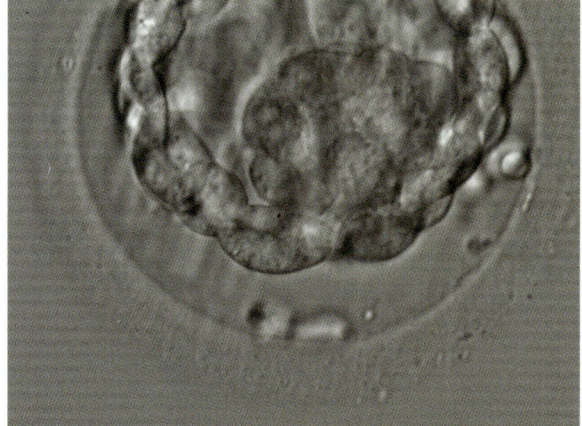

Video 1.12 Assisted hatching of an expanded blastocyst (Gardner score 4aa) showing partial collapse.

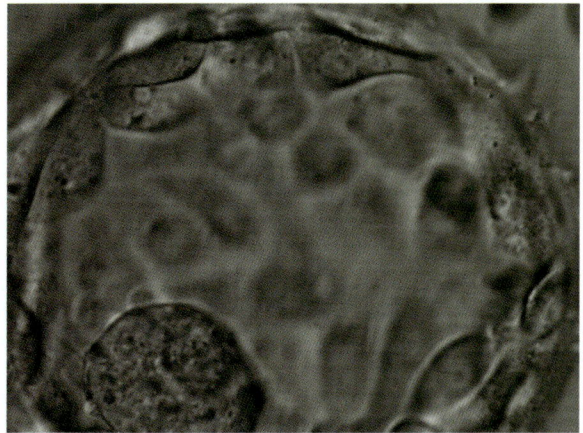

Video 1.13 Vitrification of a hatching blastocyst (Gardner Score 5aa) using a sequence of media containing ethylene glycol and DMSO (GM501 VitriStore, Gynemed, Lensahn, Germany).

5–10 μl) and the presence of extensive air bubbles is avoided [61].

In order to improve outcome among patients with a history of multiple implantation failures special hyaluronan-enriched transfer media are available [62]. Assisted hatching (Video 1.12) of the candidate embryos in cases of suspicious zona pellucidae, increased female age, or repeated implantation failure is an alternative technique to optimize embryo transfer [63, 64].

Ultrasound guidance allows the clinician to control optimal placement of the embryos [65], a technique which was shown to be of benefit in terms of outcome. After transfer, luteal phase support has been clearly demonstrated to improve pregnancy rates in women undergoing ART. Because of the increased risk of ovarian hyperstimulation syndrome associated with the use of hCG, progesterone has become the treatment of choice for luteal phase support [66].

1.9. Cryopreservation

For ethical reasons surplus zygotes, embryos, and blastocysts of adequate quality are cryopreserved; thus, saving them for usage in later treatment cycles. In principle, two cryopreservation techniques are available: slow freezing and vitrification. Due to a remarkable improvement in the efficiency of vitrification techniques as an alternative to the classical slow-freezing procedure [67], the former approach is

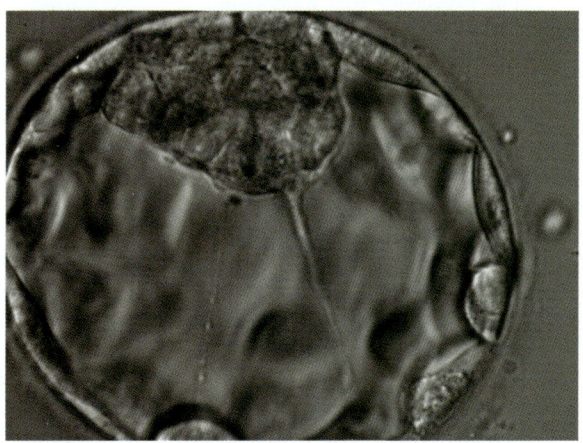

Video 1.14 Artificial shrinkage of the blastocoel by means of an ICSI pipette.

generally accepted as the most efficient technique [68], although all developmental stages, particularly zygotes, can successfully be frozen in a slower mode.

The physical process by which a viscous solution supercools to very low temperatures and finally solidifies into a stable glassy phase, without undergoing crystallization, at a practical cooling rate is called vitrification. According to its definition, with vitrification formation of ice crystals is theoretically impossible both in the intra-cellular as well as the extra-cellular spaces. The fundamental issue in vitrification is to achieve and maintain conditions within the cells which guarantee an amorphous state throughout the cooling as well as during the warming process. Independent of the carrier device that determines the cooling and/or the warming rate, the key to success in order to achieve a "glass-like" state depends on an optimal balance between the speed of cooling–rewarming (time and temperature) and the optimal cell dehydration and penetration of cryoprotectant when they are exposed to concentrated hypertonic solutions (Video 1.13). Particularly, in blastocysts the amount of water in the blastocoel is highest, bearing the highest risk of cryodamage. In order to avoid this, blastocoel fluid can be removed artificially [69] (Video 1.14), thus increasing survival rate and finally cumulative pregnancy rate.

References

1. Steptoe PC, Edwards RG. Birth after the reimplantation of a human embryo. *Lancet* 1978; 2: 366.

2. Cohen J, Malter H, Fehilly C, et al. Implantation of embryos after partial opening of oocyte zona pellucida to facilitate sperm penetration. *Lancet* 1988; 2: 162.

3. Ng SC, Bongso A, Ratnam SS, et al. Pregnancy after transfer of sperm under zona. *Lancet* 1988; 2: 790.

4. Palermo G, Joris H, Devroey P, et al. Pregnancies after intracytoplasmic injection of single spermatozoon into an oocyte. *Lancet* 1992; 340: 17–18.

5. Patrizio P, Silber S, Ord T, et al. Two births after microsurgical sperm aspiration in congenital absence of vas deferens. *Lancet* 1988; 2: 1364.

6. Silber SJ, Nagy ZP, Liu J, et al. Conventional in-vitro fertilization versus intracytoplasmic sperm injection for patients requiring microsurgical sperm aspiration. *Hum Reprod* 1994; 9: 1705–9.

7. Gleicher N, Friberg J, Fullan N, et al. Egg retrieval for in vitro fertilisation by sonographically controlled vaginal culdocentesis. *Lancet* 1983; 2: 508–9.

8. Feichtinger W, Kemeter P. Transvaginal sector scan sonography for needle-guided transvaginal follicle aspiration and other applications in gynecologic routine and research. *Fertil Steril* 1986; 45: 722–5.

9. Trounson A, Leeton JF, Wood C, et al. Pregnancies in humans by fertilization in vitro and embryo transfer in the controlled ovulatory cycle. *Science* 1981; 212: 681–2.

10. Porter RN, Smith W, Craft IL, et al. Induction of ovulation for in-vitro fertilisation using buserelin and gonadotropins. *Lancet* 1984; 2: 1284–5.

11. Itskovitz-Eldor J, Kol S, Mannaerts B. Use of a single bolus of GnRH agonist triptorelin to trigger ovulation after GnRH antagonist ganirelix treatment in women undergoing ovarian stimulation for assisted reproduction, with special reference to the prevention of ovarian hyperstimulation syndrome: preliminary report. *Hum Reprod* 2000; 9: 1965–8.

12. Veeck LL. The human oocyte. In Veck LL, ed. *An Atlas of Human Gametes and Conceptuses.* 1st ed. New York, London: Parthenon Publishing; 1999: pp. 19–24.

13. Rattanachaiyanont M, Leader A, Léveillé MC. Lack of correlation between oocyte-corona-cumulus complex morphology and nuclear maturity of oocytes collected in stimulated cycles for intracytoplasmic sperm injection. *Fertil Steril* 1999; 71: 937–40.

14. Ebner T, Moser M, Shebl O, et al. Blood clots in the cumulus-oocyte complex predict poor oocyte quality and post-fertilization development. *Reprod Biomed Online* 2008; 16: 801–7.

15. Daya S, Kohut J, Gunby J, et al. Influence of blood clots in the cumulus complex on oocyte fertilization and cleavage. *Hum Reprod* 1990; 5: 744–6.

16. Stanger JD, Stevenson K, Lakmaker A, Woolcott R. Pregnancy following fertilization of zona free, coronal cell intact human ova. *Hum Reprod* 2001; 16: 164–7.

17. Ebner T, Moser M, Sommergruber M, et al. Incomplete denudation of oocytes prior to ICSI enhances embryo quality and blastocyst development. *Hum Reprod* 2006; 21: 2972–7.

18. Van de Velde H, Nagy ZP, Joris H, et al. Effects of different hyaluronidase concentrations and mechanical procedures for cumulus cell removal on the outcome of intracytoplasmic sperm injection. *Hum Reprod* 1997; 12: 2246–50.

19. Parinaud J, Vieitez G, Milhet P, et al. Use of a plant enzyme preparation (Coronase) instead of hyaluronidase for cumulus cell removal before intracytoplasmic sperm injection. *Hum Reprod* 1998; 13: 1933–5.

20. Evison M, Pretty C, Taylor E, et al. Human recombinant hyaluronidase (Cumulase) improves intracytoplasmic sperm injection survival and fertilization rates. *Reprod Biomed Online* 2009; 18: 811–14.

21. Van den Bergh M, Bertrand E, Biramane J, et al. Importance of breaking a spermatozoon's tail before intracytoplasmic injection: a prospective randomized trial. *Hum Reprod* 1995; 10: 2819–20.

22. Dozortsev D, Qian C, Ermilov A, et al. Sperm-associated oocyte-activating factor is released from the spermatozoon within 30 minutes after injection as a result of the sperm-oocyte interaction. *Hum Reprod* 1997; 12: 2792–6.

23. Yong HY, Pyo BS, Hong JY, et al. A modified method for ICSI in the pig: injection of head membrane-damaged sperm using a 3–4 μm diameter injection pipette. *Hum Reprod* 2003, 18: 2390–6.

24. Palermo GD, Schlegel P, Colombero LT, et al. Aggressive sperm immobilization prior to intracytoplasmic sperm injection with immature spermatozoa improves fertilization and pregnancy rates. *Hum Reprod* 1996; 11: 1023–9.

25. Yanagida K, Katayose H, Yazawa H, et al. The usefulness of a piezo-micromanipulator in intracytoplasmic sperm injection in humans. *Hum Reprod* 1999; 14: 448–53.

26. Yanagida K, Katayose H, Hirata, et al. Influence of sperm immobilization on onset of Ca^{2+} oscillations after ICSI. *Hum Reprod* 2001; 16: 148–52.

27. Montag M, Rink K, Delacrétaz G, van der Ven H. Laser-induced immobilization and plasma membrane permeabilization in human spermatozoa. *Hum Reprod* 2000; 15: 546–52.

28. Ebner T, Yaman C, Moser M, et al. Laser assisted immobilization of spermatozoa prior to intracytoplasmic sperm injection in humans. *Hum Reprod* 2001; 16: 2628–31.

29. Ebner T, Moser M, Yaman C, et al. Successful birth after laser assisted immobilization of spermatozoa before intracytoplasmic injection. *Fertil Steril* 2002; 78: 417–18.

30. Vanderzwalmen P, Bertin G, Lejeune B, et al. Two essential steps for a successful intracytoplasmic injection: Injection of immobilized spermatozoa after rupture of the oolemma. *Hum Reprod* 1996; 11: 540–7.

31. De Oliveira NM, Sanchez R, Fiesta S, et al. Pregnancy with frozen–thawed and fresh testicular biopsy after motile and immotile sperm microinjection, using the mechanical touch technique to assess viability. *Hum Reprod* 2004; 19: 262–5.

32. Casper RF, Meriano JS, Jarvi KA, et al. The hypo-osmotic swelling test for selection of viable sperm for intracytoplasmic sperm injection in men with complete asthenozoospermia. *Fertil Steril* 1996; 65: 972–6.

33. Aktan TM, Montag M, Duman S, et al. Use of a laser to detect viable but immotile spermatozoa. *Andrologia* 2004; 36: 366–9.

34. Ebner T, Tews G, Mayer RB, et al. Pharmacological stimulation of sperm motility in frozen and thawed testicular sperm using the dimethylxanthine theophylline. *Fertil Steril* 2011; 96: 1331–6.

35. Van de Velde H, De Vos A, Joris H, et al. Effect of timing of oocyte denudation and micro-injection on survival, fertilization and embryo quality after intracytoplasmic sperm injection. *Hum Reprod* 1998; 13: 3160–4.

36. Dozortsev D, Nagy P, Abdelmassih S, et al. The optimal time for intracytoplasmic sperm injection in the human is from 37 to 41 hours after administration of human chorionic gonadotropin. *Fertil Steril* 2004; 82: 1492–6.

37. Biggers JD, Summers MC. Choosing a culture medium: making informed choices. *Fertil Steril* 2008; 90: 473–83.

38. Ebner T, Shebl O, Moser M, et al. Group culture of human zygotes is superior to individual culture in terms of blastulation, implantation and live birth. *Reprod Biomed Online* 2010; 21: 762–8.

39. O'Neill C. The potential roles for embryotrophic ligands in preimplantation embryo development. *Hum Reprod Update* 2008; 14: 275–88.

40. Gilchrist RB, Lane M, Thompson JG. Oocyte-secreted factors: regulators of cumulus cell function and oocyte quality. *Hum Reprod Update* 2008; 14: 159–77.

41. Corn CM, Hauser-Kronberger C, Moser M, et al. Predictive value of cumulus cell apoptosis with regard to blastocyst development of corresponding gametes. *Fertil Steril* 2005; 84: 627–33.

42. Ebner T, Shebl O, Holzer S, et al. Viability of cumulus cells is associated with basal AMH levels in assisted reproduction. *Europ J Obstet Gynecol Reprod Biol* 2014; 183: 59–63.

43. Cheng EH, Chen SU, Lee TH, et al. Evaluation of telomere length in cumulus cells as a potential biomarker of oocyte and embryo quality. *Hum Reprod* 2013; 28: 926–36.

44. Wathlet S, Adriaenssens T, Segers I, et al. New candidate genes to predict pregnancy outcome in single embryo transfer cycles when using cumulus cell gene expression. *Fertil Steril* 2012; 98: 432–9.

45. Ebner T, Moser M, Sommergruber M, et al. Selection based on morphological assessment of oocytes and embryos at different stages of preimplantation development: a review. *Hum Reprod Update* 2003; 9: 251–62.

46. Meseguer M, Herrero J, Tejera A, et al. The use of morphokinetics as a predictor of embryo implantation. *Hum Reprod* 2011; 26: 2658–71.

47. Bielanska M, Tan SL, Ao A. Chromosomal mosaicism

throughout human preimplantation development in vitro: incidence, type, and relevance to embryo outcome. *Hum Reprod* 2002; 17: 413–19.

48. Rosenbusch B, Schneider M, Gläser B, et al. Cytogenetic analysis of giant oocytes and zygotes to assess their relevance for the development of digynic triploidy. *Hum Reprod* 2002; 17: 2388–93.

49. Akarsu C, Cağlar G, Vicdan K, et al. Smooth endoplasmic reticulum aggregations in all retrieved oocytes causing recurrent multiple anomalies: case report. *Fertil Steril* 2009; 92: 1496. e1–3.

50. Ebner T, Moser M, Shebl O, et al. Prognosis of oocytes showing aggregation of smooth endoplasmic reticulum. *Reprod Biomed Online* 2008; 16: 113–18.

51. Otsuki J, Okada A, Morimoto K, et al. The relationship between pregnancy outcome and smooth endoplasmic reticulum clusters in MII human oocytes. *Hum Reprod* 2004; 19: 1591–7.

52. Sadowy S, Tomkin G, Munné S, et al. Impaired development of zygotes with uneven pronuclear size. *Zygote* 1998; 63: 137–41.

53. Scott L, Alvero R, Leondires M, et al. The morphology of human pronuclear embryos is positively related to blastocyst development and implantation. *Hum Reprod* 2000; 15: 2394–403.

54. Gianaroli L, Magli MC, Ferraretti AP, et al. Pronuclear morphology and chromosomal abnormalities as scoring criteria for embryo selection. *Fertil Steril* 2003; 80: 341–9.

55. Hardarson T., Hanson C., Sjögren A, et al. Human embryos with

unevenly sized blastomeres have lower pregnancy and implantation rates: indications for aneuploidy and multinucleation. *Hum Reprod* 2001; 16: 313–18.

56. Ebner T, Maurer M, Shebl O, et al. Planar embryos have poor prognosis in terms of blastocyst formation and implantation. *Reprod Biomed Online* 2012; 25: 267–72.

57. Gardner DK, Lane M, Stevens J, et al. Blastocyst score affects implantation and pregnancy outcome: towards a single blastocyst transfer. *Fertil Steril* 2000; 73: 1155–8.

58. Ahlström A, Westin C, Reismer E, et al. Trophectoderm morphology: an important parameter for predicting live birth after single blastocyst transfer. *Hum Reprod* 2011; 26: 3289–96.

59. Rovei V, Dalmasso P, Gennarelli G, et al. IVF outcome is optimized when embryos are replaced between 5 and 15 mm from the fundal endometrial surface: a prospective analysis on 1184 IVF cycles. *Reprod Biol Endocrinol* 2013; 16: 114.

60. Ebner T, Yaman C, Moser M, et al. The ineffective loading process of the embryo transfer catheter alters implantation and pregnancy rates. *Fertil Steril* 2001; 76: 630–2.

61. Eytan O, Elad D, Zaretsky U, et al. A glance into the uterus during in vitro simulation of embryo transfer. *Hum Reprod* 2004; 19: 562–9.

62. Nakagawa K, Takahashi C, Nishi Y, et al. Hyaluronan-enriched transfer medium improves outcome in patients with multiple embryo transfer failures. *J Assist Reprod Genetics* 2012; 29: 679–85.

63. Ebner T, Moser M, Tews G. Possible applications of a non-contact 1.48 micron wavelength diode laser in assisted reproduction technologies. *Hum Reprod Update* 2005; 11: 425–35.

64. Primi MP, Senn A, Montag M, et al. A European multicentre prospective randomized study to assess the use of assisted hatching with a diode laser and the benefit of an immunosuppressive/ antibiotic treatment in different patient populations. *Hum Reprod* 2004; 19: 2325–33.

65. Rinaldi L, Floccari A, Selman H. Ultrasound guidance of embryo transfer: a role for midwife. *Sex Reprod Healthc* 2014; 5: 47–9.

66. Baker VL, Jones CA, Doody K, et al. A randomized, controlled trial comparing the efficacy and safety of aqueous subcutaneous progesterone with vaginal progesterone for luteal phase support of in vitro fertilization. *Hum Reprod* 2014; 29: 2212–20.

67. Vanderzwalmen P, Bertin G, Debauche CH, et al. Vitrification of human blastocysts with the Hemi-Straw carrier: application of assisted hatching after thawing. *Hum Reprod* 2003; 18: 1504–11.

68. ALPHA Scientists in Reproductive Medicine, ESHRE Special Interest Group Embryology. Istanbul consensus workshop on embryo assessment: proceedings of an expert meeting. *Reprod BioMed Online* 2011; 22: 632–46.

69. Vanderzwalmen P, Bertin G, Debauche CH, et al. Births after vitrification at morula and blastocyst stages: effect of artificial reduction of the blastocoelic cavity before vitrification. *Hum Reprod* 2002; 17: 744–51.

2

Description of time-lapse systems: EmbryoScope™

Markus Montag, Tine Q. Kajhøj, and Inge E. Agerholm

2.1. History of the EmbryoScope® time-lapse system

The current EmbryoScope time-lapse system is the fourth version and a result of a continuing evolution of the system. The original idea was to use oxygen consumption as an additional parameter for bovine embryo quality [1]. In connection to the measurement of oxygen consumption image acquisition was established. Soon it became apparent that the timing of the embryo development also was a strong parameter for embryo quality and Unisense FertiliTech A/S (now Vitrolife AB) decided to focus on the image acquisition. In 2009 the EmbryoScope time-lapse system was CE marked as a Class IIa medical device and soon after the first EmbryoScope was introduced to a clinic in Spain. In 2011 the device was FDA 510(k) cleared and has since then been used in more than 49 countries. Vitrolife now offers the system.

2.2. The EmbryoScope® time-lapse system

The EmbryoScope time-lapse system presents a unique integration of bench-top incubation, time-lapse monitoring software for embryo development analysis, and communication tools.

2.2.1. The EmbryoScope™ time-lapse incubator

The heart of the system is the EmbryoScope time-lapse incubator (Figure 2.1), providing a stable incubation environment and automated high-quality image acquisition. Images are continuously collected and stored in a database for constant access and analysis through the EmbryoViewer software and the ES server. Despite

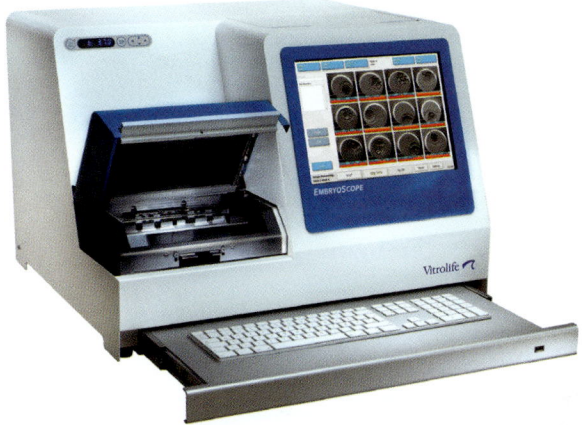

Figure 2.1 The EmbryoScope® time-lapse incubator. The EmbryoScope time-lapse incubator has access to the incubation chamber through the door on the left side. The EmbryoSlide culture dish is placed on a slide holder that moves over the position of the camera and the light path for image acquisition. The atmosphere in the incubation chamber is under constant control and the air is recirculated through an air filtration system with UV-light, Hepa, and VOC filtering units. (Figure has been provided by Vitrolife A/S Denmark. Copyright: Vitrolife A/S Denmark.)

the continuous acquisition of images in intervals as short as 10 minutes and with custom-defined numbers of focal planes, the total exposure to light is less than in standard observation conditions [2].

The EmbryoScope time-lapse system provides a completely flexible way of culturing and logging of embryos during development, which is adaptable to a clinic's own workflow.

2.2.2. The EmbryoSlide® culture dish

The EmbryoSlide culture dish (Figure 2.2a/b) is a single-use, sterile culture dish especially designed for culture of embryos in the EmbryoScope time-lapse incubator. The culture dish is embryo-toxicity and

(a)

Figure 2.2a/b The EmbryoSlide® culture dish. The EmbryoSlide culture dish has a grip for ease of handling (a). The dish can hold up to 12 embryos, which are located in separate microwells (b). (Figure has been provided by Vitrolife A/S Denmark. Copyright: Vitrolife A/S Denmark.)

(b)

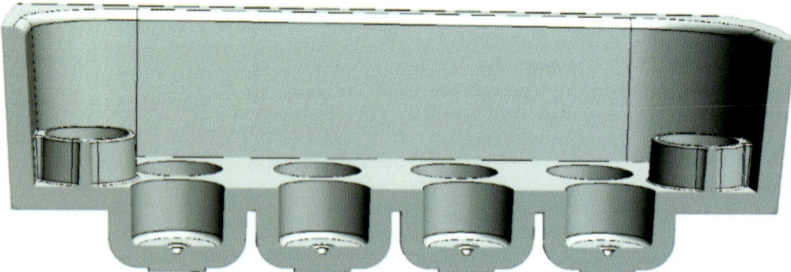

cytotoxicity tested, CE marked (Class IIa) and FDA-cleared 510(k), and compatible with standard and inverted microscopes.

Each EmbryoSlide culture dish holds up to 12 embryos, each cultured individually in droplets of 25 µl media. The embryo is placed within a microwell at the bottom of the culture well. Patented design and culture well numeration ensures concordant identification of embryos between all the products within the EmbryoScope time-lapse system.

In addition to the 12 culture wells, the Embryo-Slide culture dish includes four individual rinsing wells for optional use during the embryo loading process or for transient collection of embryos for transfer or freezing upon embryo selection. All wells (culture wells and rinsing wells) are integrated into a common oil reservoir.

2.2.3. The EmbryoViewer™ software

The EmbryoViewer software is at first sight a documentation tool to assess embryo development (Figure 2.3). However, the software provides numerous other features including the comparison of embryos, generation of videos, export of data, generation of a patient report, access to incubation conditions, and instrument parameters and the patient and treatment characteristics.

2.2.4. The ES™ server

The ES server enables storage and access to data on a common database as multiple EmbryoScope time-lapse incubator systems can be connected to the ES server (Figure 2.4). In order to facilitate image analysis and annotation, the EmbryoViewer software can be implemented on several clients,

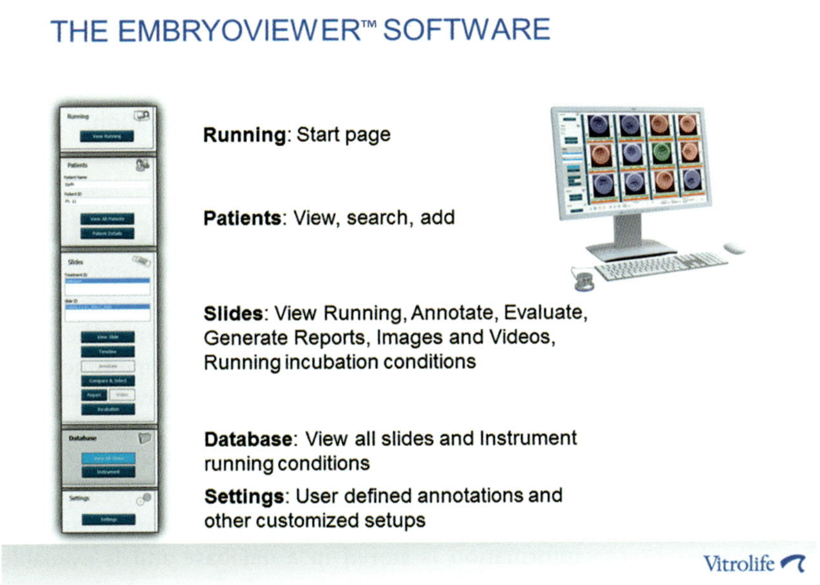

THE EMBRYOVIEWER™ SOFTWARE

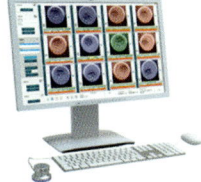

Running: Start page

Patients: View, search, add

Slides: View Running, Annotate, Evaluate, Generate Reports, Images and Videos, Running incubation conditions

Database: View all slides and Instrument running conditions

Settings: User defined annotations and other customized setups

Vitrolife

Figure 2.3 The EmbryoViewer™ software. The EmbryoViewer software allows for a user friendly and intuitive interaction. The different parts of the software for entering details regarding the patient, the treatment details, the embryos, the incubation parameters, and the settings are separate; and, an export generates a defined database structure. (Figure has been provided by Vitrolife A/S Denmark. Copyright: Vitrolife A/S Denmark.)

LABORATORY

EMBRYOVIEWER®
SOFTWARE

EMBRYOSCOPE™
TIME-LAPSE INCUBATOR

ES
SERVER

DOCTOR'S OFFICE

REMOTE OFFICE

EMBRYOSCOPE™
COUNSELING APP

EMBRYOVIEWER®
SOFTWARE

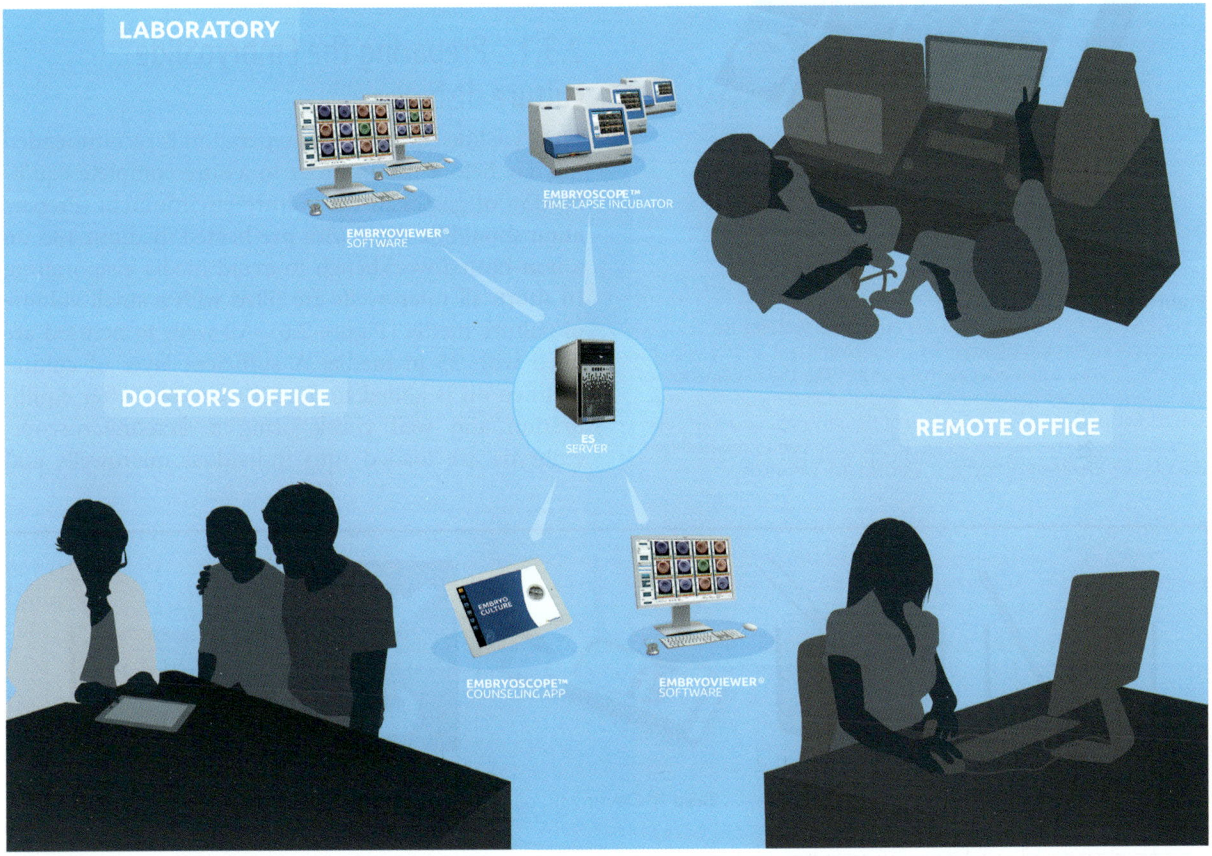

Figure 2.4 The ES server. The ES server connects to several EmbryoScope time-lapse incubators, so that all data are stored in one database. This can be assessed through several clients, which can be placed in the laboratory and/or in different locations within a clinic or within a clinic network. (Figure has been provided by Vitrolife A/S Denmark. Copyright: Vitrolife A/S Denmark.)

which can all be used in parallel. This enables interaction with the database in the laboratory as well as in a remote office or even a location outside of the clinic if secure internet connectivity is possible.

2.2.5. Additional features of the EmbryoScope™ time-lapse system

Another additional feature (where local rules permit) is the EmbryoScope Counseling App, which is available in the App store. This application comes with educational images and movies that explain the difference between standard incubation and incubation in a time-lapse system. The app can be used to access

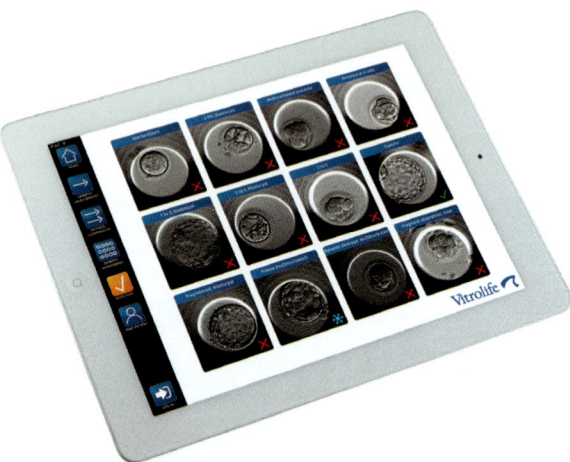

Figure 2.5 The EmbryoScope® Counselling App. The EmbryoScope Counselling App has numerous features for interaction within the clinic (embryologist/clinician) and/or with the patients. Besides some educational features like the explanation of the differences between time-lapse and standard incubation, the app allows one to show the videos of the embryos of a particular patient using secure access codes. (Figure has been provided by Vitrolife A/S Denmark. Copyright: Vitrolife A/S Denmark.)

patient and treatment cycles on the ES server (Figure 2.5). This gives the medical professional the ability to show patients their own embryos that have been chosen for transfer or cryopreservation.

2.3. Getting started with EmbryoScope™ time-lapse incubation

Incubation in the EmbryoScope time-lapse system requires setting up the EmbryoSlide culture dish in advance, loading oocytes or embryos into the slide, and placing the slide in the EmbryoScope time-lapse incubator. Once the required basic parameters have been entered into the software, the system will automatically start collecting images at pre-defined time-intervals and with the desired number of focal planes. All embryo development and instrument running information is stored in a database and is available for future reference. At the end of an incubation cycle, the embryos are removed from the EmbryoSlide culture dish for transfer or cryopreservation.

2.3.1. Preparing the EmbryoSlide® culture dish

EmbryoSlide culture dish preparation is recommended the day before use in order to assure complete equilibration of gasses in the culture environment. Preparation should be done with pre-heated medium and on a non-heated workbench to avoid media evaporation. In short: all microwells are filled with a small volume of culture media (Figure 2.6). All wells to be used are filled with 25 µl media. A sufficient layer of culture certified oil is added to the oil reservoir. After equilibration and well check using a stereomicroscope, embryos are loaded into individual microwells and

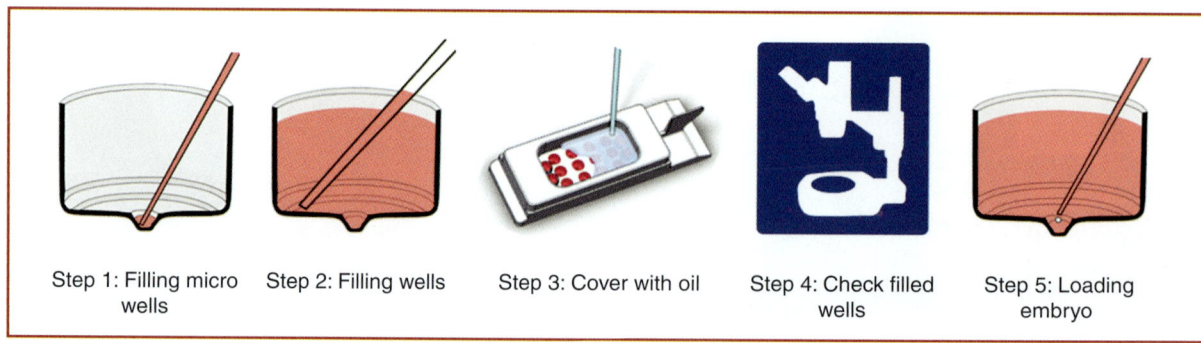

Step 1: Filling micro wells

Step 2: Filling wells

Step 3: Cover with oil

Step 4: Check filled wells

Step 5: Loading embryo

Figure 2.6 Preparation of the EmbryoSlide® culture dish. The preparation of the EmbryoSlide culture dish is based on an easy strategy using microcapillaries for priming the microwells with medium, followed by complete filling with standard pipettes or multidispension pipettes. The oil overlay is crucial as the EmbryoScope time-lapse incubator is a non-humidified device. (Figure has been provided by Vitrolife A/S Denmark. Copyright: Vitrolife A/S Denmark.)

Table 2.1 Timings and definitions of the individual events that can be observed and recognized at a specific time

Timings/parameter	Definition of the event
t0	Time of IVF or mid-time of micro/injection (ICSI/IMSI)
tPB2	The second polar body is completely detached from the oolemma
tPN	Fertilization status is confirmed
tPNa	Appearance of individual pronuclei; tPN1a, tPN2a; tPN3a
tPNf	Time of pronuclei disappearance ; tPN1f; tPN2f...
tZ	Time of PN scoring
t2	Time for two discrete cells
tRoll (i) tRoll (e)	Initiation and end of blastomere rolling at the two-cell stage
tTM	Trichotomous mitosis
t3 to t9	Three to nine discrete cells
tSC	First evidence of compaction
tM	Initation of morulation
tSB	Initiation of blastulation
tB	Full blastocyst (last frame before zona starts to thin)
tE or tEB	Initiation of expansion; first frame of zona thinning
tHN	Herniation; end of expansion phase and initiation of hatching process
tHD or tHB	Fully hatched blastocyst

the dish is immediately transferred to the Embryo-Scope time-lapse incubator.

2.3.1.1. Using the EmbryoSlide™ culture dish with sequential or single-step media

The EmbryoSlide culture dish can be used with both sequential and single-step media. While single-step media without change [3] fits well the concept of continuous undisturbed monitoring by time-lapse technology, clinic procedures might prescribe media change during the culture period.

For changing media in the EmbryoSlide culture dish, it is recommended to remove 20 μl of media from each well of the dish and add another 20 μl of fresh, pre-equilibrated medium to each well.

2.3.2. The EmbryoViewer™ software in clinical use

2.3.2.1. Time-lapse parameters of embryo development

The continuous monitoring of embryos at pre-set time intervals and at a defined number of focal planes enables observation of embryo development from the time point of insemination (e.g., by intra-cytoplasmic

sperm injection, ICSI) up to the last assessment prior to removal of the embryos from the EmbryoSlide for transfer, freezing or for discarding. This increase in information has also led to a new nomenclature that characterizes distinct developmental events. An overview of the most relevant time-lapse events is given in Table 2.1 and an abbreviation list as it is used in the EmbryoViewer software is given in Table 2.2.

Another new term that has evolved is "annotation," which defines the process of characterizing the embryo throughout its development by assigning the definition given by the nomenclature to a specific time-point. This results in an overview of the course of development, which can be linked to morphology. The combination of morphology over time is also referred to as "morphokinetics."

In order to have a common language among time-lapse users, a guideline paper on the nomenclature has recently been published [4].

2.3.2.2. Annotation strategies

In standard incubation the dish containing the embryos is removed from the incubator usually once a day and the corresponding stage of development as

15

Table 2.2 Abbreviation list as used in the EmbryoViewer™ software

Variable	Description	Values
NOT2PN	Maximum number of pronuclei differs from two	TRUE/FALSE
UNEVEN2	Uneven size of blastomeres at the two-cell stage	TRUE/FALSE
UNEVEN4	Uneven size of blastomeres at the four-cell stage	TRUE/FALSE
MN2	Multinuclearity occurs at the two-cell stage	TRUE/FALSE
MN4	Multinuclearity occurs at the four-cell stage	TRUE/FALSE
tPB2	Time from insemination until second polar body is extruded	Hours
tPNa	Time from insemination until pronuclei have appeared	Hours
tPNf	Time from insemination until pronuclei have faded	Hours
t2	Time from insemination to complete division to two cells	Hours
t3	Time from insemination to complete division to three cells	Hours
t4	Time from insemination to complete division to four cells	Hours
t5	Time from insemination to complete division to five cells	Hours
t6	Time from insemination to complete division to six cells	Hours
t7	Time from insemination to complete division to seven cells	Hours
t8	Time from insemination to complete division to eight cells	Hours
t9+	Time from insemination to complete division to nine or more cells	Hours
tM	Time from insemination to formation of morula	Hours
tSB	Time from insemination to start of blastulation	Hours
tB	Time from insemination to formation of blastocyst	Hours
tEB	Time from insemination to formation of expanded blastocyst	Hours
tHB	Time from insemination to hatching blastocyst	Hours

well as the quality of the embryo is documented in the patient file. A common strategy has been described by Alpha scientists in reproductive biology and the special interest group of ESHRE and is known as the Alpha/ESHRE consensus [5]. This consensus paper describes standard incubation defined time-points for embryo assessment. The aim of this approach was to enable a fair comparison between embryo classifications in publications from different laboratories and to create a standard for the optimal time when embryos should be assessed.

As time-lapse imaging records the embryo development over the entire incubation period, one can also assess the time-points that are defined by the ALPHA/ESHRE consensus paper. However, one can re-wind the video and thus assess any developmental event in a retrospective way during the actual treatment cycle as well as at a later time-point, e.g., when a patient comes

for another treatment cycle and it is helpful to observe how the embryos developed in the previous cycle.

The freedom in assessing embryo development at any time opens up completely new annotation strategies. Usually the morning is the busiest time in the laboratory:

-Oocytes that have been inseminated on the day before need to be checked for fertilization
-Embryos that are considered for transfer or freezing need to be assessed, selected, and cryopreservation needs to get started
-Embryos that need to be biopsied have to be looked up for the proper time and stage
-Embryos in culture on the days in between fertilization and transfer need to be checked constantly on a daily base
-Ovum pick-up needs to be prepared and started

Using the EmbryoScope time-lapse system allows flexibility for embryo assessment as every time-point that has passed can be assessed retrospectively. Fertilization does not need to be checked in the morning at 16–18 h post-insemination, instead one can sit down at any time after insemination and re-wind the video to the time-period of 16–18 h.

In short, annotation can be done on every day, but at the time which best suits the laboratory. This also means that weekend embryo evaluations are unnecessary as the annotation and scoring can be performed on Monday.

2.3.2.2.1. Daily annotation strategy

A very efficient annotation strategy is to screen on days 1, 3, and 5 and omit day 2 and day 4 (Table 2.3). The first check is on day 1 for correct fertilization at 16–18 h after insemination. During this step all oocytes that are not fertilized or that are not correctly fertilized (1PN, $\geq$ 3PN) can be deselected by marking the wells with the red annotation mark, which indicates in the software that this cell has been or will be non-selected. For all oocytes with 2PN the presence of 2PN is marked in the software.

The next check is on day 3 and during this check all embryos that are not considered to be in an acceptable developmental stage will be marked red (discarded). If there is time, one can then look already back into previous events that occurred on day 2, such as the time of division to two-cell (t2), three-cell (t3),

four-cell (t4), and five-cell (t5) and annotate these in the software. These divisions are easy to spot and after annotation a cell division chart appears, which reveals unusual cleavage behavior that is known to result in low implantation potential (also see 2.4.1.).

The next check is on day 5 and again the focus is on proper development up to that day. So all embryos that have not reached a stage which is considered to be adequate will be marked red. This will finally leave only a few embryos for which annotation has to be made. If these have been annotated up to t5 already, then only t8, tM, tSB, tB, and tEB (tHB) are required. This will allow assessment of the full morphokinetic development of these embryos and support the decision about which one(s) to transfer.

2.4. Morphokinetics, embryo development, and outcome

2.4.1. Cell division charts and their significance

Cell division charts give an immediate overview of the development of an embryo (Figure 2.7). Looking at a cohort of embryos in one EmbryoSlide culture dish from one patient thus enables a very fast classification. Especially, embryos with a direct cleavage can be easily identified. Based on current reports in the literature, these embryos present a very low implantation potential. So the cell division charts allow

Table 2.3 Assessment strategy

Assessment days	Suggested assessment strategy	Annotation remark
Day 1	PN assessment at 16–18 h Annotation of PN fading Deselection of all unfertilized or not correctly fertilized oocytes	2PN tPNf Mark embryo RED (which indicates: deselect)
Day 2	No assessment	No annotation
Day 3	Deselect embryos that did not reach a proper cell stage Annotate remaining embryos	Mark embryos RED t2, t3, t4, t5, t8
Day 4	No assessment	No annotation
Day 5	Deselect embryos that did not reach a proper cell stage Annotate remaining embryos Select embryos for transfer/cryopreservation	Mark embryos RED tM, tSB, tB, tEB, tHB Mark embryo(s) GREEN for transfer and BLUE for cryopreservation

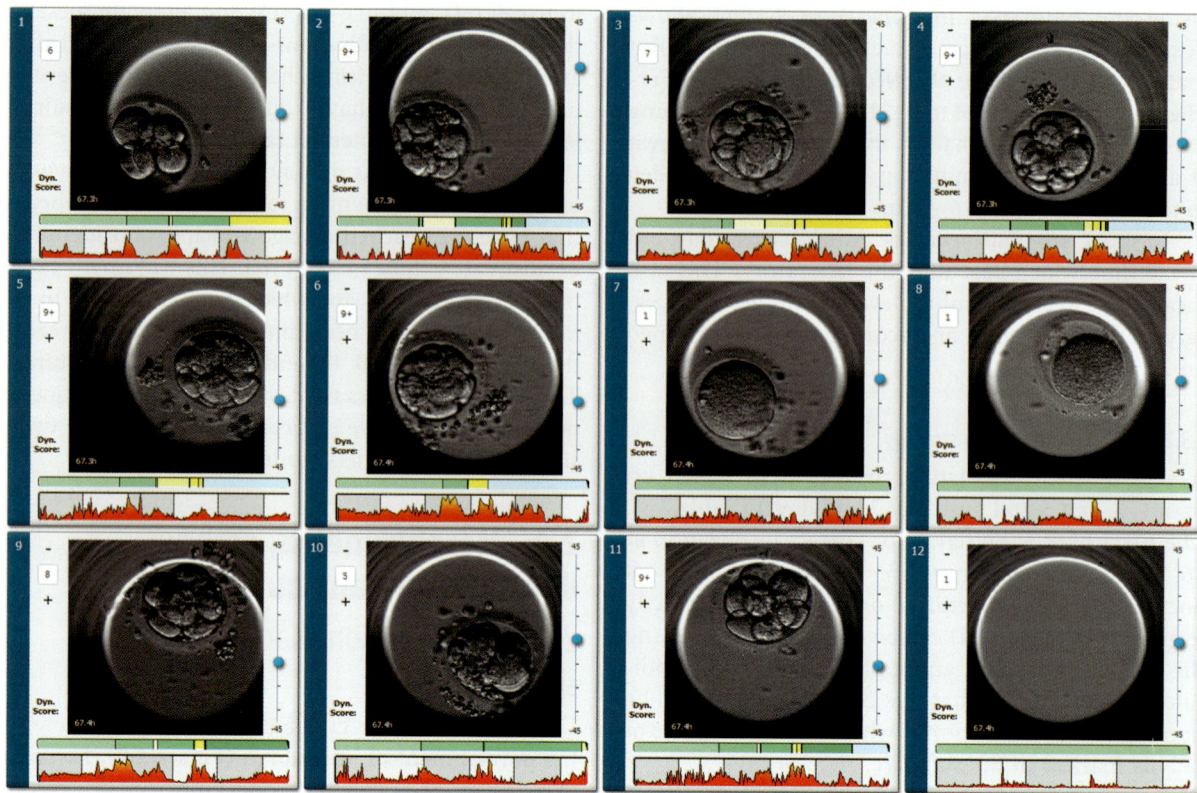

Figure 2.7 Embryo implantation potential according to cell division charts. After annotation of the key characteristics of embryo development, a division chart appears under the image of the embryo. This allows for a fast classification of the developmental pace and helps to identify direct cleavage or unexpected duration of unequal cleavage stages like the three-cell, five-cell, six-cell, and seven-cell stage by using different color codes. (Figure has been provided by Vitrolife A/S Denmark. Copyright: Vitrolife A/S Denmark.)

deselection of such embryos, allowing efforts to be concentrated on the remaining embryos.

2.4.2. Known implantation data

The concept of Known Implantation Data (KID) describes information of embryos including their specific status of implantation (Figure 2.8). The KID ratio differs from the traditional implantation rate in that only embryos for which it is specifically known whether they implanted or not are counted. For instance in the case of a double embryo transfer, only in cases where both embryos implanted or cases where none of the embryos implanted are counted as KID. For a double embryo transfer with one implanting embryo there is no KID because it is not known which of the transferred embryos was the one to implant.

Known Implantation Data are particularly useful for correlating morphokinetic traits during embryo development to implantation potential because implanting embryos with their associated morphokinetic information can be analyzed as a group to differentiate morphokinetic patterns for implanting embryos from those of non-implanting embryos.

2.4.3. Calculated variables

Variables describing embryo morphokinetics can be presented in different forms to reflect specific trends during embryo development.

In the EmbryoViewer software the specific timings of developmental events can be registered as described above e.g., "t2" describes the time at which the embryo divided to a two-cell embryo.

Duration of developmental stages can be calculated from the specific timing variables by calculating the difference between two annotated cell stage timings.

KNOWN IMPLANTATION DATA (KID)
(By fetal heart beat)

Only embryos from transfers in which **all** transferred embryos **do** implant or in which **all** transferred embryos **do not** implant are counted in the KID ratio.

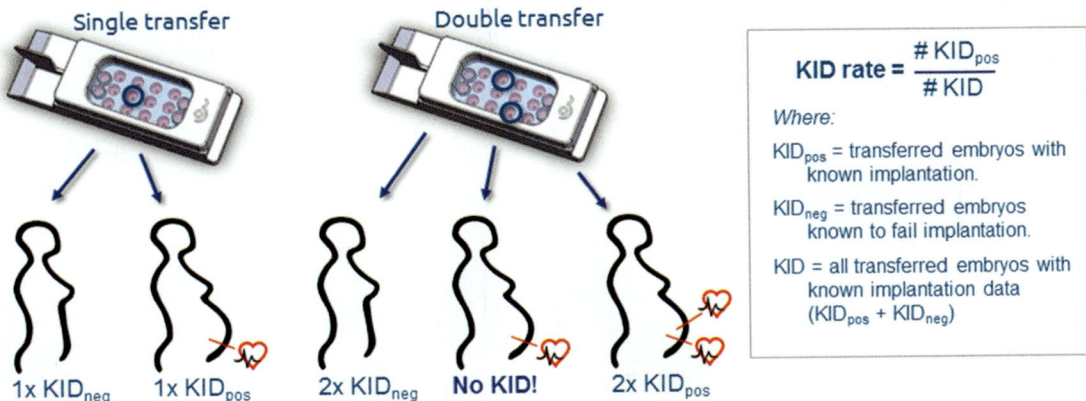

$$\text{KID rate} = \frac{\#\,\text{KID}_{pos}}{\#\,\text{KID}}$$

Where:

KID_{pos} = transferred embryos with known implantation.

KID_{neg} = transferred embryos known to fail implantation.

KID = all transferred embryos with known implantation data $(\text{KID}_{pos} + \text{KID}_{neg})$

- KID describes whether or not a specific transferred embryo has implanted.
- KID is important for relating implantation success to a specific embryo.
- KID ratio is a measure of positive implantation of a cohort of embryos with Known Implantation Data

Figure 2.8 The concept of Known Implantation Data (KID). Known Implantation Data are important to identify a direct link between a specific timing or characteristic of an embryo with implantation or live birth. As shown, a single embryo transfer will yield 100% KID results, whereas after a double embryo transfer only some embryos will generate KID results. (Figure has been provided by Vitrolife A/S Denmark. Copyright: Vitrolife A/S Denmark.)

Ratios of developmental stages can be calculated by mathematical expressions using specific timings and have been shown to be indicative of embryo potential [6].

2.5. Examples for new findings and aspects of embryo development

The application of time-lapse imaging in clinical embryology has within a short time period confronted embryologists with aspects of embryo development which they have never seen before. Some events or aspects of embryo development that could previously only be explained by untested hypotheses are being verified by time-lapse technology, which provides direct proof of the underlying mechanism.

Using the EmbryoScope time-lapse system, videos can be generated from individual embryos or from up to three embryos side by side. The video generator allows export of the entire sequence or only a defined time-interval of the embryo development. The speed of the video can be adjusted and it is also possible to add custom-defined logos or descriptions to the video. The time of the sequence shown in relation to insemination is displayed in the lower right corner of the video.

2.5.1. Cytoplasmic waves

Cytoplasmic waves are linked to the fertilization event and indicate that the oocyte has been activated through the sperm cell and that the initiation of the fertilization cascade has occurred. In Video 2.1 a cytoplasmic wave can be spotted just prior to the appearance of the pronuclei. It requires a decent image quality to spot cytoplasmic waves.

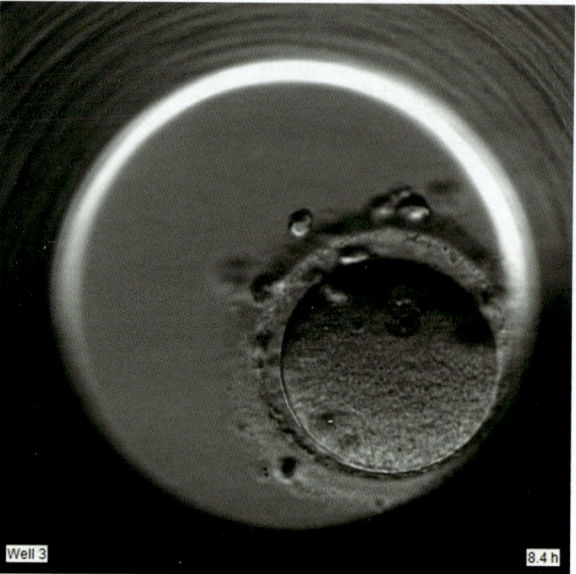

Well 3   8.4 h

Video 2.1 Cytoplasmic waves in an oocyte prior to appearance of pronuclei. A cytoplasmic wave can be spotted between 4.4 and 6.4 hours – just prior to appearance of the pronuclei – travelling over the surface of the oocyte from the upper right to the lower left side.

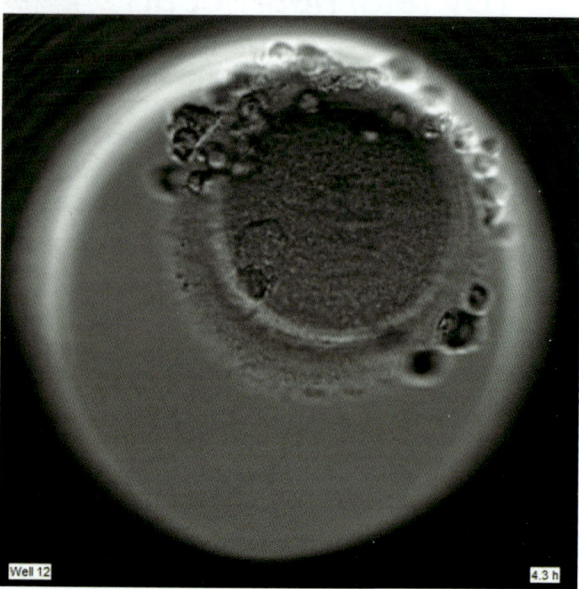

Well 12   4.3 h

Video 2.2 Extrusion of the second polar body. The first polar body can be spotted from the beginning of the video at a position close to 8 pm on the oocyte. The second polar body is extruded at 3.1 hours and changes shape and splits into two fragments at 6.7 hours.

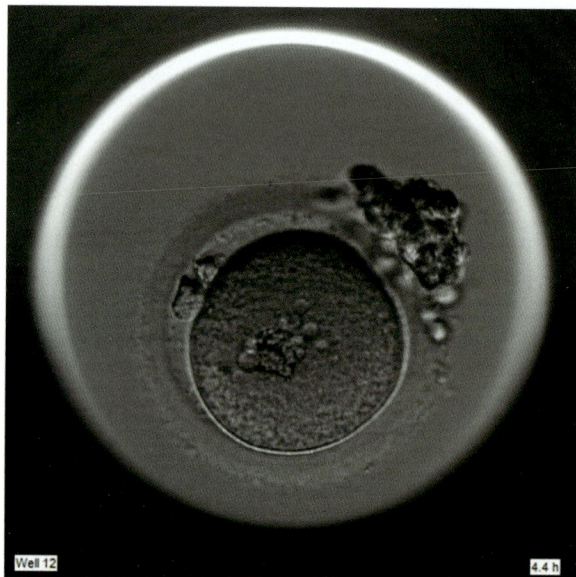

Well 12   4.4 h

Video 2.3 Appearance of the female and male pronucleus (PN). Following extrusion of the second polar body at 3.2–3.6 hours, the female pronucleus can be seen at 7.6 hours on a virtual line from the second polar body towards the center of the oocyte. The male pronucleus can be spotted at 8.0 hours close to the inclusion body towards the center of the oocyte. Both pronuclei get closer to each other and grow in diameter over time. Also note the change of the nucleolar precursor bodies within the pronuclei.

2.5.2. Extrusion of the second polar body

Oocytes that have been placed in the EmbryoScope time-lapse system immediately after ICSI allow for the visualization of the extrusion of the first polar body [4]. This event is called tPB2 and usually occurs about 3 hours after insemination, as shown in Video 2.2 [7].

2.5.3. Formation of the pronuclei

The appearance of the female and male pronuclei may occur as early as 6 hours after insemination (tPNa) [7]. With the high resolution and image quality of the EmbryoScope time-lapse system one can distinguish the female and the male pronucleus, as the female PN appears close to the side of the second polar body extrusion (Video 2.3).

2.5.4. Incorrect ICSI

In case the oolemma has not been penetrated properly during the injection of a human spermatozoon, the sperm cell will remain within a membrane-enclosed

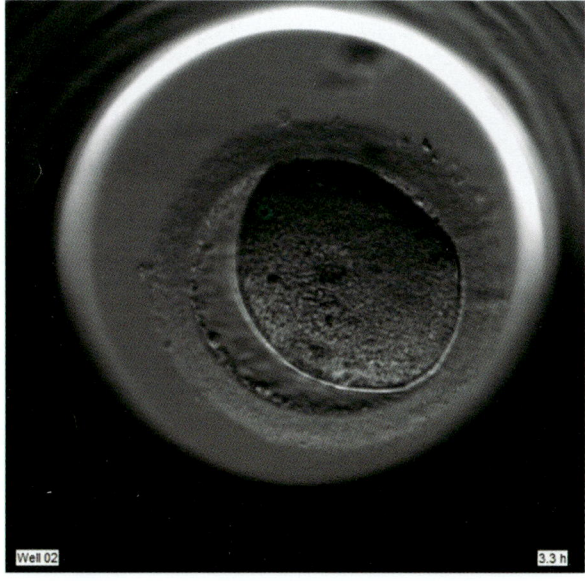

Well 02   3.3 h

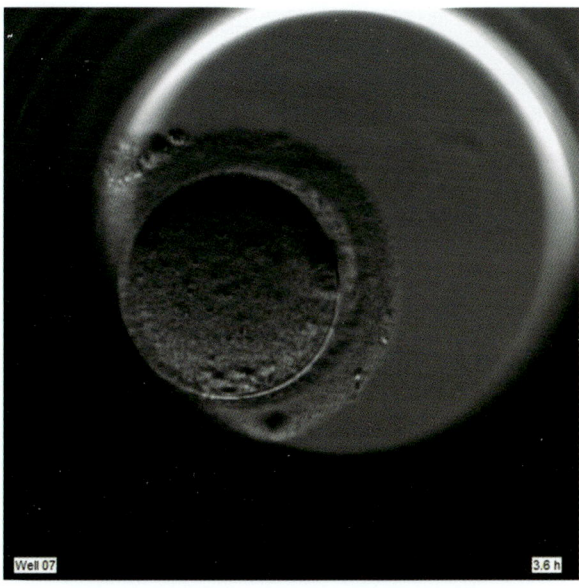

Well 07   3.6 h

Video 2.4 Sperm in vacuole after incorrect ICSI. This video shows a sperm cell that is captured in a membrane-enclosed vacuole as a result of an improper breakage of the oolemma during ICSI. The sperm cell can be seen best at a sequence at 10 hours.

Video 2.5 Fusion of 2PN to 1PN. The presence of 2PN shows that this oocyte has been fertilized correctly. However, both pronuclei fuse at around 12–14 hours and the result is a larger single pronucleus.

vacuole (Video 2.4). This usually results in failed fertilization as the sperm cell cannot interact with the ooplasm and activate the oocyte properly.

2.5.5. Fusion of 2PN to generate a 1PN

The potential use of an oocyte that exhibits only 1PN at 16–18 hours after IVF or after ICSI has been intensively discussed among embryologists in the past. Provided that no other embryo is available, most embryologists do consider transferring an embryo derived from a 1PN oocyte after IVF, but not after ICSI. Video 2.5 shows an example, where 2PN initially formed after ICSI and later fused to a single, slightly larger 1PN. This example shows that time-lapse imaging from the very beginning can help identifying the nature of a 1PN oocyte. In the case shown, the oocyte although showing only 1PN has responded to oocyte activation and presumably contains normal chromatin content. As fusion of 2PN to 1PN is not the normal course of early development, would normally not consider an embryo derived from such an oocyte for transfer. However, knowing the history of the fusion of the 2PN, some laboratories might consider transfer if no other embryo is available.

2.5.6. Presence of smooth endoplasmic reticulum (SER)

The identification of smooth endoplasmic reticulum (SER) is important given the low success rates, high miscarriage rates, and the risk for imprinting defects linked with oocytes containing SER. Time-lapse imaging helps to better identify SER in oocytes as the course of SER formation – appearance and disappearance – can be followed over time (Video 2.6).

2.5.7. Development of an embryo up to blastocyst stage

Embryos that develop normally to the blastocyst stage show a regular developmental pace (Video 2.7).

Embryos that arrested during development and do not proceed properly are usually considered by most embryologists as developmentally incompetent and discarded (Video 2.8).

Embryos that degenerate during development can be marked accordingly and further annotation is not required. Some embryos may also degenerate at the blastocyst stage (Video 2.9).

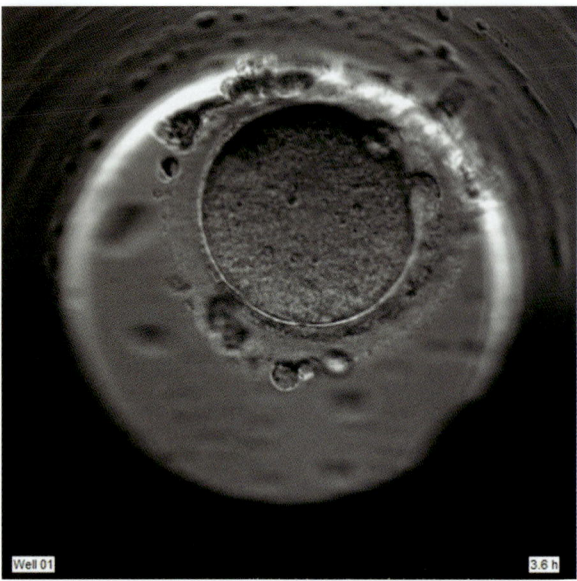

Video 2.6 Appearance and disappearance of SER in an oocyte. A disc of smooth endoplasmic reticulum can be spotted in the center of this oocyte. The SER seems to disappear around 8 hours, which would be impossible to see with standard observation. It shows that SER is not necessarily present for a longer time period.

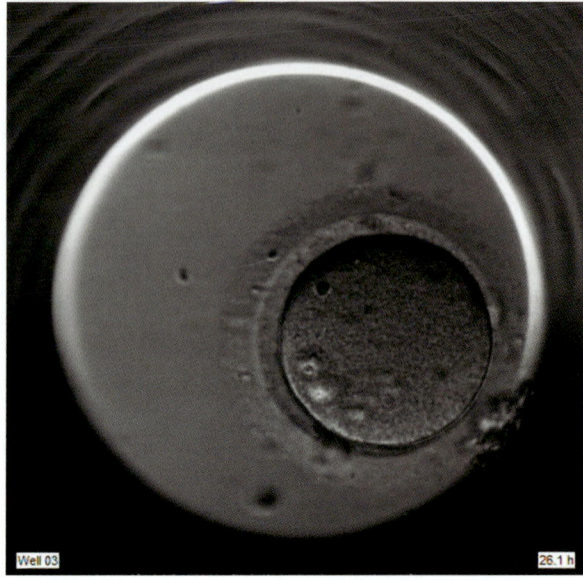

Video 2.7 Normal developing blastocyst. This video shows a normal developing embryo with regular cleavage cycles and short intermediate cell stages, which give rise to a nice blastocyst.

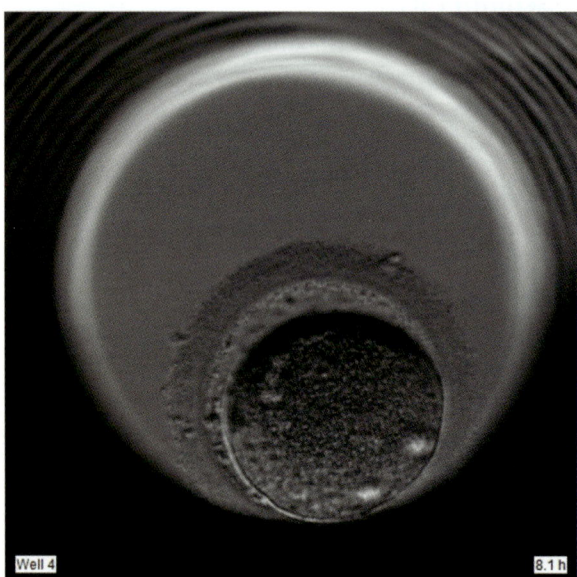

Video 2.8 Arrested embryo. The embryo shown in this video already has a later time of division to the two-cell stage. The further development is very irregular and the embryo does not proceed to become a blastocyst, instead it arrests in development.

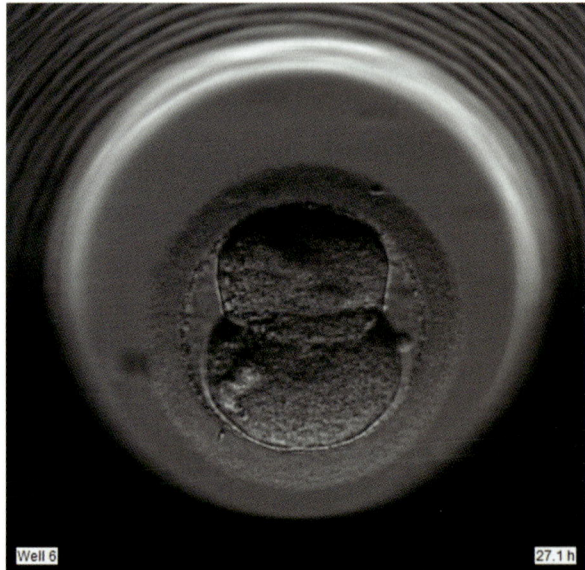

Video 2.9 Degenerating blastocyst. Following a quite moderate development to the eight-cell stage, this embryo presents a blastomere that degenerates at the morula stage. During further development the blastocyst is unable to expand completely and further cells degenerate, leading to a completely degenerated blastocyst at the end.

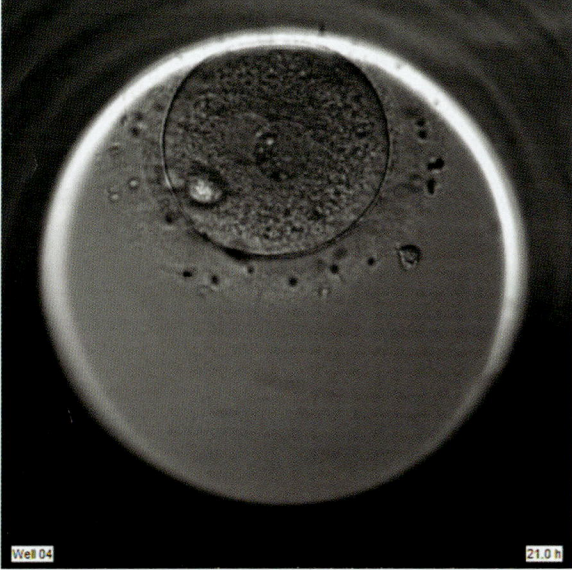

Well 04   21.0 h

Video 2.10 Development of a blastocyst from one blastomere of a two-cell stage. Following the division to the two-cell stage, one of the two blastomeres stops further development. The developing embryo reaches the blastocyst stage from one blastomere only. In standard incubation and observation such a development is difficult to identify.

Using time-lapse imaging allows visualization of abnormal embryo development, which is not usually detectable with standard observation.

Video 2.10 shows an embryo in which at the two-cell stage one of the two blastomeres did not continue division. Consequently the embryo developed only from one blastomere and eventually reached the blastocyst stage. If no time-lapse information was available, only very careful observation and recording of blastomere sizes would uncover such a developmental event. Although the embryo most likely will have a chance to implant it would in many cases not be the first choice for transfer.

2.5.8. Embryo cleavage to the two-cell stage

The first cleavage of embryos is a very important step. It shows that the embryo is on its way to normal development and that the mitotic process is re-initiated. The normal cleavage pattern results in a two-cell embryo with even blastomeres (Video 2.11).

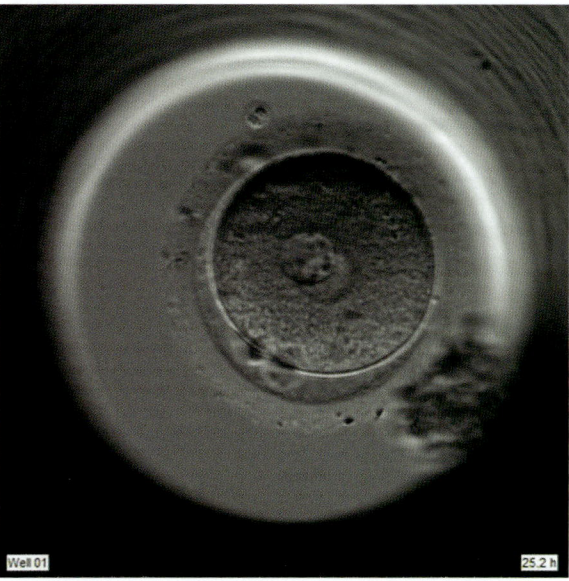

Well 01   25.2 h

Video 2.11 Cleavage at the two-cell stage to an even embryo. A fertilized oocyte dividing into two blastomeres of equal size with one nucleus in each blastomere is shown in this video.

In case of an uneven cleavage to the two-cell stage, the embryo most likely will also show an uneven blastomere size at the four-cell stage (Video 2.12). Unevenness at the two-cell stage is generally considered to characterize a low-quality embryo.

During division from one cell stage to another, embryos often show a very unusual wobbling effect, which is characterized by a continuous change in the shape of the resulting blastomeres (Video 2.13). If one observes embryos at this time-period in a standard observation approach and without time-lapse information, there is a high chance that the embryo will be characterized as a low-quality grade and with unevenly sized blastomeres.

2.5.9. Direct division

Time-lapse imaging has revealed that a substantial number of embryos exhibit a very fast transition from the one-cell to the three-cell stage (Video 2.14) or from the two-cell to the five-cell stage. If the embryos remain in the two-cell or four-cell stage for too short time to be able to duplicate DNA, this process is called direct cleavage. Retrospective analyses have shown that such embryos are characterized by a low

23

Well 1   8.4 h

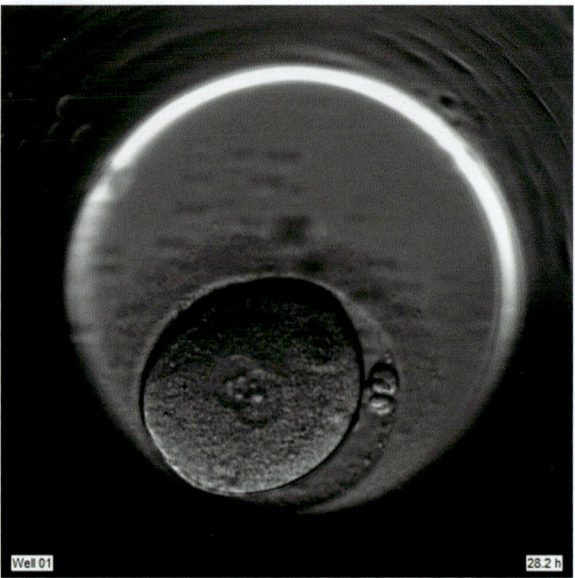

Well 01   28.2 h

Video 2.12 Cleavage at the two-cell stage to an uneven embryo. The embryo in this video shows the formation of the pronuclei close to the membrane. Although the pronuclei move closer towards the center, the cleavage results in an uneven two-cell stage embryo. In the following division to the four-cell stage both blastomeres give rise to equal cells, with two smaller and two larger blastomeres as a final result.

Video 2.13 Wobbling at the division to the two-cell stage. At the early cleavages to the two-cell and four-cell stage one can sometimes see that the blastomeres shortly after cleavage appear uneven and that they do not stay in the position defined by the cleavage. The blastomeres seem to wobble and they usually change shape and size during this short period. Note a small fragment that is re-absorbed at around 33 hours.

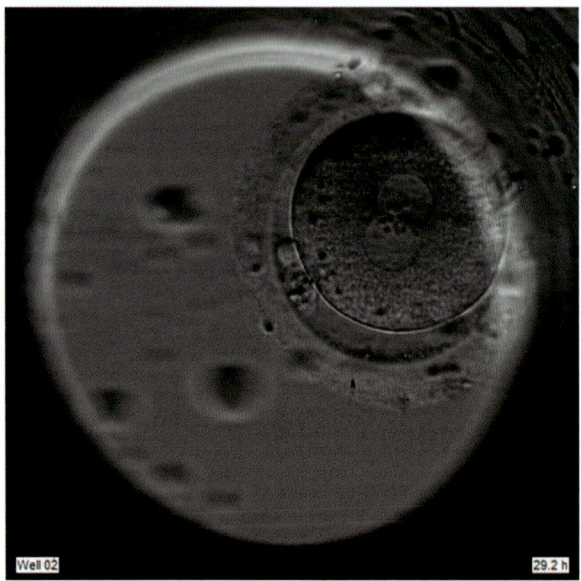

Well 02   29.2 h

Video 2.14 Direct cleavage from the one-cell stage to the three-cell stage. This video illustrates the direct cleavage from one to three cells. The resulting blastomeres in this example are unevenly sized, but all contain a nucleus. Direct cleavage has been described to be associated with low implantation rates.

implantation rate, a high rate of chromosomal aberrations, and, in case of implantation, by a high miscarriage rate [8, 9]. This phenomenon is considered as one of the most robust deselection criteria. Without time-lapse imaging, direct cleavage is overlooked in the majority of embryos that show this phenomenon.

2.5.10. Reverse cleavage

For the first time, time-lapse has revealed the phenomenon that a blastomere, after division into two daughter blastomeres, can re-form within a short time period by fusion of these two blastomeres to one blastomere (Video 2.15). This process has been named reverse cleavage and several authors have studied the possible impact of reverse cleavage on implantation potential. It is obvious and logical that a division of a blastomere followed by a fusion will result in a doubling of the genetic material and consequently give rise to one blastomere with a double ploidy status compared to all other blastomeres of the same embryo. Although a few live births have been reported after transfer of

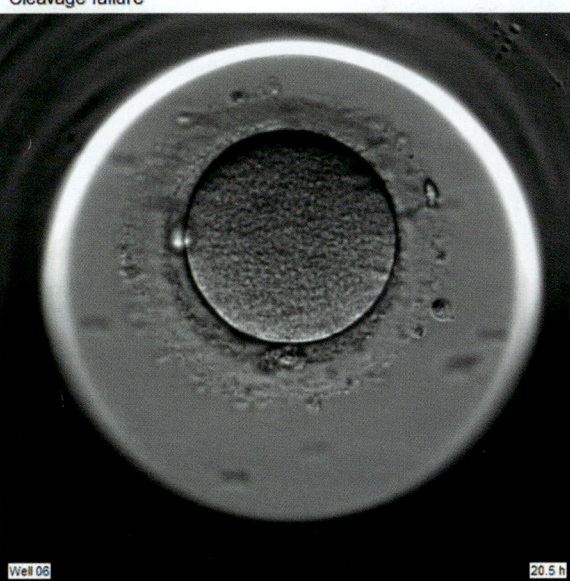

Video 2.15 Reverse cleavage from four to five to four blastomeres. This video on reverse cleavage shows an embryo that at the four-cell stage undergoes cleavage of one blastomere to a five-cell stage embryo (around 48 hours). However, the divided blastomeres fuse again and the result is again a four-cell stage embryo.

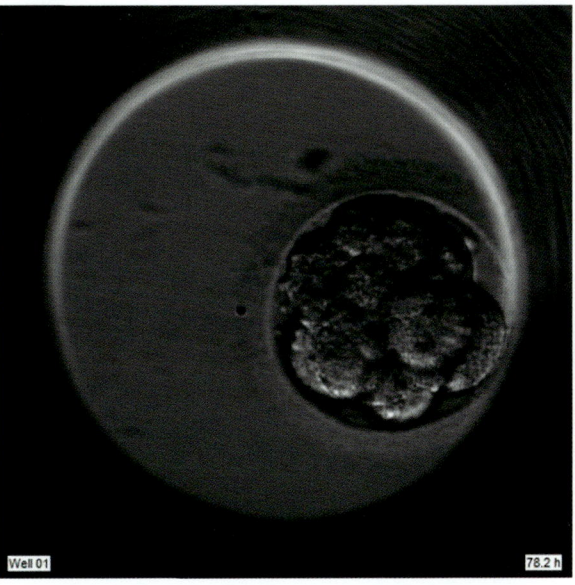

Video 2.16 Extrusion of blastomeres at the morula stage. This embryo shows the initiation of compaction and formation of a morula. During this process, some blastomeres are extruded from the compacted mass and eventually re-absorbed, whereas others are extruded. One extruded blastomere can be seen at the 11 am position, which degenerates. During expansion such cells are pushed towards the zonal pellucida and eventually lyse.

reverse cleaved embryos, the etiology and the potential impact on the child is still unknown. However, most embryologists consider reverse cleavage as a sign of a disturbed course of cell division and are cautious in selecting these embryos for transfer. A recent publication confirmed the low implantation potential of embryos showing reverse cleavage [10].

2.5.11. Exclusion of blastomeres at the morula stage

In standard observation it is quite common to see parts of the trophectoderm that look very granular at the blastocyst stage, which usually leads to downgrading the quality of the TE. Video 2.16 shows an example in which seemingly low-quality TE is actually not related to the TE, but to blastomeres that were not integrated into the developing embryo or were actively excluded at the morula stage. During further expansion of the blastocysts, these blastomeres are compressed and usually degenerate or lyse, which leaves a granular mass between the trophectoderm

and the zona pellucida. The impaired visibility of the TE cells in this area is the reason for the impression of a low-quality TE.

The exclusion of such blastomeres is currently being discussed as a potential way for the embryo to get rid of eventually aneuploid or abnormal cells.

2.5.12. Blastocyst expansion with collapse

Blastocysts need to leave the zona pellucida in order to be able to implant. The hatching process in vivo and in vitro is different. Whereas in vivo uterine enzymes support the dissolution of the zona and facilitate the escape from the zona, the in vitro situation is different. Usually the blastocyst needs to expand, which causes a compression and visual thinning of the zona pellucida. Eventually trophectoderm cells protrude through the zona and initiate the hatching process. Depending on the resistance and thickness of the zona pellucida as well as the number of cells of the blastocyst, the embryo undergoes several rounds of expansion, collapse, and re-expansion

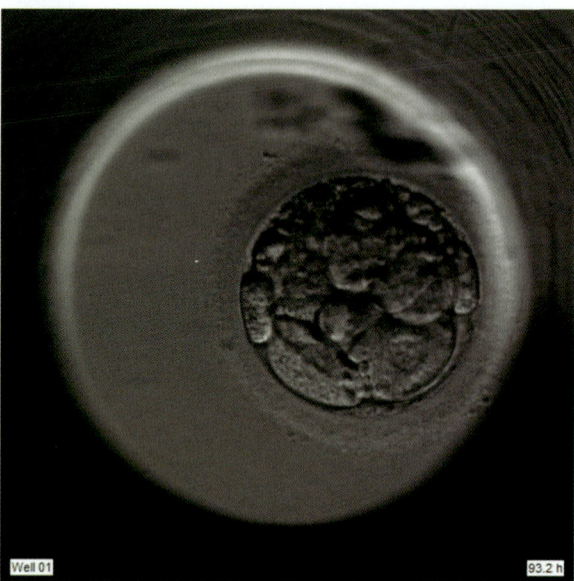

Well 01   93.2 h

Video 2.17 Blastocyst pulsing. The blastocyst in this example has only a limited number of cells. During formation of the blastocoel and the expansion period, the blastocyst starts pulsing, which is a series of collapses followed by re-expansion. It has been discussed that extensive pulsing results in embryos with a lower implantation potential.

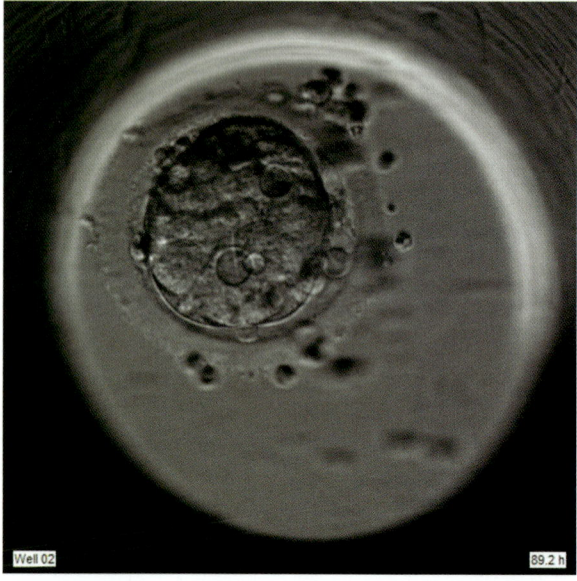

Well 02   89.2 h

Video 2.19 Normal hatching blastocyst. This video shows a blastocyst that starts hatching at the 5 pm position by extrusion of few cells of the trophectoderm through the zona pellucida. Note that the zona pellucida is compressed during the expansion process due to the resistance of the zona structure. Hatching only starts at a certain degree of compression.

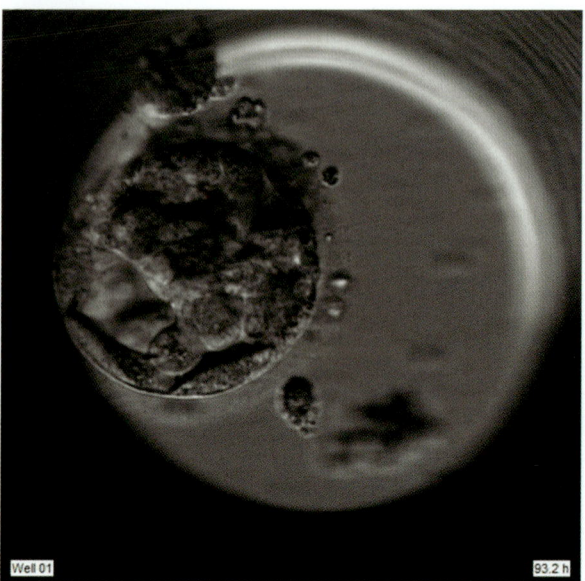

Well 01   93.2 h

Video 2.18 Strings in blastocyst. During the expansion this blastocyst seems to have the inner cell mass still connected on both sides to trophectoderm cells. The ICM eventually moves towards the left side, leaving a cytoplasmic string/connection to the other side in place. One can also spot that the blastocyst collapses shortly and that only after the following re-expansion is the cytoplasmic string withdrawn.

(Video 2.17). The presence of several cycles of pulsing has been reported to result in or be associated with a lower implantation potential [11, 12].

2.5.13. Cytoplasmic strings: a new way to observe this feature

The presence of cytoplasmic strings in blastocysts has been long recognized. It has been suggested that the presence of such strings during blastocyst collapse and re-expansion may lead to an indirect splitting of the ICM. This can occur when ICM cells are attached to strings and are pulled to the other side during pulsing or during blastocyst expansion (Video 2.18). Such blastocysts are prone to generate monozygotic twins after implantation.

2.5.14. Zona pellucida characteristics and the hatching process

Normal hatching is a process that is initiated in vitro by the trophectoderm cells (Video 2.19).

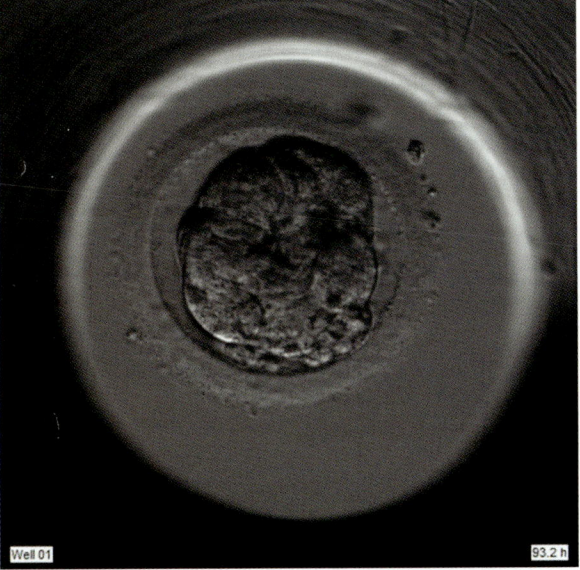

Video 2.20 Blastocyst hatching after polar body biopsy. The embryo in this video was subjected to polar body biopsy using a laser that left a permanent slit behind in the zona pellucida. During the expansion of the blastocyst the zona thus has no resistance and the hatching process starts as soon as trophectoderm cells are pushed into the laser-created opening in the zona. Note that the thickness of the zona remains almost unchanged.

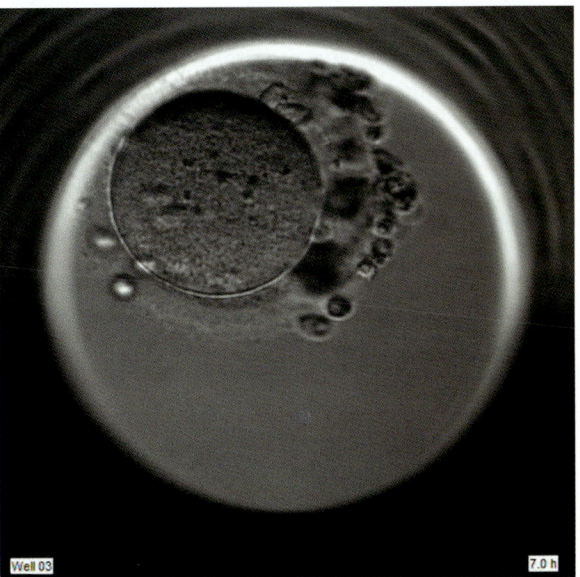

Video 2.21 Blastocyst hatching after partial zona thinning. The embryo in this video has been subjected to partial zona thinning with a laser at the eight-cell stage. The thin part of the zona can be seen at the lower left side of the embryo. During the expansion process the embryo needs less force to induce a rupture of the zona and hatching starts over a larger area and with less resistance by the zona pellucida.

In the case that the zona pellucida has a permanent opening, e.g., due to a laser-induced opening for biopsy or due to complete assisted hatching, the blastocyst starts hatching in parallel to the expansion process (Video 2.20). This is mainly due to the fact that the resistance, which is normally imposed by the zona pellucida, is absent.

In contrast to complete zona openings, weaker parts of the zona allow for a controlled initiation of hatching (Video 2.21). Such areas are created by zona thinning and mark the hatching side.

2.6. Benefits of using the EmbryoScope® time-lapse system

The public discussions about time-lapse imaging are very much focused on clinical benefits. However, an integrated time-lapse imaging system such as the EmbryoScope will most likely change the routine practice in the laboratory and result in a more flexible workflow.

2.6.1. Benefits for the laboratory

A continuous undisturbed culture condition without the need to remove a dish from the incubator can be considered as an optimal way of culturing human embryos. Less physical handling of culture dishes clearly reduces the overall risks that are associated with such an intervention such as mixing up patient embryos or dropping a dish.

One of the key benefits of using the EmbryoScope time-lapse system is the high degree of user flexibility offered by the system. Using the system implies that no analytical process (patient registrations, annotation, evaluation, instrument readings, etc.) is required at specific time-points. Instead, such processes can be scheduled at a convenient time for the laboratory because all data are stored by the system for continuous accessibility.

2.6.1.1. Documentation support

The EmbryoScope time-lapse system offers a range of features to ease the clinic administration and monitoring. Reports containing patient information, laboratory data, instrument running data (for example incubation conditions), embryo annotation, and evaluation details can be generated from the Embryo-Viewer workstation for monitoring purposes or to share within the clinic or with patients. These reports cover some of the documentation that is required by the European Tissue Directive. Also, specific images or video sequences can be generated and saved in file-format for internal or distribution purposes.

2.6.1.2. Decision support

Embryos can be annotated on a daily basis or on the day of transfer depending on the preferred approach in the clinic. Embryos that are morphologically abnormal can be marked in the "view slide" overview to avoid unnecessary annotations of obviously unviable embryos. Following this procedure will decrease the time spent on evaluating each embryo, which can be carried out at the most convenient time. In addition, the possibility to discuss the appearance of an embryo among several embryologists within one laboratory enables the training and education of staff. New or unknown behavior of embryo development can easily be shown and discussed with colleagues in other laboratories.

To ease the decision process of which embryos to transfer or freeze, Vitrolife A/S has developed the KID-Score™ decision support tool. The KIDScore D3 Basic is a safe, robust, and easy-to-use model, which will provide the immediate benefit of using time-lapse for embryo evaluation because it will help reduce the number of embryos a clinic needs to consider for transfer or freezing. The model is based on morphokinetic traits associated with implantation from a very large data set and is validated for day 3 transfers in a wide range of clinics. To use the model as intended, it is necessary to annotate a fixed set of seven variables upon which the KIDScore D3 Basic can be applied in Compare & Select feature of the EmbryoViewer. When applied, KID-Score D3 Basic will provide a score reflecting the implantation potential of each embryo for a specific patient. KIDScore is based on statistical probabilities of implantation and specifically designed to support the avoidance of embryos with low implantation potential.

2.6.2. Benefits for patient communication

The EmbryoScope time-lapse system includes several features for enhanced patient communication. The EmbryoScope Counseling App has been described above as a patient-oriented communication tool, which facilitates improved possibilities for patient comprehension of the background behind embryo selection and embryo morphokinetics and also includes the option to show the patients their own embryos by a video playback.

Furthermore, the EmbryoViewer software includes the function of saving specific images or whole videos of embryos during development while cultured in the EmbryoScope time-lapse incubator. Such videos and images can be shown to patients or given to them.

The EmbryoViewer software also provides the option to generate a variety of reports, one of which is intended for the patient.

2.6.3. Clinical benefit

The EmbryoScope time-lapse system is a widely documented time-lapse system in the international literature and is backed by peer-reviewed articles in the most prominent journals in the field. Since its introduction in 2009 into clinical practice, numerous articles have focused on the application of the EmbryoScope time-lapse system in the IVF laboratory in regard to basic embryology, research, and improvement of clinical outcome.

The immediate clinical benefit becomes obvious when one considers that certain characteristics like direct cleavage and reverse cleavage, which are both reported to be associated with low implantation rates [8–10], are mostly overlooked in standard observation. Being able to identify these events, and exclude these embryos from the primary choice for transfer, will already result in a better outcome.

The only randomized controlled trial that has been reported so far on time-lapse was conducted using the EmbryoScope time-lapse system. The base of this RCT was an algorithm which identified implantation-competent embryos for transfer than standard morphology for cycles that were incubated only in the EmbryoScope time-lapse system [13]. This algorithm was first tested in a retrospective study and resulted in higher implantation and pregnancy rates in combination with the EmbryoScope time-lapse system compared to standard incubation and morphology selection [14]. Based on this, a prospective randomized controlled trial was performed that confirmed the results of the previous retrospective study [15]. In addition to the higher outcome results, it was shown that the early pregnancy loss rate was significantly lower after incubation in the EmbryoScope time-lapse system compared to standard incubation. Similar findings have been reported by a center in the UK [16].

Another recent study showed that the presence of multinucleation at the two-cell stage could have a negative effect on implantation and pregnancy rates [17]. As multinucleation can only be spotted within a certain period of the cell cycle, it is mandatory to identify the cell cycle of a particular cleavage stage, which can only be done by time-lapse imaging with good image quality.

2.7. Summary and conclusions

The EmbryoScope time-lapse system is a benchtop-like integrated incubator with a built in camera for time-lapse imaging that offers a platform solution for the routine embryology laboratory. Besides the EmbryoViewer software the ES server allows for new opportunities to access data including use of an app for patient counseling. The system has been integrated in the routine of several hundreds of clinics around the globe and has substantially impacted the current discussion on embryo analysis using time-lapse imaging in human assisted reproduction. Direct as well as indirect evidence for the clinical benefit of time-lapse imaging using the EmbryoScope time-lapse system has been shown in a number of peer-reviewed publications and conference abstracts. The impact of time-lapse in IVF is of obvious benefit and most of the clinics who have implemented this fascinating technology strongly believe that standard incubation and daily observation at limited time-points will be obsolete in a couple of years.

References

1. Ramsing NB, Callesen H. Detecting timing and duration of cell divisions by automatic image analysis may improve selection of viable embryos. *Fertility and Sterility*. 2006;86(3):S189.

2. Li R, Pedersen KS, Liu Y, Pedersen HS, Laegdsmand M, Rickelt LF, et al. Effect of red light on the development and quality of mammalian embryos. *Journal of Assisted Reproduction and Genetics*. 2014. Epub 2014/05/24.

3. Ciray HN, Aksay T, Goktas C, Ozturk B, Bahceci M. Time-lapse evaluation of human embryo development in single versus sequential culture media: a sibling oocyte study. *Assisted Reproduction and Genetics*. 2012;29(9):891–900.

4. Ciray HN, Campbell A, Agerholm IE, Aguilar J, Chamayou S, Esbert M, et al. Proposed guidelines on the nomenclature and annotation of dynamic human embryo monitoring by a time-lapse user group. *Human Reproduction*. 2014. Epub 2014/10/26.

5. Alpha/ESHRE. The Istanbul consensus workshop on embryo assessment: proceedings of an expert meeting. *Human Reproduction*. 2011;26(6): 1270–83.

6. Cetinkaya M, Pirkevi C, Yelke H, Colakoglu YK, Atayurt Z, Kahraman S. Relative kinetic expressions defining cleavage synchronicity are better predictors of blastocyst formation and quality than absolute time points. *Journal of Assisted Reproduction and Genetics*. 2015;32(1):27–35.

7. Aguilar J, Motato Y, Escriba MJ, Ojeda M, Munoz E, Meseguer M. The human first cell cycle: impact on implantation. *Reproductive Biomedicine Online*. 2014;28(4): 475-84. Epub 2014/03/04.

8. Rubio I, Kuhlmann R, Agerholm I, Kirk J, Herrero J, Escriba MJ, et al. Limited implantation success of direct-cleaved human zygotes: a time-lapse study. *Fertility and Sterility*. 2012;98(6):1458–63. Epub 2012/08/29.

9. Zaninovic N, Ye Z, Zhan Q, Clarke R, Rosenwaks Z. Cell stage onsets, embryo developmental potential and chromosomal abnormalities in embryos exhibiting direct unequal cleavages (DUCs). *Fertility and Sterility*. 2013;100(3):S242.

10. Liu Y, Chapple V, Roberts P, Matson P. Prevalence, consequence, and significance of reverse cleavage by human embryos viewed with the use of the Embryoscope time-lapse video system. *Fertility and Sterility*. 2014;102(5):1295–300 e2. Epub 2014/09/17.

11. Watanabe S, Kamihata M, Matsunaga R, Kuwahata A, Ochi M, Horiuchi T. Contractions during the expanded blastocyst stage decrease the success rate of frozen-thawed blastocyst transfer: time-lapse video analysis. *Fertility and Sterility*. 2013;100(3):S245.

12. Marcos Alises J, Perez S, Gumbao D, Silva MM, Meseguer M, Landeras Gutierrez J. Contraction of blastocyst is strongly related with lower implantation success: a time-lapse study. *Fertility and Sterility*. 2013;100(3):S501.

13. Meseguer M, Herrero J, Tejera A, Hilligsoe KM, Ramsing NB, Remohi J. The use of morphokinetics as a predictor of embryo implantation. *Human Reproduction*. 2011;26(10):2658–71. Epub 2011/08/11.

14. Meseguer M, Rubio I, Cruz M, Basile N, Marcos J, Requena A.

Embryo incubation and selection in a time-lapse monitoring system improves pregnancy outcome compared with a standard incubator: a retrospective cohort study. *Fertility and Sterility.* 2012;98(6):1481–9 e10.

15. Rubio I, Galan A, Larreategui Z, Ayerdi F, Bellver J, Herrero J, et al. Clinical validation of embryo culture and selection by morphokinetic analysis: a randomized, controlled trial of the EmbryoScope. *Fertility and Sterility.* 2014;102(5): 1287–94.

16. Barrie A, Schnauffer K, Kingsland C, Troup S. Treatment outcome and early pregnancy loss: a comparison of conventional and EmbryoScope® systems. *Fertility and Sterility.* 2013;100(3):S248.

17. Ergin EG, Caliskan E, Yalcinkaya E, Oztel Z, Cokelez K, Ozay A, et al. Frequency of embryo multinucleation detected by time-lapse system and its impact on pregnancy outcome. *Fertility and Sterility.* 2014;102(4): 1029–33.

Description of time-lapse systems: Primo Vision™

Csaba Pribenszky

3.1. Introduction

The Primo Vision time-lapse embryo monitoring system was originally aiming to serve our own research group's primary interest: cell stress studies, in the early 2000s. We wanted to image cells that had been exposed to stress before, without any additional stress factor. In order to achieve this, a compact, inert digital inverted microscope was designed that can be fitted into classic incubators. The microscope provides high-resolution, Hoffmann contrasted bright field images, in a way that a single frame contains all the embryos of a patient. As a consequence, neither the dish nor the embryos are moved to get into the field of view in order to avoid the physical stresses accompanying the movements, heat accumulation due to friction, and harmful effects from lubricants. Microscopes are connected to a controlling unit outside of the incubator that switches them on only for the image acquisition to avoid the negative effects of electromagnetic field. In order to identify the embryos individually, a microwell culture dish was designed that keeps the embryos in individual microwells that are close to each other to harvest the group effect as well, thus reaching an enhanced embryo quality during in vitro development. The system is modular in a way that a controlling unit can handle up to six microscopes independently. Connecting more controlling units into the network can further expand the system, providing the possibility of starting time-lapse monitoring only for a selected group of patients then proceeding gradually to being able to offer time-lapse follow-up for all patients. The software provides a practical interface to easily start time-lapse projects then to review, annotate, and analyze them. Classic morphology together with the timing of kinetic events are all taken into consideration in assessing embryo quality.

3.1.1. What does not kill me will make me stronger

Why did we need a stress-free, completely physiological embryo monitoring system? It is the thoughts of Friedrich Nietzsche, quoted from *Twilight of the Idols* (or *How One Philosophizes with a Hammer*), that would be the best quotation for our initial studies aiming to improve cells' resistance, viability, and developmental competence in various ART procedures by a "stress treatment." The phenomenon that a sublethal stress induces a response with a temporary increase in a general, rather non-specific resistance to various further stresses, has been observed in almost all levels of life, from bacteria to multicellular organisms including humans. The first reaction to stress in humans or animals is the "fight-or-flight" or the "acute stress response," described by Cannon [1]. This response was later recognized as the first stage of a general adaptation syndrome (GAS) that regulates stress responses among vertebrates and other organisms [2]. On the cellular level the stress reaction incorporates sensing, assessing, and then counteracting stress-induced damage, consequently temporarily increasing tolerance to such damage [3].

Our concept was to utilize controlled environmental impact as a treatment for cells and tissues to condition, to improve the cells themselves [4]. The emphasis is on the definition "controlled." For an example, although uncontrolled sheer stress during moving embryo culture dishes or vigorous pipetting may harm cells, controlled mechanical stress applied at the right

Primo Vision EVO⁺

Video 3.1 Summary of embryo kinetics as seen on time-lapse.

time of embryo development might precondition cells, enabling them to perform better [5, 6].

A series of procedures and devices have been developed to train cells so that they perform better in forthcoming processes, such as cryopreservation, in vitro culture, in vitro maturation or enucleation, and somatic cell nuclear transfer (for reviews see [4, 7, 8]).

But how did these stress studies come into embryo monitoring?

Testing the viability of porcine embryos after oocyte enucleation and somatic cell nuclear transfer (SCNT) during their cleavage and differentiation without impairing their development in in vitro culture was a difficult task. Moreover, the procedure was involving the stress pre-conditioning of the recipient oocytes before SCNT [9]. So either we were waiting for 7 to 8 days at every experimental round to see the outcome of the oocyte treatment or we find a way to peek into the incubator, without causing any further stress for the developing embryos. Moreover, we also wanted to see the effect of the controlled stress treatment that we utilized for the oocytes without adding any further stresses. This was the need that planted the seed of a unique miniaturized inverted microscope, that could be placed inside of an existing incubator and that does imaging with frequent intervals

(every 5 to 10 minutes) during in vitro culture of extremely sensitive embryos without adding new stress factors. By saying no stress we meant that:

- embryos or culture dishes should not be moved during culture to avoid shear stress or heat accumulation,
- there should not be any electric current, even very low voltage, to avoid the possible negative effects of the electromagnetic field,
- light intensity and wavelength should be optimized to avoid photo-toxicity,
- inert materials are to be used to cut off volatile organic compounds (VOC).

Furthermore, culturing embryos in groups was an important need as group culture gives clear benefits over single embryo culture in animal systems we used at that time, such as porcine, bovine or murine [24, 25, 26].

3.2. The concept of a Primo Vision embryo monitoring system: Primum nil nocere (First, do no harm!)

The idea to apply time-lapse techniques in mammalian embryology dates back to the beginning of the last century with the prestigious publication of Lewis and Gregory in *Science* in 1929 [10], stating that differences in the time of division of rabbit eggs may significantly indicate differences in the cells. Many time-lapse related studies have been conducted since this early publication. All of the early studies have used individually designed systems; commercial production of various time-lapse equipment has only started recently. However, the principle is the same i.e., either constructing an incubator over the stage of a commercially available inverted microscope, or putting a modified microscope into a commercially available incubator.

There were several approaches for time-lapse imaging of in vitro developing embryos by the time when the first and second versions of Primo Vision time-lapse embryo monitoring systems were launched. As of 2008, devices with the following concepts were available for research use in embryology:

- Classic inverted microscope equipped with a plastic environmental chamber.
- Small environmental chamber (stage-top incubator) that is placed onto the microscopic stage of an inverted or stereomicroscope.
- Integrated systems were also available, in several embodiments:
  - A robotic carousel sample holder, integrated into a regular cell-culture incubator, rotating the Petri dishes into the field of view of a regular inverted microscope that is placed into the incubator.
  - An inverted microscope is set up in a regular incubator, and is fed by a robotic arm that takes the samples (culture dishes) from the shelf-system in the inner side of the incubator.

None of the above mentioned approaches satisfied our need for a stress-free environment during monitoring, as potential sources of damage using the discussed principles may impair severely embryo development:

- *Suboptimal culture conditions.* In classic environmental chambers and stage-top incubators suboptimal distribution of heat, light from laboratory neon, suboptimal gas, and humidity control can hamper embryo development.
- *Frequent exposure to visible light during imaging.* Light has been published as an important stress factor that may affect embryo development

negatively. High intensity of light (e.g., that of the sun) can directly kill embryos, while wavelengths in the blue zone or below also have a direct detrimental effect on embryo development.

- *Uncontrolled physical, chemical, and biological damage* related to the applied instruments. Automatic moving and focusing systems have been installed in both incubators and inverted microscopes. However, these systems require complicated mechanical arrangements and electronic regulation. Additional heat from motors, electromagnetic field, volatile compounds of lubricants, seals, and other applied materials may all cause detectable or hidden but existing damage to embryos.
- *Sterilization* and especially maintenance of sterility of these instruments is difficult or impossible, thus increasing the risk of microbiological growth, which in turn may hamper further investigations for longer periods.
- *Shear stress.* The required constant movement of dishes, and the consequent shear stress, may compromise the development [5].
- *Potential risk of malfunction or breakdown.* Optimally, routine application of real-time monitoring systems for human embryos would need the unachievable 100% stability or warranty that an accident occurring in the mechanical or electronic part does not cause any harm in the development of the observed embryos. The probability of such accidents increases in parallel with the complexity of instruments, number of cable contacts, motors, and moving parts.

According to our approach, a real-time light microscopic system to monitor pre-implantation stage embryo development (especially of the human embryo) should meet the following criteria:

- Monitoring should be free of any measurable or potential harmful elements that may impair embryonic development compared to routine embryo culture procedures. Moreover, real-time monitoring should even minimize the detrimental effects of routine morphological investigations of the present-day practice.
- Provide an optically satisfactory picture suitable for safe diagnosis of key milestones and important features of pre-implantation development.
- Provide pictures at any time point about the actual state of all embryos and all previous events

33

without disturbing the safe incubation and the monitoring process.

- Enable individual identification of embryos, while monitoring in parallel at least nine to 16 embryos.
- Take pictures with a reasonable frequency (5 to 10 minutes) during the whole pre-implantation period (0 to 6 days).
- The monitoring system should enable medium changes and safe repositioning of the Petri dish without mixing up embryos. The removal and replacement of dishes as well as the medium change should not be more time-consuming than the present routine in ART laboratories.
- Assemble pictures into time-lapse videos, enable morphometric analysis of individual pictures, and archive files with the documentation of patients.
- The software should support the user to easily and straightforwardly make decisions about embryo quality.

The first point, the "nil nocere" (do no harm) principle, is the basis of all medical interventions and shall also be a key condition when new methods, such as continuous embryo monitoring, are introduced [11].

3.3. Description of the Primo Vision Embryo Monitoring System

The Primo Vision Embryo Monitoring System is an automated time-lapse equipment for in vitro monitoring of embryos developing inside of the incubator. The device is used for presenting real-time images of the embryos developing inside of the incubator, displaying a video of the developmental history of the embryos, providing support for embryo evaluation, and digitally archiving the images of embryo development for quality control and quality assurance purposes.

By this system, embryos can stay in the incubator completely undisturbed, but still under the most thorough control during the whole period of in vitro development, providing a maximum amount of information, in order to achieve optimal embryo selection while accurately recording all moments of embryo development.

The Primo Vision Embryo Monitoring System includes:

- An inert, compact, sealed, digital inverted microscope (Primo Vision Digital Microscope)

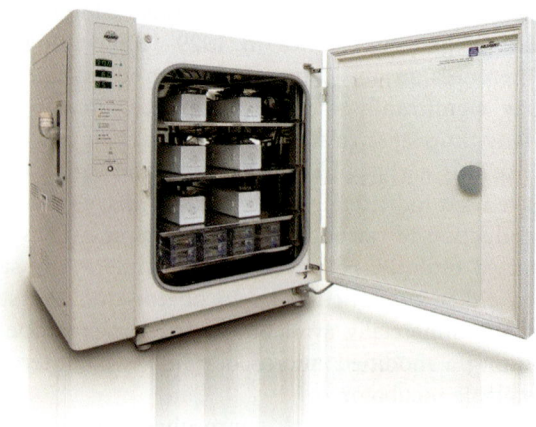

Figure 3.1 Primo Vision microscopes placed into an incubator, with Primo Vision microwell group culture dishes on their sample holders.

that is to be installed inside of any regular incubator.

- A controlling unit that controls the microscope(s) and runs the Primo Vision software. The controlling unit is located outside of the incubator and is connected to the microscope(s) with special USB cable(s) through the factory-made access port on the side or back of the incubator. One controlling unit is able to control and follow-up up to six microscopes independently.
- Microwell group culture dishes (Primo Vision Embryo Culture Dish) that hold up to nine or 16 embryos individually in microwells for identification in a setup that supports paracrine and autocrine communication between the embryos.

During the procedure, the custom-made Petri dish, with individually identified microwells for each embryo, is placed into the incubator in the dish holder of the Primo Vision Digital Microscope. The microscope takes images of the embryos during their in vitro development by user-defined frequency. All the embryos are seen on the image at the same time. Focusing, initiating the time-lapse project, and imaging in multiple focal planes is simply executed from outside the incubator, using the software running on the external controlling unit. The photos taken by the microscopes are presented on the screen connected to the controlling unit and are saved in a dedicated database where they are used for:

1. Supporting decision on embryo quality,
2. Creating different types of time-lapse movies, reports of the developing embryos,
3. Specific and precise manual analysis of embryo development using both morphometric and kinetic markers (detection of cleavage times, multinucleation, vacuolization, fragmentation, ploidy, PB extrusion, classic morphology, etc.).

3.3.1. The Primo Vision Microscope

The Primo Vision Microscope is a special, compact, airtight, inert, digital inverted microscope, designed for safe and comfortable in-incubator use. External dimensions are $24 \times 8 \times 12$ cm, weight is ~ 2 kg. The microscope's field of view is ~ 3×2 mm and fits all the embryos of a patient. The optics itself is a proprietary, custom-made, high-resolution system with Hoffman modulation contrast. The microscope's housing incorporates the custom-made, high-precision optical system, a camera and fine mechanics for focusing and imaging in multiple focal planes. Being an inverted setup, the lamp console is fixed to the top of the microscope's case above the dish holder. On the back panel of the housing, a sealed mini connector is found that connects the microscope to the controlling unit outside of the incubator through a USB cable.

The lamp of the Primo Vision Microscope is designed to minimize the light intensity, limited to a safe wavelength. The optical system, the LED, and the CCD are all designed and harmonized to provide the maximal resolution at the field of view of the microscope. The approximate light intensity in the time-lapse imaging system is 6 μW/cm^2 compared with the 80 mW/cm^2 measured in a general inverted microscope, so even with imaging every 10 min, the total light energy to which the embryos are exposed for 5 days is approximately 10% of that which an embryo would be exposed to during a single routine daily check. Other potentially harmful factors, including the light of the laboratory itself, are also eliminated with the time-lapse follow-up.

Microscopes are controlled by the external controlling unit, which connects the microscope only for milliseconds to capture images, or for a maximum of 2 minutes when the live mode is active. The microscope is powered through the provided USB cable connected to the mini plug on the back panel of the microscope. Six of these microscopes can be connected to a controlling unit that can further be connected to each other, creating a network.

Figure 3.2 The Primo Vision Microscope, a compact, sealed, inert, inverted digital microscope to be used inside of incubators with either a humid or dry environment.

The optical system of the microscope inside of the housing is capable of moving in ~ 1 μm steps, providing a possibility of fine imaging in multiple focal planes. These focal planes/layers can then be visualized in the viewer/analyzer software.

The quality of the optics makes it possible to recognize morphokinetic events as well as qualifying classic morphology of the embryos at different stages of development, without the need of taking them out from the incubator.

As an example, Figure 3.3 shows the scan layers of an expanded blastocyst. The proper qualification of the inner cell mass (ICM) and the trophectoderm (TE) cells is routinely possible.

3.3.2. The Primo Vision Embryo Culture Dish

The Primo Vision Time-Lapse Embryo Monitoring System uses a microwell group culture dish developed for two reasons:

a. to provide a tool that ensures accurate individual identification of each embryo growing in a group and
b. to provide a culture environment that ensures enhanced embryo development based on better exploitation of autocrine and paracrine factors.

Both the volume and morphology of the microwells and the microwell-to-microwell distance are crucial in achieving improved embryo development, and are utilized in the Primo Vision dishes.

(a)

(b)

(c)

(d)

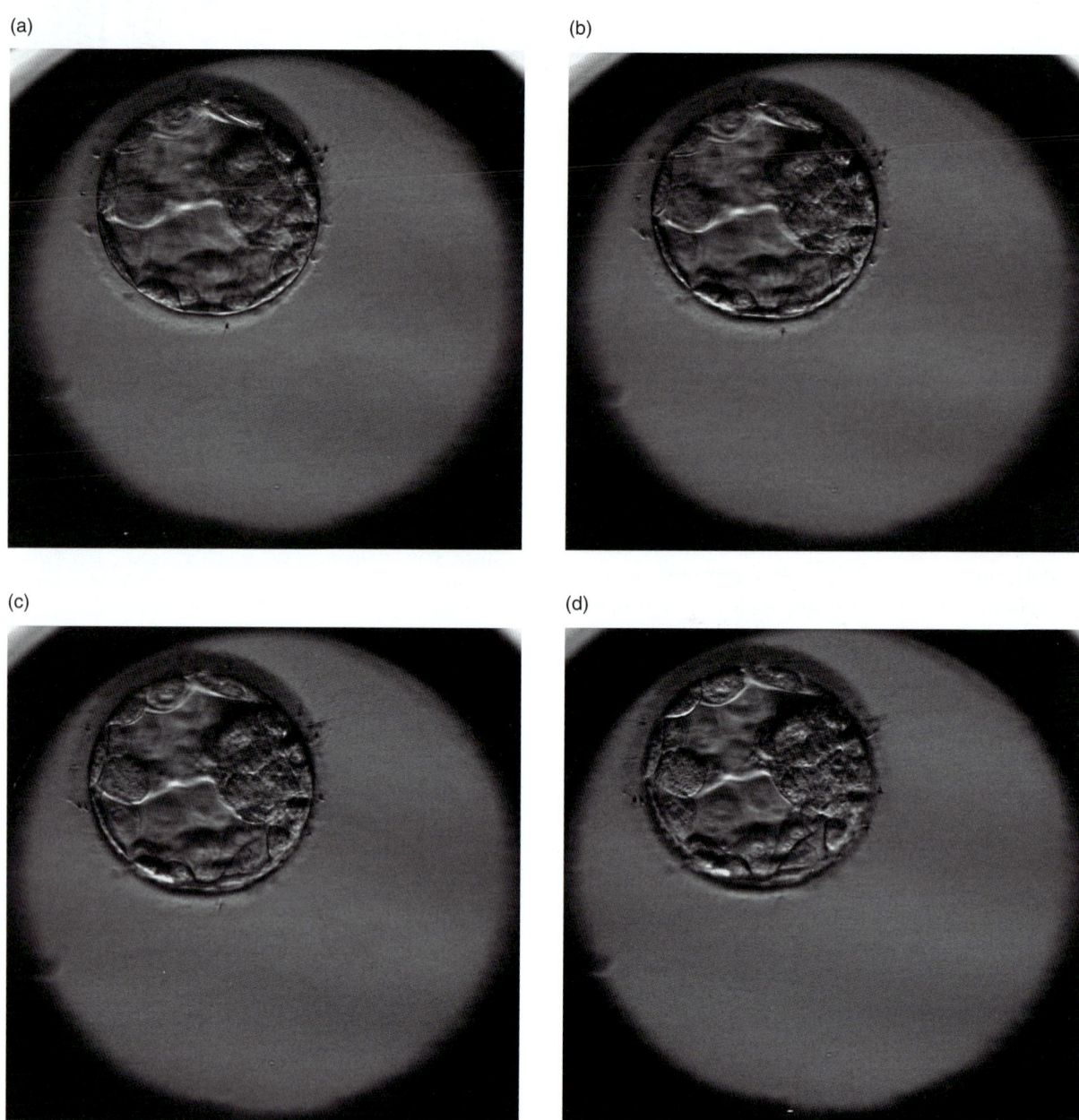

Figure 3.3 a–h Image quality as seen on an expanded blastocyst imaged by the Primo Vision Microscope, in multiple focal planes. Note the nuclei in the trophectoderm cells. Images were made in 10 μm increments in nine focal planes.

The arrangement of the microwells allows easy adjusting, tracking, and identifying of the embryos and also confers improved culture conditions, as described before [12].

The microwells form a matrix of four rows with four wells each (or 3 × 3 in other dish types). The rows are identified by numbers from 1 to 4 and the columns by letters from A to D (Figure 3.4). The positioning of the Primo Vision Embryo Culture Dish into the dish holder of the microscope is facilitated by

an incision on the edge of the dish. The letters and numbers together with the incision on the edge of the dish ensure the reliable identification of the embryos either under the stereomicroscope or in the microscope unit of the Primo Vision System.

The microwell culture system is based on a method used for short-term co-culture of zona-free mouse embryos to produce chimerae [13, 14]. A further, early purpose of the culture system was to establish a system for culturing zona-free embryos

(e)

(f)

(g)

(h)

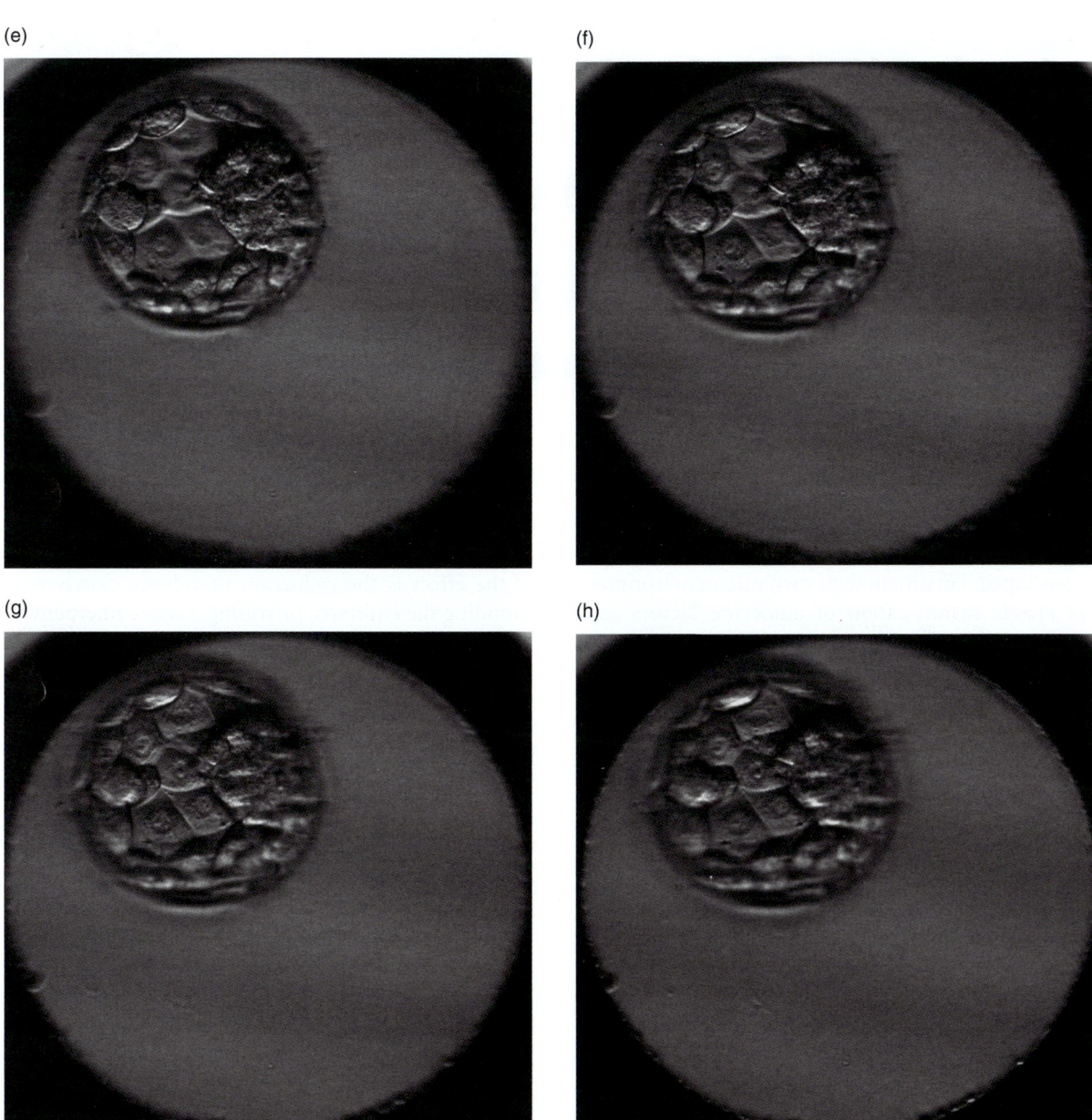

Figure 3.3 a–h (*cont.*)

[15, 16]. The microwells were used to prevent both aggregation of multiple zona-free embryos and dissociation of single embryos to individual blastomeres before compaction. The system has proved its value for culture of cattle, sheep, pig, and mouse zona-free somatic cell cloned embryos by improving in vitro developmental rates and resulting in full term development after transfer to recipients [17, 18, 19, 20, 21, 22]. A similar system was used to support the cleavage of biopsied blastomeres from human eight-cell stage embryos to obtain embryonic stem cells [23].

It has been hypothesized that the microwell group culture system provides a suitable macroenvironment (nutritional support) and microenvironment (accumulation of autocrine and/or paracrine factors from the embryos themselves) for the embryo, resulting in improved development to the blastocyst

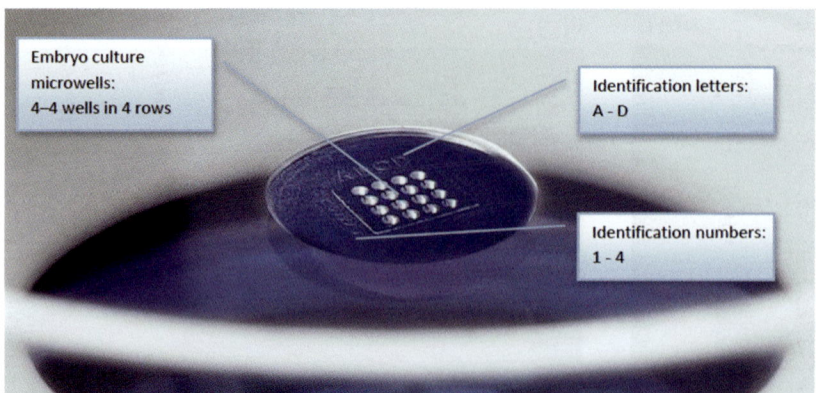

Embryo culture
microwells:
4–4 wells in 4 rows

Identification letters:
A - D

Identification numbers:
1 - 4

Figure 3.4 The microwell matrix of a 16-well Primo Vision dish.

stage [24, 25]. The microwell culture system has been shown to be also useful for counterbalancing the missing group effect in individually cultured zona-intact bovine embryos [24]. This effect may be explained by the potential that allows the embryos to develop and maintain their own microenvironment and enable manifestation of autocrine factors supporting embryo development. In addition, the partial openness of the microwells allows the required nutrient supplementation and dilution of potential toxic factors, including ammonia and free radicals [12].

A recent study using bovine inseminated oocytes showed a clear benefit of the Primo Vision dish over individual culture or classical group culture. Blastocyst development in Primo Vision dishes was similar to classical group culture, and better than Corral dishes or individual culture. In Primo Vision dishes, a higher number of "slow" embryos developed to the blastocyst stage compared with their individually cultured counterparts, while no differences were observed for "fast" embryos. "Slow" embryos in a "standard drop" had a higher chance of becoming a blastocyst compared with individual culture (OR: 2.3), whereas blastulation of "fast" embryos was less efficient in a "delayed drop" than in individual culture (OR: 0.3). The number of non-cleaved embryos in Primo Vision dishes did not negatively influence blastocyst development. Likewise, removing non-cleaved embryos and re-grouping the cleaved embryos afterwards did not affect blastocyst development and quality compared with group culture in Primo Vision dishes. The experiments revealed that group culture of bovine embryos in Primo Vision dishes is superior to individual culture, primarily because of the higher blastocyst rate achieved by slow embryos. Non-cleaved or arrested embryos do not hamper the ability

of co-cultured bovine embryos to reach the blastocyst stage in group culture [26].

Further investigations are required to reveal the precise mechanism of the apparently improved developmental competence. One possible explanation of the effect is the reduction in volume closely surrounding the embryos, providing a stable microenvironment and allowing manifestation of autocrine effects.

3.3.2.1. Using the microwell culture dish in everyday practice

The Primo Vision Embryo Culture Dish is to be loaded with culture media on the day prior to starting the time-lapse sequence, to let the media equilibrate. The dish is recommended to be used with ~ 80 μl embryo culture media, overlaid by 2.5–3 ml of paraffin oil. The dome of the drop should be as flat as possible. The rim around the microwells does not mark the edge of the droplet, but helps to stabilize the embryos sitting within the microwells.

Primo Vision offers two different embryo culture dishes, and as the size of the microwells differs a bit, the suggested method of loading also differs.

The drop then can be enlarged to the desired volume and size by putting the given amount of culture media onto the middle of the dish, over the wells, spread over the rim to make the dome flat.

After this, approximately 2.5 to 3 ml paraffin oil shall be overlaid, to cover the whole droplet. This is important, as it will not only affect image quality, but culture conditions as well. Now the dish should pass an overnight equilibration.

As evaporation occurs during preparation, the speed is important, and it is advised to use cold

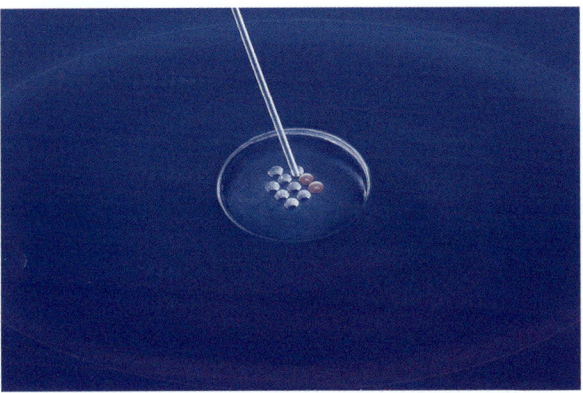

Figure 3.5 Loading the nine-well dish.

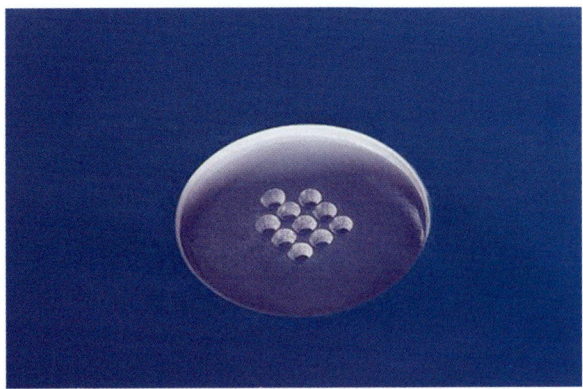

Figure 3.6 The flat dome of the culture media over the microwells.

Figure 3.7 The 16-well dish.

surfaces, cold media, and low flow in the laminar hood.

As the diameter of the microwells and the shape is different in the case of the 16-well dish, methods of dish filling and loading may differ from the nine-well dish.

3.3.3 Controlling Unit

The Primo Vision Controlling Unit functions as the central controller of the Primo Vision system: controls the connected microscopes, stores the images taken by the microscopes, and runs the provided software. A proper monitor (min. 1920 × 1080 pixels [HD] resolution) is needed for the desired operation. Inbuilt into the controlling unit, the controller card commands the microscopes, up to six, independently, at the same time.

The controller card turns off the power to the microscopes in between image acquisitions for the safety of embryos, thus electricity and light are turned on only when the microscope is activated (takes a photo or is in live mode). Accordingly, no continuous electric current or electromagnetic radiation is present around the embryos. Electromagnetic radiation has been reported to affect in vitro embryo development by altering the speed and synchrony of cleavages, gene expression, and enzymic activity of the embryos of different species and to reduce in vitro embryo survival [27, 28, 29].

3.4. The Primo Vision software

The software controls the microscopes, allows the users to set up projects: set imaging frequency, duration, imaging at multiple focal planes. It creates and manages the patient database. It also provides tools for embryo evaluation. Classic morphology can be annotated as well, together with kinetic events. It is fully customizable: patient data and annotations, both for kinetics and morphology, specific for the given clinic, can be set. Also different profiles containing different patient information, events, etc. can be set for different user levels. (e.g., the "routine profile" will be set for those embryologists who are only involved in the routine procedures in the lab; that profile will not allow any extra to be done, while at higher user levels all possibilities are open).

The software can manage up to 56 microscopes individually and can create statistical analyses of the database built up during usage.

The software is built up in a server–client architecture. Controlling units that may have up to six microscopes can be linked up to share information, and to present all microscopes on the same user interface. Consequently the controlling units have server and client software components.

The server components will start automatically when the Windows system starts and will run in the background. The tasks include handling the microscopes: switching on and off the USB ports when necessary, giving power to the microscopes, taking photos from the embryos, providing a live image to the client software, handle the different error conditions (e.g., user misuse by unplugging microscopes while projects are running). Further tasks are to run the main controlling module. It has a database, which encodes, saves, and organizes all the data about the system: data on the microscopes, the users, the projects, the patients, and the embryos themselves. This module handles the images, encodes them to the hard drive, produces video file for each embryo, or plays the time-lapse videos if necessary. A further task in the background is to keep all controlling units synchronized.

The client software: this is what the user sees; this is the surface where all controls and analysis is performed. Each controlling unit has the pre-installed client software, but this client software also can run on any other, normal Windows computer, like a Windows deskstop PC, or a laptop. It means that the user can use the system remotely, so the controlling units can run silently, behind the scenes, independently from the client PC.

The client part of the software has the user interface; the users can only work with this module. It communicates with the controlling units' server components, asking for information from them, giving commands to the microscopes, starting new projects, analyzing them, playing the video from the embryos, producing the report file, etc.

The analysis part of the software provides extra features for the manual analysis of embryo development. The images taken by the software are automatically stored and used for compiling a video suitable for processing. By the use of the program-generated time-lapse movie of the embryo development, the user can define cleavage times and the occurrence of special/suspicious/important events (uneven cleavage, fragmentation, re-absorption of cellular fragments, multinucleation, vacuolization,

etc.) together with classic morphology. The software is capable of generating graphs, creating videos, and reports, showing embryo developmental dynamics and data tables with the cleavage/event data. All these files are saved into the project's (patient's) database. The software contains reference values for selected events and durations, which are used as reference points to list the embryos in the cohort according to how far away they are from the set reference values.

The information provided by this software supports decision-making on embryo competence, and helps the user establish parameters that will aid in the selection of the embryo(s) for transfer.

3.4.1. Description of the software in more detail

Login: The user starts with entering her login name and password. Many user levels can be set starting from allowing full access to modify settings as far as visitor level. Visitors cannot put down annotations or notes. If there is a multilaboratory setup, the user with the right authority can switch among laboratories (Figure 3.8).

The home screen: Upon entering the software, the user arrives at the laboratory overview. This is basically the map of the lab, indicating the incubators of the lab and the Primo Vision microscopes within them. It also shows microscope status in brief with patient names and ID, and the actual age of the developing embryos (D0, D1, D2, D3, D4...). Microscopes can also be rearranged here within the incubators by drag and drop.

In the menu bar the user parameters can be set, one can search within projects, assign microscopes to incubators, manage and disconnect microscopes, upload projects that were not recorded using this software, run the setup in case of a superuser, and logout (Figure 3.9).

After clicking onto an incubator the user actually can enter into an incubator and check more detailed project info. The user can initiate a new time-lapse project at an available microscope, or start the analysis of a running project. The user can further proceed to the microscope view by clicking onto one of the microscopes (Figure 3.10).

Settings: The cogwheel on the home screen of the software takes us to the settings. This is the first step the user does at the first setup: specify lab needs and

Figure 3.8 Login page of the software.

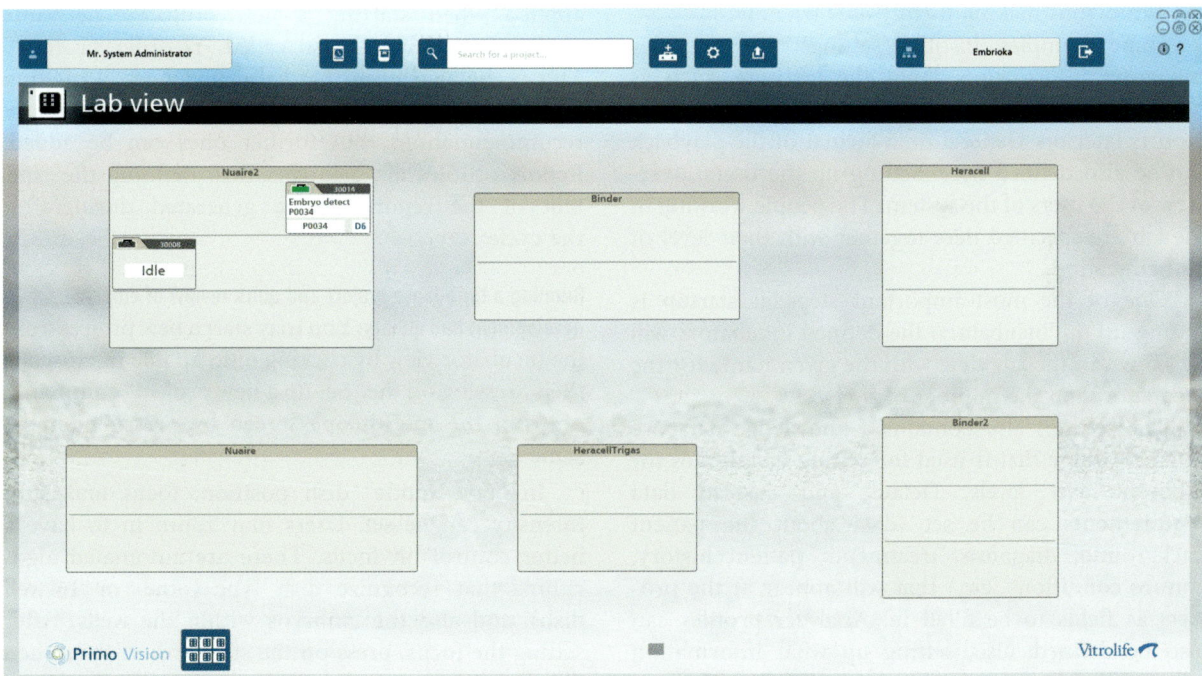

Figure 3.9 The home page, the "laboratory overview" of the software indicating the actual incubators in the lab and the position and status of the microscopes set into them.

Figure 3.10 The incubator view: shows an incubator selected in the lab view with more detailed information about its Primos.

specialties; and trim the software to the specialties of the clinic, thus making it truly user friendly!

Some examples: institute settings: the superuser can enter information about the institute, such as name, address, and logo, that will show up in the reports later on. General time format of the playback can be also defined here. Setting up the users: overview of the users of the system. The people working in the lab are specified here together with their level of authentication.

One of the most important steps at startup is setting up the incubators: the defined incubators will be shown in the lab view with the given name for the incubator, and the number of shelves.

The settings also contains a knowledge database in embryology that is used for setting up profiles for different user levels. Default and custom data requirements can be set (data about the patient background, diagnosis, treatments, patient history, culture conditions, etc.) that will appear at the projects as fields to be filled in. Analyzer profiles can also be defined; also, setting up what information will need to be annotated or scored during embryo analysis. Regular users will have access to the default profile only, while superusers and research users will

have the option of selecting the capture profile to be applied when starting a new project. The same applies for setting the so called "re-order profiles." This is the algorithm that helps to suggest a ranking among the selected embryos. This contains factory recommendations, but further ones can be added. Report profiles are also specified: defining the content of the reports to be generated during/after the cycle.

Running a time-lapse project and quick review of embryo development and quality: You may start a new project from the incubator view by clicking onto an idle microscope then pressing on the "Set up a new project" command, or from the microscope screen by pressing on the same icon.

In "Live mode" dish position, focus and light intensity can be set. Users may zoom in to have a better control on focus. There are automated algorithms that recognize dish type (nine- or 16-well dish), and also the embryos within the wells. After setting the focus, press on the start project icon, then the project info pops up (Figure 3.11).

Here the user can enter the required patient name and ID, select a "Capturing profile" (in case of

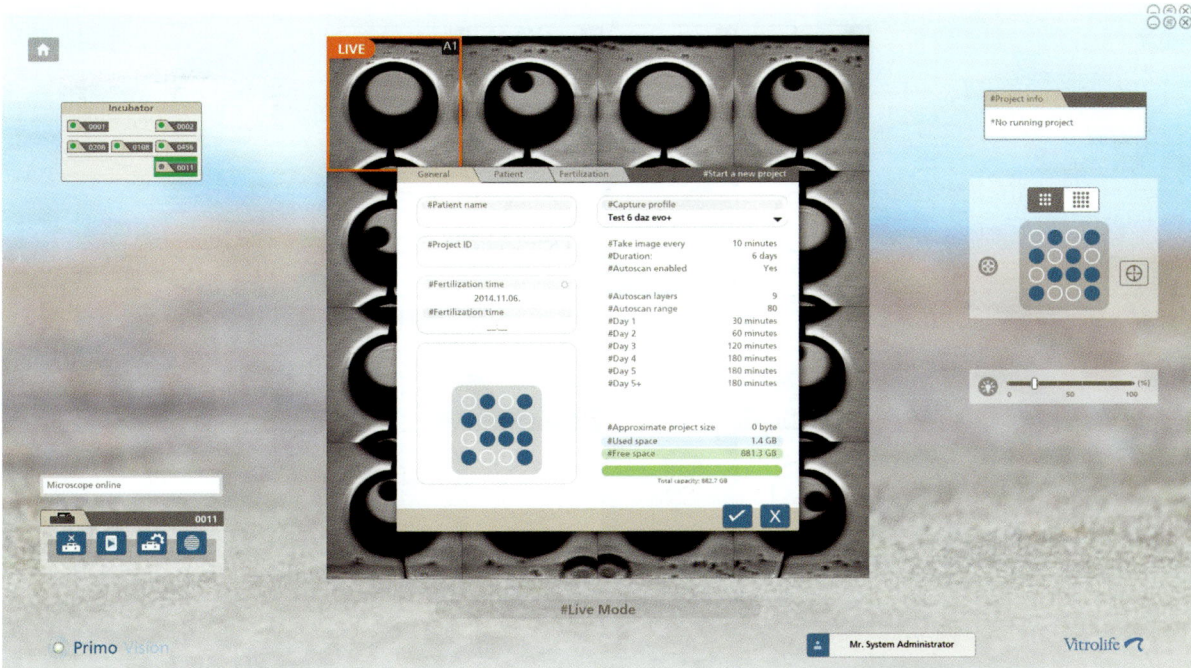

Figure 3.11 Setting up a new project.

research or a superuser). Also, basic information is presented about the project and the system. Users may start the time-lapse sequence immediately, but may also enter patient and culture data here that were previously specified at the setup as required information. This information may be entered later on as well. Further data about patient, diagnosis or treatment can also be added.

The running project screen appears after approval. In that time an algorithm finds and crops the microwells and the embryos on the images taken, and creates a compiled image that is displayed on the screen. The image sequence can then be played back, decisions can be made, and the deep analysis can be initiated from here (Figure 3.12).

The screen, on the left, also shows the actual incubator and the position of the microscope within the incubator on the shelf, information about the project on the right, and the timeline below. The timeline shows the timing for different stages for orientation, and the time elapsed from fertilization. Here you can make decisions (transfer, cryopreserve, discard, not decided) and define actions (PGD, PGS). Pausing a project for media change or stopping it is also possible here (Figure 3.12).

From here, users may enter the second level of analysis (deep analysis) by double-clicking on one of the embryos.

Analyzing a project: There are two levels of analysis. Level one is a fast but superficial one, described previously, on the "microscope view." Here the user may run the time-lapse movie back and forth, follow and compare embryo development, and mark decisions through a drop-down menu.

If the user would like to harvest all possibilities and analyze embryos based on classic morphology and kinetics as it was set in the settings, have the chance to annotate, to compare embryos, utilize algorithms in embryo evaluation, create reports of the development of embryos, create videos, provide data for statistical analysis, then double-click on one of the embryos on the overview. This action takes you to the second level of the analysis part (Figure 3.12).

The second level of analysis starts with the embryo view. Here annotations can be made that are both kinetic and morphologic. All these annotations will automatically be saved into the database of the project, from where they can be exported into an Excel macro, where all project data are compiled, giving

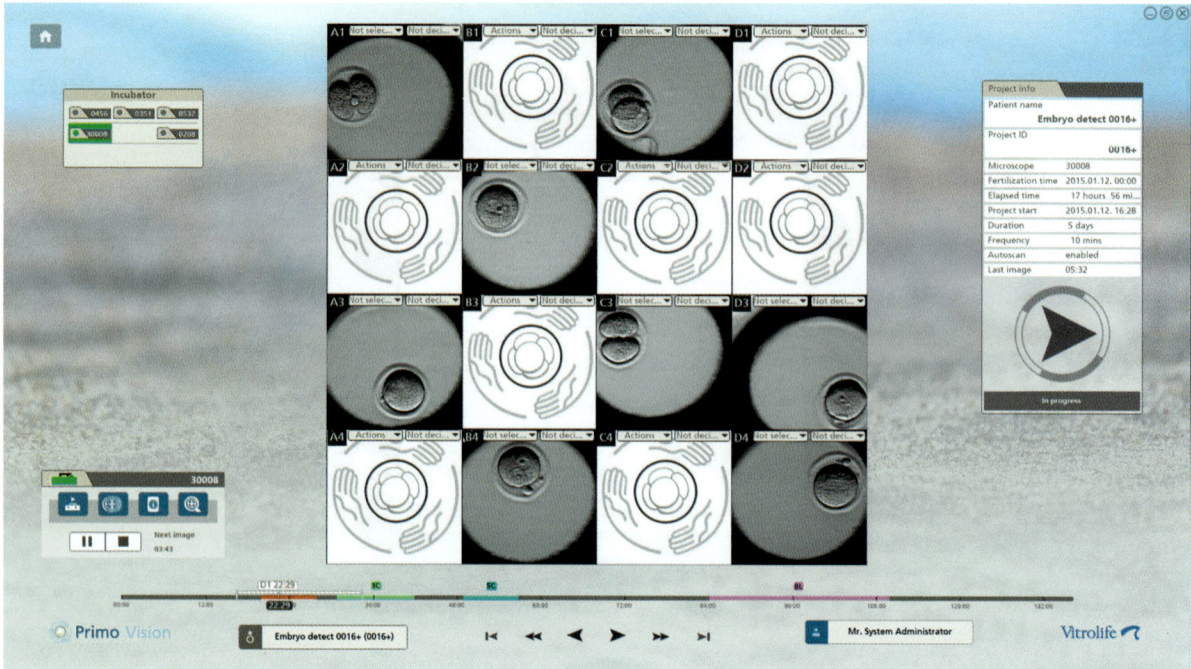

Figure 3.12 A screen with double function: real-time information about the stage of embryo development and quick diagnosis and decision-making.

options for further analysis and the possibility to feed the data into statistical software.

The timeline at the bottom shows time from fertilization, scan, and video controls. Images can be played back with scrolling with the mouse wheel, or grabbing and dragging the time marker (Figure 3.13).

The right panel of the embryo is its diary, called "Smart log." It logs all annotations and free text inputs to the actual time in the embryo's life. You may click on them, which takes you to the corresponding frame in the sequence. The annotations and comments of the different users are seen here with their login name marked (Figure 3.13).

Tools for measurements are also provided on the right panel. This is used to measure blastomere sizes and ratios, area ratios (e.g., at quantifying fragmentations), sizes, etc. These measurements are also logged in the smart log (Figure 3.14).

The actual embryos can be sent to the "Compare" window (Figure 3.16), where embryos are compared automatically based on their kinetic normality. After closing this embryo analysis window, the "working desk" appears. The left side of the desk is a graphic menu. The first panel is an "Overview" from where a

new embryo can be dragged and dropped into the middle of the desk to continue the deep analysis of the cohort.

Further embryos can be dragged and dropped to any of the four places of the desk for a quick comparison. Dragging and dropping makes analysis fast and intuitive. Having fixed spaces on the desk enables rapid workflow as there is no need for rearranging and resizing windows, there is no unnecessary moves and clicks (Figure 3.15).

The second panel is the "Compare." Embryos can be dragged and dropped here from the overview. This provides the tool to compare embryos and further support selection.

Embryos in this panel can be "re-ordered," listed according to how far away are they from the set reference values, provided that these embryos have been annotated. Note that this function takes the mathematical distance of the actual duration of events of an embryo and compares it to the corresponding values set in the reference values. Further ranking based on cleavage normality and morphology is needed for the proper evaluation and making the right decision for selection (Figure 3.16).

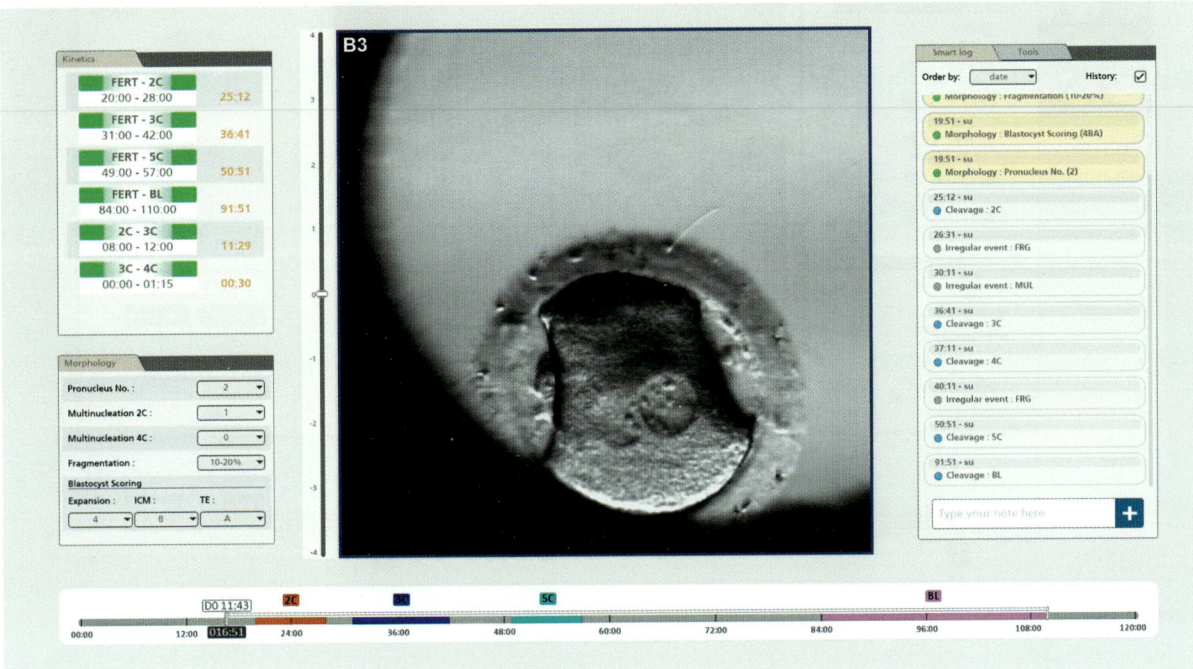

Video 3.2 Embryo development as seen in the analyzer window.

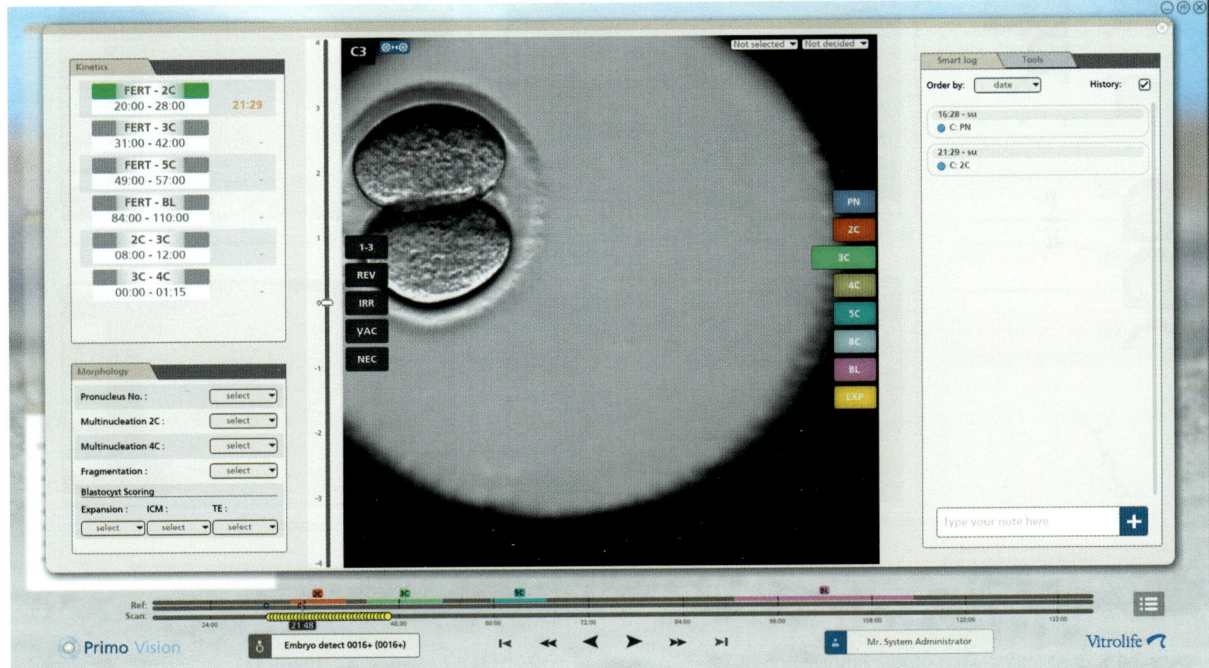

Figure 3.13 The "deep analysis" window. Embryo morphology and kinetics can be annotated simply here.

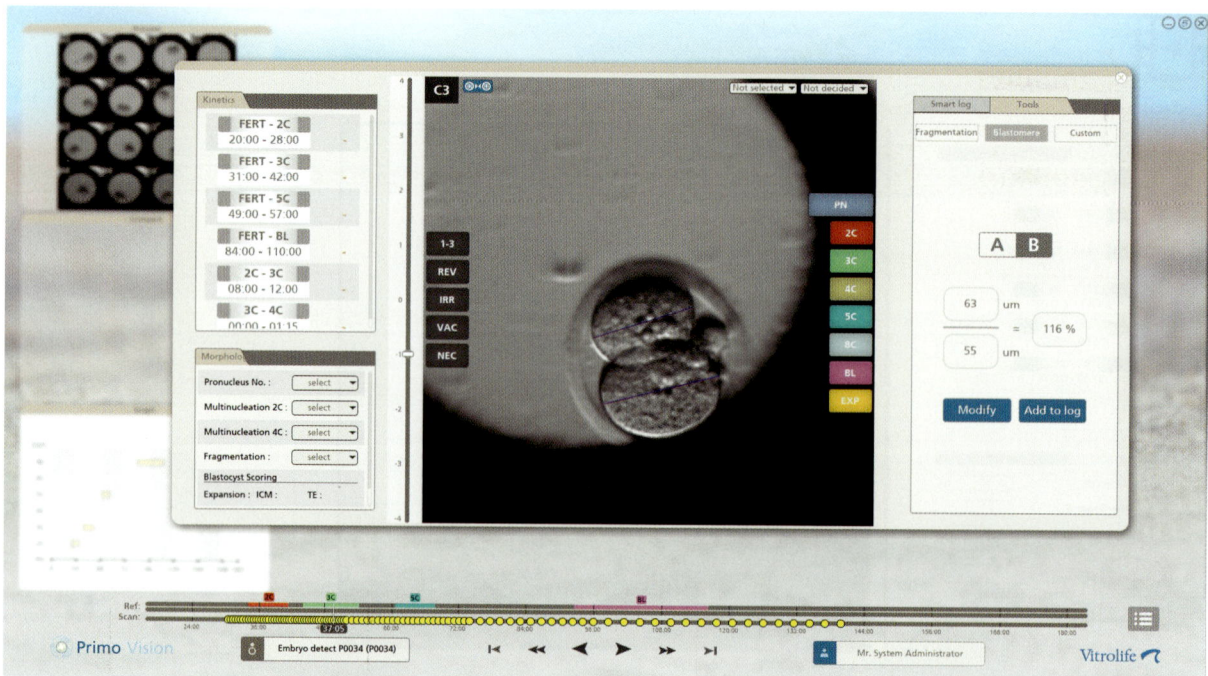

Figure 3.14 The "deep analysis" window. Further tools for embryo analysis.

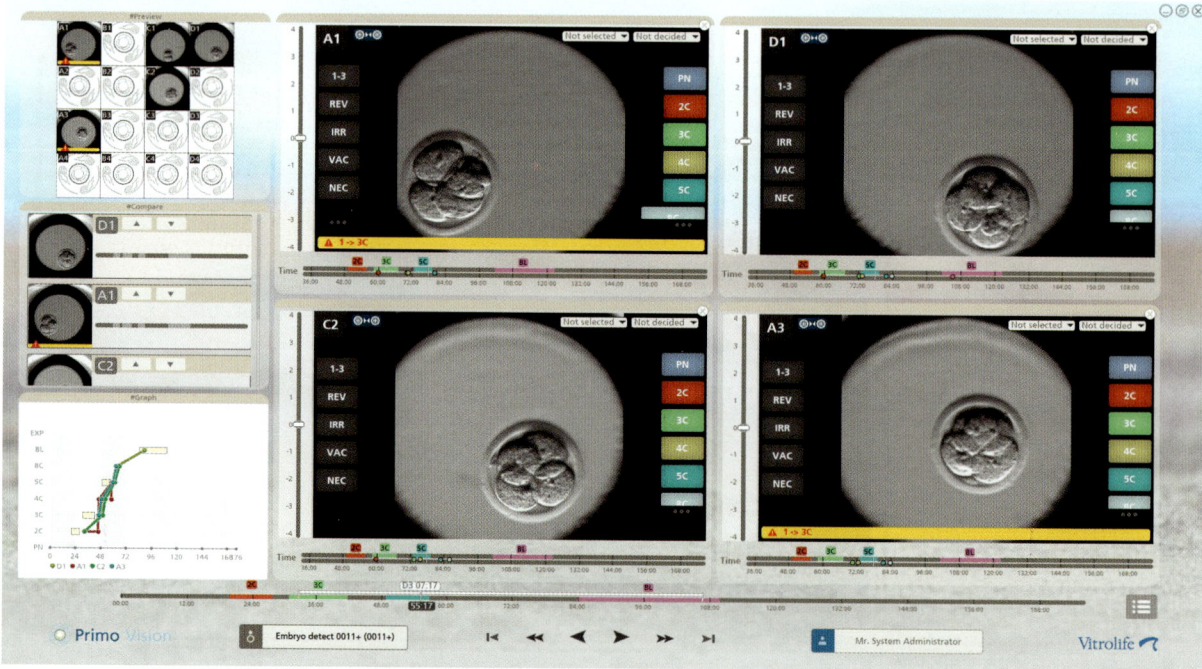

Figure 3.15 Quick comparison of four embryos on the "desk."

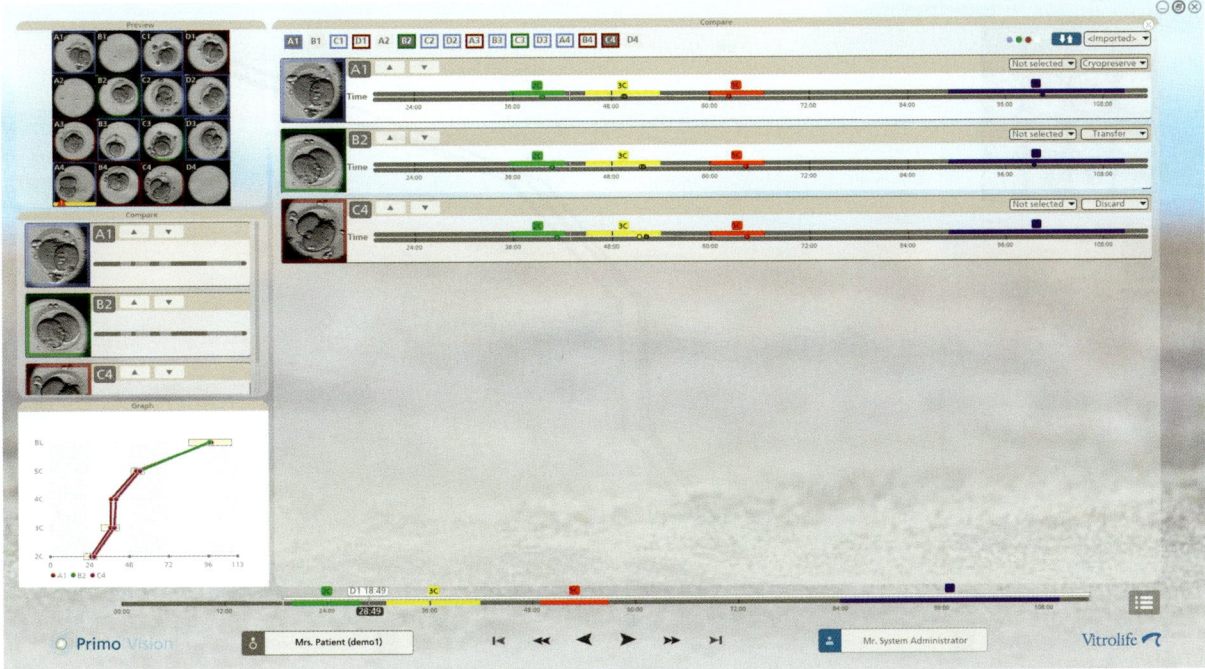

Figure 3.16 The compare panel that also runs the re-ordering algorithms.

The bottom left panel shows the "Developmental graph" of the embryos that are in the compare panel. This panel can also be dragged and dropped onto the table and provides a quick view of embryo development and the reference values.

The Menu is visualized by clicking the bottom right icon. From the Menu one can get back to the Overview screen, go to the home screen, enter the Project information and continue adding data, finalizing the analysis, and create reports or videos.

3.5. System setup

The Primo Vision System is ready for operation out of the box, after connecting the microscope(s) and the accessories to the external Controlling Unit and starting the software.

3.6. System maintenance

Since the equipment is operated in a clean environment, disinfection of the entire external surface of the microscope and cables, complying with the lab's protocol, is important.

On the other hand, the Primo Vision Time-lapse Embryo Monitoring System does not require special annual preventive maintenance.

3.7. Closing remarks

One of the advantages of using time-lapse techniques is to follow up embryo development and to obtain information on morphology. This information is provided continuously every 5 to 20 minutes, not just at distinct time-points of the day but at any time-points of the day. Techniques now available make it possible to perform embryo evaluation while embryos are inside of the incubator, thus reducing handling stress. During the course of routine time-lapse examination, hundreds of images are made, and saved and archived in digital format, enabling another fundamentally important expectation of the scientific society: proper documentation and quality control of the laboratory phase, right at its heart, inside of the incubator. Apart from quality control there is another everyday use of the digital imaging, and that is learning, teaching, and communication. The listed possibilities alone justify the use of time-lapse technology in the embryology lab.

3.7.1. What do we learn from continuous embryo follow-up?

Pronuclear scoring involves the assessment of the number and relative position of the nucleolar

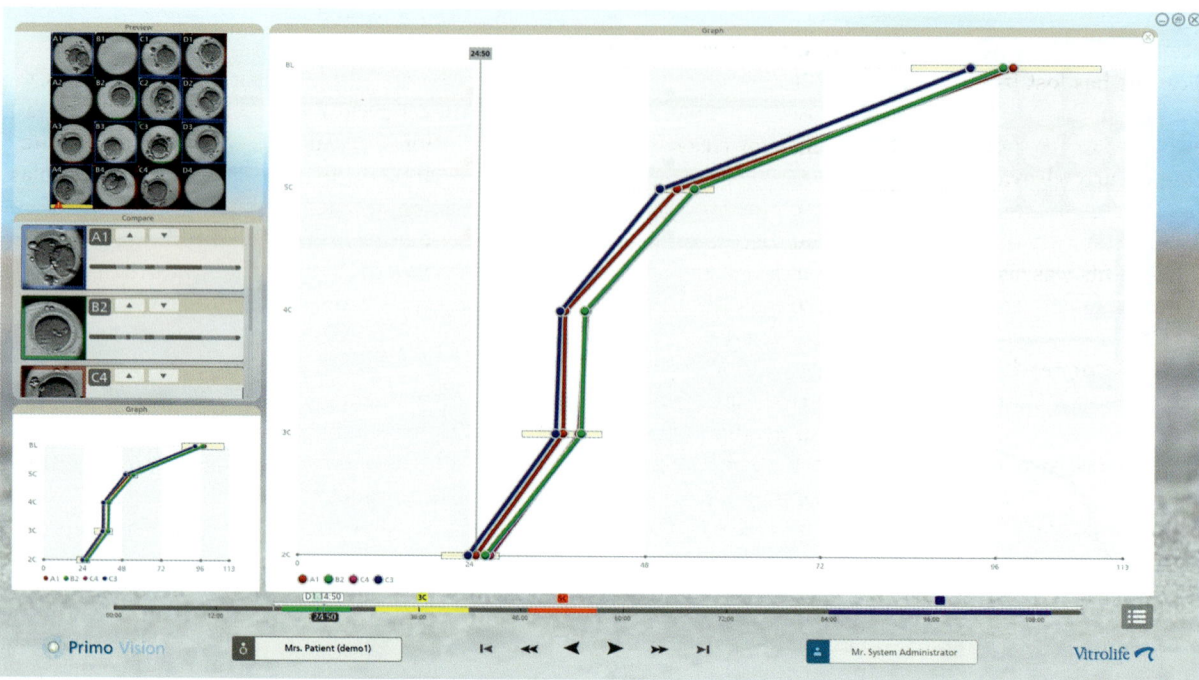

Figure 3.17 The embryo developmental graphs.

(a)

(b)

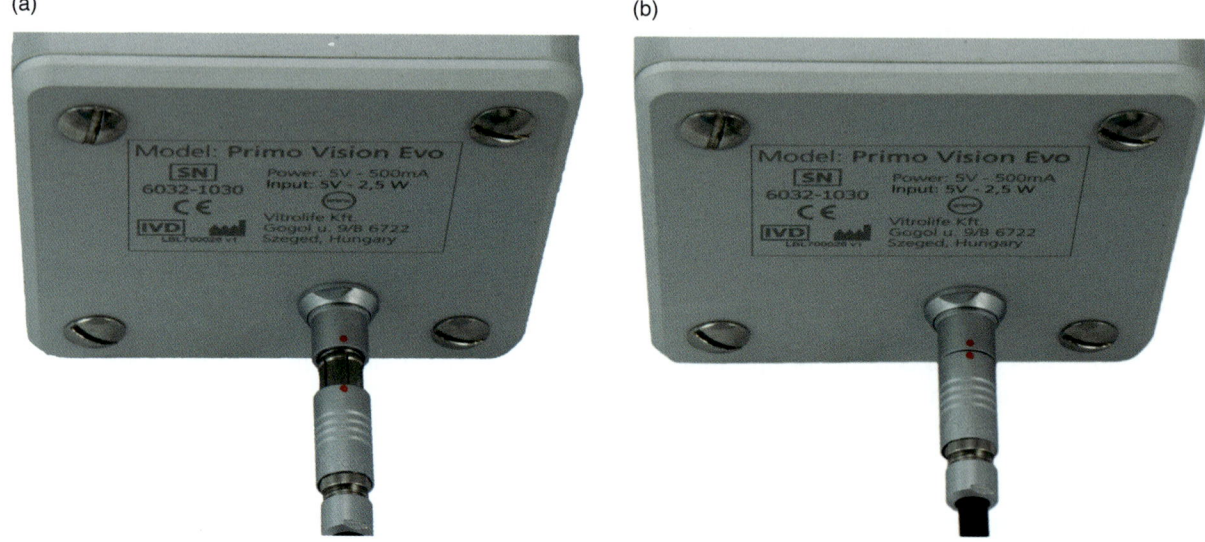

Figure 3.18 a, b Connecting the Primo Vision microscope on the back panel.

precursor bodies (NPBs), which are established in the pronuclei. Any inequality in the distribution of the NPBs within the pronuclei is considered to be abnormal, but time-lapse studies revealed that NPBs move around inside of the pronuclei, and can produce up to a two-score difference within 2 hours, making their traditional, "static evaluation" and its value questionable in the present format.

Morphologically, early cleaving embryos have been regarded as higher-quality ones. However, early cleavage has lost its classic meaning in the time-lapse environment. By checking time-lapse recordings of embryos we are looking for timeframes for cleavages, as too early cleavage can equally be an unfavorable sign of embryo quality as cleaving too late.

Such simple questions as cell number at certain time-points was also questioned, when it became possible to follow the cleavage pattern of the actual embryo. A morphologically sound five-cell stage embryo can reach the five-cell stage by normal but also abnormal cleavage paths. A five-cell stage embryo can be the result if the first cytokinesis produced three blastomeres, two of which cleaved further. After a normal first cytokinesis, one blastomere can cleave to two cells, while the second one may cleave to three daughter cells, resulting, again, in a normal looking but abnormal five-cell stage embryo. Though the morphological evaluation may reveal the same score for the given examples, their potential to implant and to develop to a healthy offspring differs significantly.

Fragmentation has also been observed as a highly dynamic process with fragments continuously rearranging around the blastomeres or being reabsorbed during the course of in vitro development. For this reason, static evaluation of fragmentation might not be absolutely correct.

A further example for the dynamic nature of morphology is the fact that blastocysts pulsate: they expand and collapse continuously. An expanded blastocyst may collapse within a short time; the blastocyst score would change, while the quality would not.

Time-lapse projects also provide insight into the timings of the cell cycle. Embryos are supposed to cleave within a definite timeframe. Which are the most important events, and are they in correlation with blastocyst formation in pregnancy? Recent studies have revealed that cleavages up until the four-cell stage are more predictive of the chance to reach the blastocyst stage, while events prone to happen after the onset of the genomic activation seem to provide information that is more relevant to pregnancy. According to our group, morphokinetics in itself is not sufficient for proper embryo evaluation; it has to be applied in combination with static morphology. Nevertheless, time-lapse is needed to qualify static morphology properly. Focusing on the importance of morphokinetics purely, our group sees its role in supporting deselection. Deselection in this context means embryos performing irregular cleavage like directly cleaving from one-cell to three-cell stages and these are ranked back in the cohort.

Today close to 600 clinics are using one of the two time-lapse solutions, and have provided evidence that this type of embryo follow-up is needed even in the everyday routine, for embryo evaluation, quality control, and service.

3.8. Disclosures

CP is also active as scientific manager of Vitrolife Kft., the manufacturer of the Primo Vision Embryo Monitoring System.

References

1. Cannon WB. *Bodily Changes in Pain, Hunger, Fear and Rage: An Account of Recent Researches into the Function of Emotional Excitement.* New York: Appleton; 1915.

2. Selye HA. A syndrome produced by diverse nocuous agents. *Nature* 1936; 138: 32.

3. Kültz D. DNA damage signals facilitate osmotic stress adaptation. *Am J Physiol Renal Physiol* 2005; 289: F504–5.

4. Pribenszky C, Vajta G, Molnar M, Du Y, Lin L, Bolund L, Yovich J. Stress for stress tolerance? A fundamentally new approach in mammalian embryology. *Biol Reprod* 2010; 83(5): 690–7.

5. Xie Y, Wang F, Zhong W, Puscheck E, Shen H, Rappolee DA. Shear stress induces preimplantation embryo death that is delayed by the zona pellucida and associated with stress-activated protein kinase-mediated apoptosis. *Biol Reprod* 2006; 75(1): 45–55.

6. Mizobe Y, Yoshida M, Miyoshi K. Enhancement of cytoplasmic maturation of in vitro-matured pig oocytes by mechanical vibration. *J Reprod Dev* 2010; 56: 285–90.

7. Pribenszky C, Horváth A, Végh L, Huang SY, Kuo YH, Szenci O. Stress preconditioning of boar spermatozoa: a new approach to enhance semen quality. *Reprod Domest Anim* 2011; 46 S2: 26–30.

8. Pribenszky C, Lin L, Du Y, Losonczi E, Dinnyes A, Vajta G. Controlled stress improves oocyte performance–cell preconditioning in assisted reproduction. *Reprod Domest Anim* 2012; 47 S4: 197–206.

9. Du Y, Lin L, Schmidt M, Bøgh IB, Kragh PM, Sørensen CB, Li J, Purup S, Pribenszky C, Molnár M, Kuwayama M, Zhang X, Yang H, Bolund L, Vajta G. High hydrostatic pressure treatment of porcine oocytes before handmade cloning improves developmental competence and cryosurvival. *Cloning Stem Cells* 2008; 10(3): 325–30.

10. Lewis WH and Gregory PW. Cinematographs of living developing rabbit eggs. *Science* 1929; 69: 226–9.

11. Vajta G, Callesen H. Establishment of an efficient somatic cell nuclear transfer system for production of transgenic pigs. *Theriogenology* 2012; 77(7): 1263–74.

12. Vajta G, Korösi T, Du Y, Nakata K, Ieda S, Kuwayama M, Nagy ZP. The Well-of-the-Well system: an efficient approach to improve embryo development. *Reprod Biomed Online* 2008; 17(1): 73–81.

13. Nagy A, Sass M, Markkula M. Systematic non-uniform distribution of parthenogenetic cells in adult mouse chimaeras. *Development* 1989; 106: 321–4.

14. Wood SA, Pascoe WS, Schmidt C, Kemler R, Evans MJ, Allen ND. Simple and efficient production of embryonic stem cell-embryo chimeras by coculture. *Proc Natl Acad Sci USA* 1993; 90: 4582–5.

15. Vajta G, Lewis IM, Hyttel P, Thouas GA, Trounson AO. Somatic cell cloning without micromanipulators. *Cloning* 2001; 3(2): 89–95

16. Vajta G, Kragh PM, Mtango NR, Callesen H. Hand-made cloning approach: potentials and limitations. *Reprod Fertil Dev* 2005; 17(1–2): 97–112.

17. Peura TT, Vajta G. A comparison of established and new approaches in ovine and bovine nuclear transfer. *Cloning Stem Cells* 2003; 5(4): 257–77.

18. Tecirlioglu RT, French AJ, Lewis IM, Vajta G, Korfiatis NA, Hall VJ, Ruddock NT, Cooney MA, Trounson AO. Birth of a cloned calf derived from a vitrified hand-made cloned embryo. *Reprod Fertil Dev* 2003; 15(7–8): 361–6.

19. Kragh PM, Vajta G, Corydon TJ, Purup S, Bolund L, Callesen H. Production of transgenic porcine blastocysts by hand-made cloning. *Reprod Fertil Dev* 2004; 16(3): 315–18.

20. Vajta G, Bartels P, Joubert J, de la Rey M, Treadwell R, Callesen H. Production of a healthy calf by somatic cell nuclear transfer without micromanipulators and carbon dioxide incubators using the Handmade Cloning (HMC) and the Submarine Incubation System (SIS). *Theriogenology* 2004; 62(8): 1465–72.

21. Ribas R, Oback B, Ritchie W, Chebotareva T, Taylor J, Maurício AC, Sousa M, Wilmut I. Modifications to improve the efficiency of zona-free mouse nuclear transfer. *Cloning Stem Cells* 2006; 8(1): 10–15.

22. Lagutina I, Lazzari G, Galli C. Birth of cloned pigs from zona-free nuclear transfer blastocysts developed in vitro before transfer. *Cloning Stem Cells* 2006; 8(4): 283–93.

23. Klimanskaya I, Chung Y, Becker S, Lu SJ, Lanza R. Human embryonic stem cell lines derived from single blastomeres. *Nature* 2006; 444(7118): 481–5.

24. Vajta G, Peura TT, Holm P, Paldi A, Greve T, Trounson AO, Callesen H. New method for culture of zona-included or zona-free embryos: the Well of the Well (WOW) system. *Mol Reprod Dev* 2000; 55: 256–64.

25. Hoelker M, Rings F, Lund Q, Phatsara C, Schellander K, Tesfaye D. Effect of embryo density on in vitro developmental characteristics of bovine preimplantative embryos with respect to micro and macroenvironments. *Reprod Domest Anim* 2010; 45: e138–45.

26. Wydooghe E, Vandaele L, Piepers S, Dewulf J, Van den Abbeel E, De Sutter P, Van Soom A. Individual commitment to a group effect: strengths and weaknesses of bovine embryo group culture. *Reproduction* 2014; 148(5): 519–29.

27. Beraldi R, Sciamanna I, Mangiacasale R, Lorenzini R, Spadafora C. Mouse early embryos obtained by natural breeding or in vitro fertilization display a differential sensitivity to extremely low-frequency electromagnetic fields. *Mutat Res* 2003; 538(1–2): 163–70.

28. Cameron IL, Hardman WE, Winters WD, Zimmerman S, Zimmerman AM. Environmental magnetic fields: influences on early embryogenesis. *J Cell Biochem* 1993; 51(4): 417–25.

29. Ravera S, Falugi C, Calzia D, Pepe IM, Panfoli I, Morelli A. First cell cycles of sea urchin *Paracentrotus lividus* are dramatically impaired by exposure to extremely low-frequency electromagnetic field. *Biol Reprod* 2006; 75(6): 948–53.

4

Description of time-lapse systems: the Eeva™ Test

Lei Tan, Shehua Shen, and Alice A. Chen

4.1. Introduction

The need to improve pregnancy rates resulting from in vitro fertilization (IVF) while reducing multiple birth rates has spawned significant research in optimizing embryo selection. Currently, the most widely used embryo selection methodology involves static morphological assessment of the embryo [1], which has limited success and is subject to high intra- and inter-observer variability [2]. To achieve high pregnancy rates while reducing multiple embryo transfer, improved embryo selection methods are needed.

Recently, emerging time-lapse technology has provided embryologists a tool to continuously monitor embryo development and an opportunity to improve embryo selection by incorporating embryo development timing parameters into potential selection models (reviews in [3–6]). However, for wide adoption of time-lapse technology in IVF practice, there are two key considerations. First, validation of the time-lapse technology is needed to demonstrate its safety and effectiveness by prospectively designed clinical trials in a multiple-clinic setting [7]. Second, time-lapse technology must be compatible with the busy workflow of any standard IVF laboratory and practical to use for any embryologist.

The objective of this chapter is to describe the development, validation, and practical application of a first-of-its-kind prognostic test (the Eeva™ Test) for embryo selection. The Eeva Test provides automated time-lapse image analysis capabilities together with a validated embryo classification model, which reduces the requirement for manual video review and eliminates the need to develop and validate one's own model. This test has been validated in a prospective multicenter clinical trial that demonstrated improved decision-making for clinical embryologists. In describing this novel test, we discuss the scientific underpinning of prediction using time-lapse imaging, describe how the Eeva Test was developed and validated, provide case examples of the Eeva Test in clinical use, and finally, review recent research studies performed using the Eeva System.

4.2. The Eeva Test

The Eeva Test (Auxogyn; Menlo Park, CA, USA) is a first-of-its-kind prognostic clinical test enabled by non-invasive time-lapse imaging and automated image analysis. Consistent with its purpose as a clinical test [8], the Eeva Test provides quantitative and objective information regarding embryo development to aid embryologists and physicians in their selection and treatment decisions. Time-lapse imaging with automated image analysis is performed by the Eeva System, the first-in-class, FDA-de novo cleared device with prognostic assessment capability that is intended to be used adjunctively with traditional morphology to aid embryo selection. The Eeva System's de novo clearance sets it apart from traditional time-lapse systems, because it is the first time the FDA has cleared a prognostic prediction capability of a time-lapse system in the IVF field. As a first-in-class device, the Eeva System established a novel device category that makes it the benchmark for FDA to evaluate all future devices in this category. The rigorous requirements and high standards held by FDA in its de novo

Note: The *Eeva*™ System has been CE Marked in the European Union and received marketing clearance in the United States by the FDA.

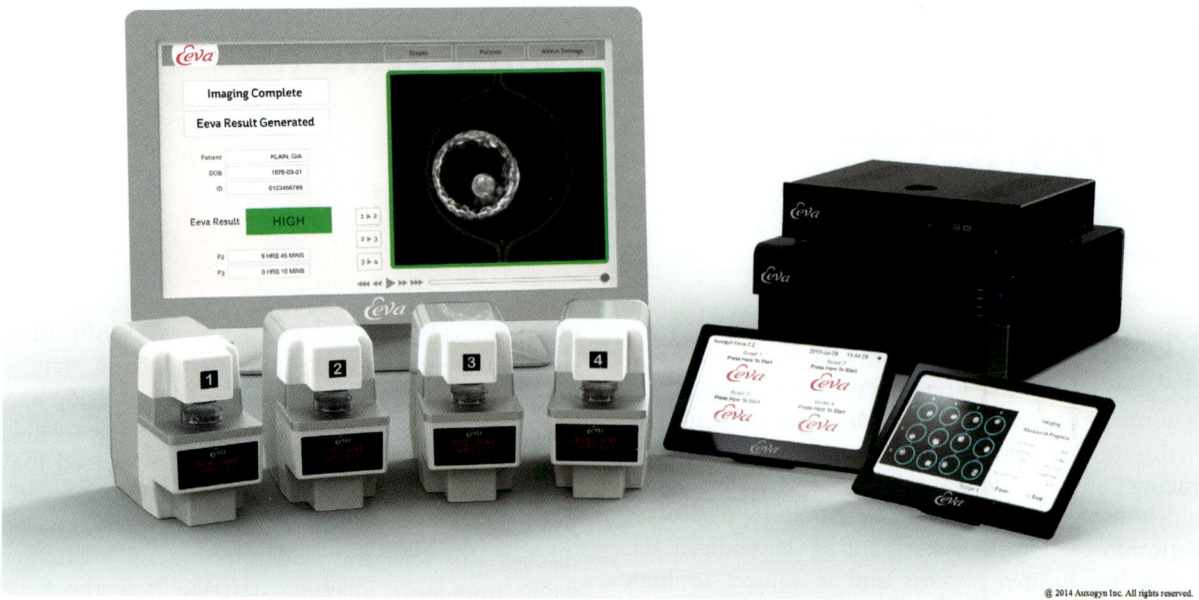

Figure 4.1 Image of hardware components for the Eeva System. Reproduced with permission from Auxogyn Inc.

pathway provide the highest level of regulatory assurance for physicians, embryologists, and patients.

To achieve clinical test results, the Eeva Test leverages a combination of unique software and hardware components (Figure 4.1). In this section, we describe the components of the Eeva Test and provide practical information for using these components collectively in the IVF laboratory workflow.

4.2.1. Software

At the core of the Eeva Test lies software that first utilizes state-of-the-art computer vision technology to automate the measurement of key cell division timings, and then generates predictive information about embryo developmental potential. By convention, time-lapse measurements require substantial time and labor, and these requirements hinder the routine use of time-lapse technology for clinical embryo selection. The software underlying the Eeva Test was based on predictive time-lapse parameters shown to aid embryo selection and, further, it was designed to address the impractical challenges of measuring hundreds of images per embryo.

4.2.1.1. Predictive information enabled by time-lapse

Information that is predictive of embryo development can potentially aid embryo selection. Recently,

increasing availability and usage of time-lapse imaging systems – both homemade and commercial variations – have enabled researchers to identify potentially predictive time-lapse markers during human embryo culture in a safe and non-invasive manner. These time-lapse research studies can be broken into two phases: an observational phase and a predictive phase. Beginning in 1997, time-lapse imaging had been focused on observing new aspects of embryo development, such as polar body extrusion, fertilization, pronuclear formation and abuttal, cytoplasmic flares, embryo hatching and so on [9–12]. Since 2010, time-lapse imaging has shifted focus to identifying predictive parameters that can help assess developmental outcomes of the embryo, such as blastocyst development, implantation, and, most recently, ploidy [13–15]. Numerous publications since 2010 have reported statistically significant correlations between time durations of cell stages and embryo outcomes. These studies, together with studies that have demonstrated that time-lapse monitoring of human embryos is safe for embryo culture, are reviewed in detail elsewhere [3, 5, 16, 17].

Various time-lapse parameters have been shown to be predictive of embryo quality. An analysis of the studies that defined a precise time window for embryo development prediction showed that the time from 2-to-3 cell (P2) and the time from 3-to-4 cell

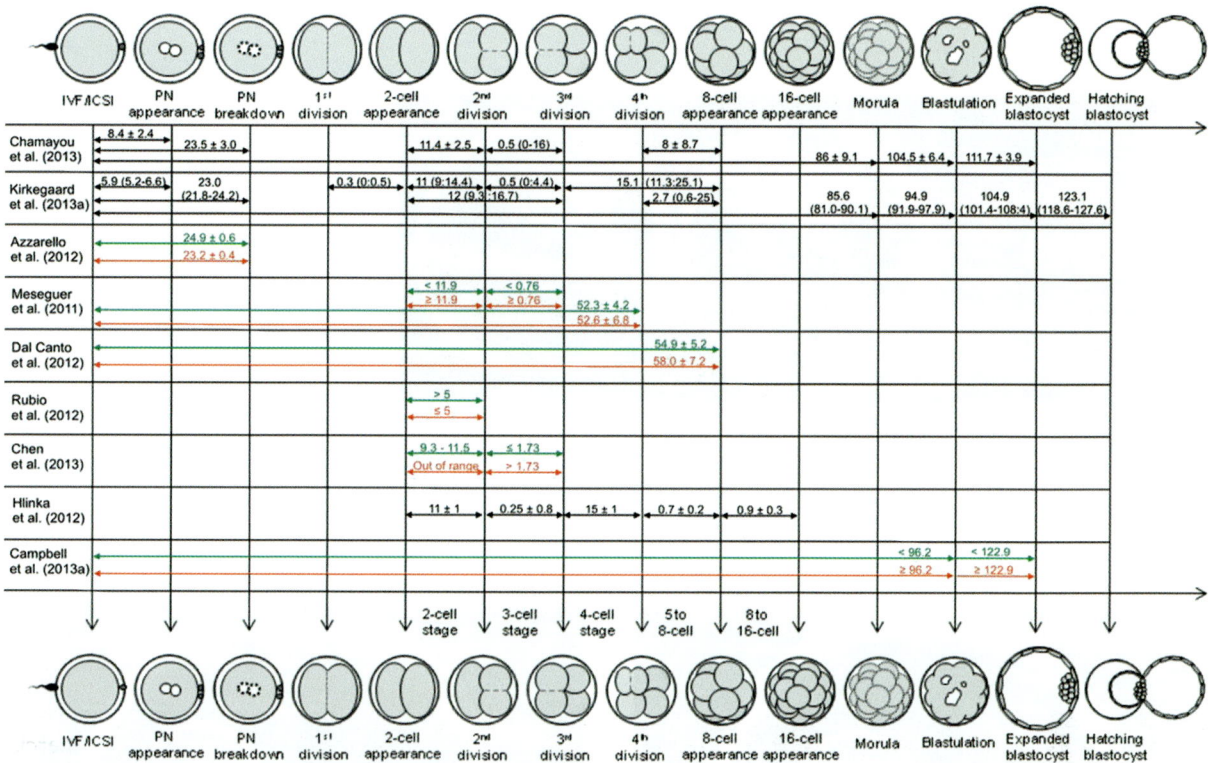

Figure 4.2 Schematic of pre-implantation embryo development with corresponding time-lapse markers. When there was no significant difference observed between "implanters" and "non-implanters," only the value for the implanted embryos is shown (in black). When significant differences were reported, the "implanter" values are shown in green, and "non-implanters" are in red. All values are expressed in hours, as mean ± standard deviation or mean (95% confidence interval) for normally distributed variables, and median (minimum: maximum) for non-normally distributed variables. PN, pronuclei. Reproduced from Kaser and Racowsky, Clinical outcomes following selection of human pre-implantation embryos with time-lapse monitoring: a systematic review, *Human Reproduction Update*, 2014, 20 [5]: 61731, by permission of Oxford University Press and the European Society of Human Reproduction and Embryology.

(P3) were invariably included in six out of seven of the studies that investigated which embryos are most likely to become a blastocyst, implant, or be euploid [5]. This was further demonstrated by a recent review of studies examining the relationship between time-lapse parameters and embryo implantation; in this recent analysis, P2 and P3 were included in five out of six of the studies that reported that implanted embryos had different developmental kinetics compared to non-implanted ones (Figure 4.2) [17]. These P2 and P3 timings were consistent predictors of embryo developmental potential, even when stimulation protocols [18], fertilization methods [19], culture media [20, 21], and culture environments [22] were varied. In addition to clinical reproducibility, mechanistic studies have also demonstrated that embryos with abnormal P2 and P3 timings exhibit distinct gene expression profiles [13], aneuploidy probabilities [15], and micronuclei patterns [15], compared to embryos with normal P2 and P3 timings. This collection of evidence supports the conclusion that time-lapse parameters P2 and P3 are reliable across independent data sets, and moreover, are grounded in basic scientific understanding. Recently, a randomized controlled trial further confirmed that incorporating P2 and P3 timings in an embryo selection protocol significantly improves pregnancy rates while lowering early pregnancy loss [23]. Overall, a vast collection of studies supports the idea that these key, predictive time-lapse parameters could be used to aid embryo selection. Manual measurement of even only a few time-lapse markers, however, requires review of hundreds of images for each embryo, which is laborious, time-consuming, subjective, and therefore impractical for broad clinical application.

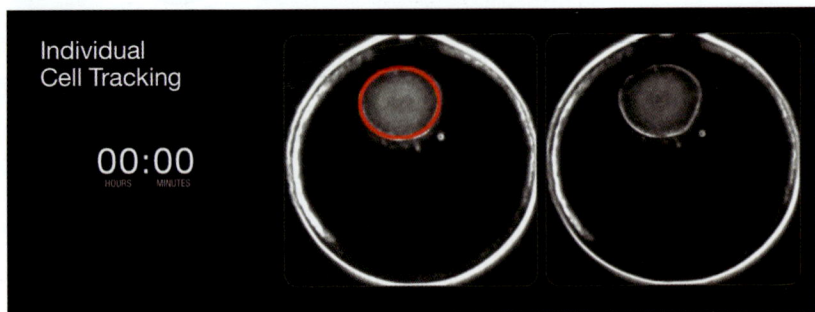

Video 4.1 Automated image analysis in the Eeva Test. Reproduced with permission from Auxogyn Inc.

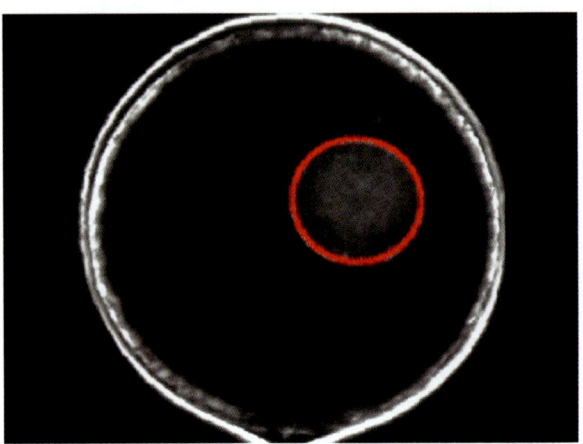

Video 4.2 Example of an Eeva High embryo with automated image analysis that tracks the number of cells. Reproduced with permission from Auxogyn Inc.

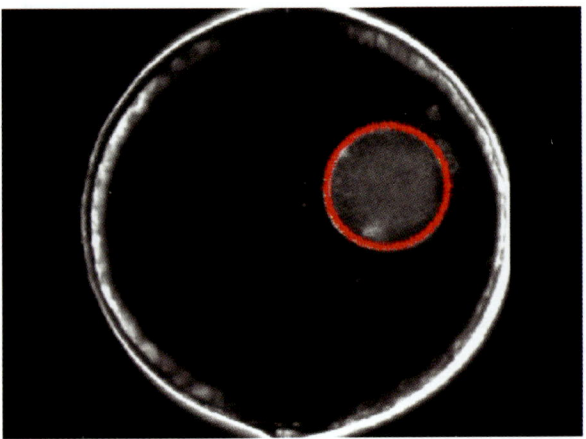

Video 4.3 Example of an Eeva Low embryo with automated image analysis that tracks the number of cells in real-time. Reproduced with permission from Auxogyn Inc.

4.2.1.2. Automated image analysis software in the Eeva Test

To reduce the burden that manual annotation and review of hundreds of images per embryo places on embryologists, the Eeva Test incorporates 1. automated image analysis software that automatically measures P2 and P3 cell division timings using state-of-the-art computer vision technology; and 2. a statistical classification model that generates a predictive result regarding embryo developmental potential. The automated image analysis software used in the Eeva Test leverages a validated, data-driven probabilistic framework that infers cell membranes, identifies cell division events, and subsequently calculates timing intervals P2 and P3 (Video 4.1). In a prospective validation study, Eeva software predictions were shown to have good (> 90%) agreement with manual predictions made by human observers [24]. In addition to the automatic quantification of P2 and P3, the software automatically inputs the calculated P2 and P3 timings into a clinically validated embryo classification algorithm. The result of the fully integrated and fully automated assessment is generation of a predictive score of "High" or "Low" developmental potential (Video 4.2, Video 4.3).

4.2.2. Hardware

Coupled with the Eeva Test software, the hardware underlying the Eeva Test enables non-invasive imaging, image data collection, and embryo culture in an undisturbed fashion. The hardware portion of the Eeva Test includes the Eeva Scope and Eeva Dish; here we discuss each component and how each fits seamlessly into the IVF clinic and routine workflow.

4.2.2.1. The Eeva Scope

The Eeva Scope is a digital microscope that uses a light-emitting diode at 625 nm wavelength with short

Video 4.4 The Eeva Scope. Reproduced with permission from Auxogyn Inc.

illumination times to minimize embryo exposure to light and to avoid damaging short wavelength light (Video 4.4). The scope automatically focuses on the embryos on the dish and captures a single, high-resolution image of all of the microwells in the Eeva Dish once every 5 minutes (Figure 4.3). Although hundreds of images are recorded during embryo culture, the total exposure time in the Eeva Scope during culture and acquisition of images is significantly lower than the microscope light exposure time reported for a standard IVF treatment [24].

The Eeva Scope employs optics that resembles those in an inverted microscope. However, the scope has a compact design that fits into conventional incubators, which allows continuous monitoring of embryo development without interrupting the culture environment by taking embryos out of the incubator. The scope is compatible with incubators from different manufacturers, including incubators commonly used in IVF laboratories (Figure 4.4). The Eeva Scope can also be integrated with a tri-gas incubator specially designed to house the system with a small footprint and low gas consumption, manufactured by IKS International (Figure 4.5).

4.2.2.2. The Eeva Dish

The Eeva Dish is a multiwell dish that provides individual identification of embryos while also enabling group embryo culture within the same media drop. The Eeva Dish is designed for oocyte or embryo handling and culture; it is therefore comprised of standard tissue culture plastic (polystyrene) and resembles a standard 38 mm Petri dish at a macroscopic level. The Eeva Dish at a microscopic level consists of a single large central ring (10 mm) that is

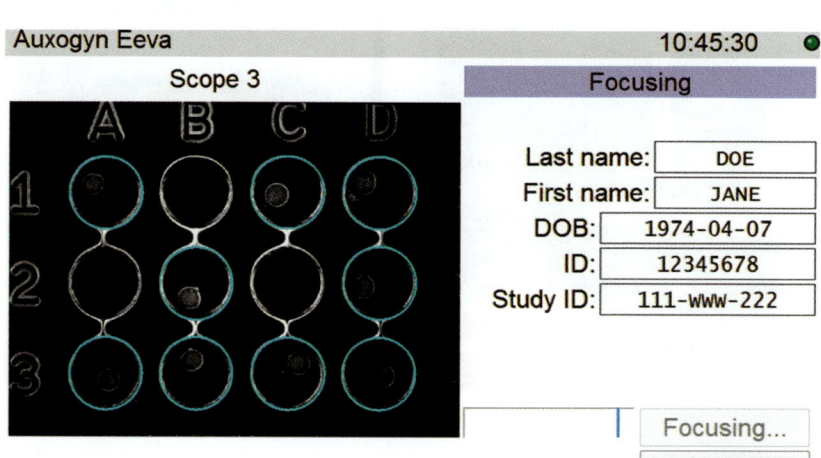

Figure 4.3 Image of a screenshot during an auto-focusing step by the Eeva Scope. Reproduced with permission from Auxogyn Inc.

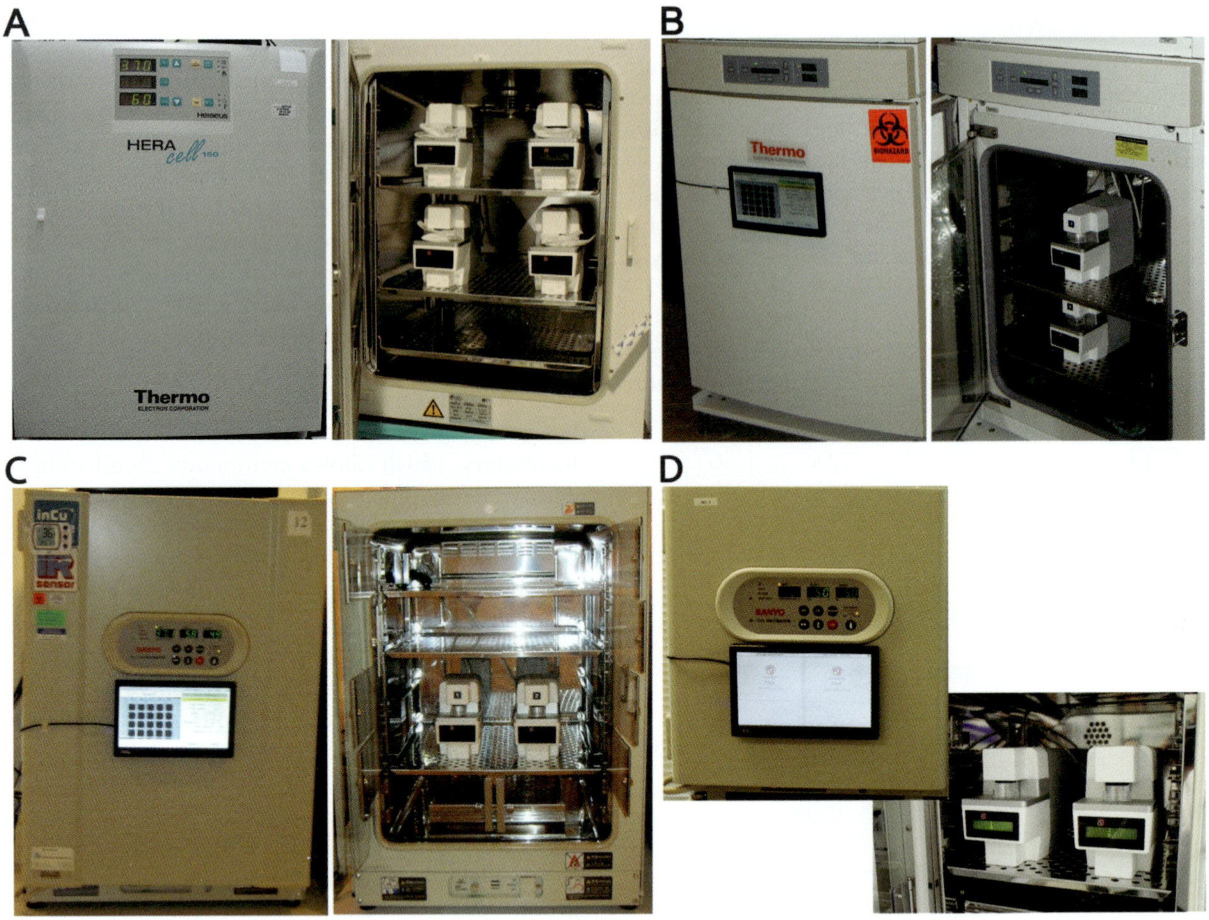

A

B

C

D

Figure 4.4 Images of Eeva Scopes in different incubators. (A) Heracell 150i. (B) Thermo Scientific Forma 3110. (C) SANYO MCO 18M. (D) SANYO MCO 5M.

Figure 4.5 Images of the Eeva System with integrated triple gas incubators and XiltriX monitoring by IKS International (image courtesy of IKS International).

Figure 4.7 Image of microwells on an Eeva Dish. Reproduced with permission from Auxogyn Inc.

Figure 4.6 Image of an Eeva Dish. Reproduced with permission from Auxogyn Inc.

Introducing the Eeva™ System

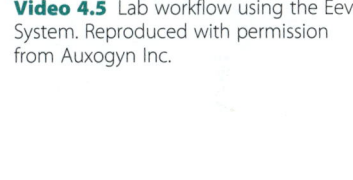

Video 4.5 Lab workflow using the Eeva System. Reproduced with permission from Auxogyn Inc.

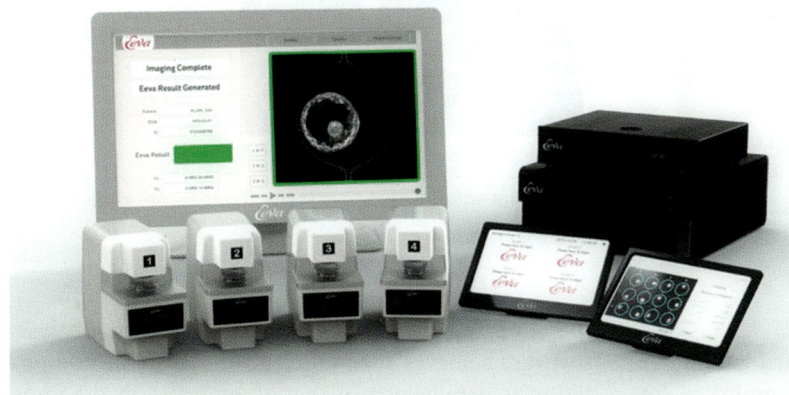

surrounded by three smaller outer rings (8 mm) (Figure 4.6). The three smaller outer rings are intended to hold media drops for rinsing embryos. The central ring contains 12 smaller microwells, and one oocyte or embryo can be placed in each individual microwell (Figure 4.7). Embryos settle at the bottom of each microwell and are group-cultured in a 100 µl media droplet under oil overlay. The group culture design of the Eeva Dish is based on the scientific finding that group culture is superior to individual culture and improves clinical outcome [25]. The

letters and numbers next to the wells are designed to be used for embryo identification and are outputted in the Eeva Test reports (Figure 4.7).

4.2.3. Practical use

The Eeva Test fits seamlessly into the IVF clinic laboratory workflow (Video 4.5). Dishes with culture media and oil overlay are placed in the incubator overnight or allowed appropriate time for media

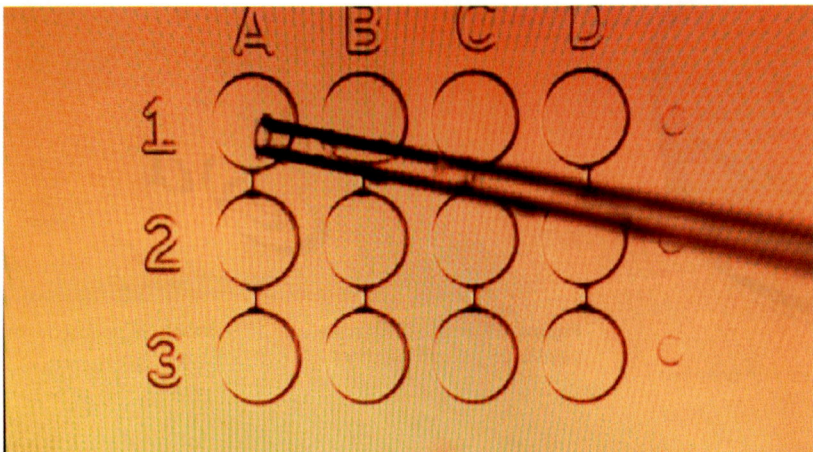

Video 4.6 Placing embryos into microwells on the Eeva Dish (mouse embryos were used in the video for illustration). Reproduced with permission from Auxogyn Inc.

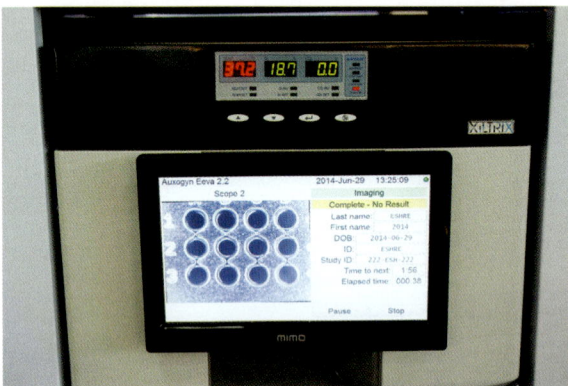

Figure 4.8 Image of the Eeva Scope Screen mounted on the incubator door (image courtesy of IKS International).

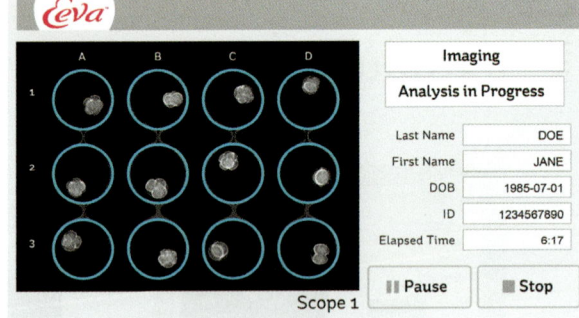

Figure 4.9 Image of the Eeva Scope Screen: close-up view. Reproduced with permission from Auxogyn Inc.

equilibration based on the user's standard laboratory protocol. After media equilibration, embryos or oocytes are loaded one at a time into the microwells of the dish (Video 4.6).

Imaging sessions in the scopes are managed via the Eeva Scope Screen, a touchscreen monitor mounted on the incubator door (Figure 4.8). The touchscreen, which can be connected to up to four scopes, allows the user to select a scope, input patient data, manage the session, and review the most recent image captured (Figure 4.9).

Once the imaging session is started by the user, image analysis software automatically processes the images taken by the scope in real-time and provides cell division measurements, timing calculations, and

the final test results by day 3. The image analysis software achieves such efficiency because the computation is parallelized so that all embryos in the same dish are analyzed at the same time just after the image is taken (Video 4.7).

After the imaging is completed, users have ready access to images, videos, and other data via the Eeva Station (Figure 4.10). Users have the option of selecting an active patient session to view the last image of the entire dish (Figure 4.11A) or a single image of a particular embryo (Figure 4.11B). Users may also review entire dish videos or individual well videos (Figure 4.12). The Eeva Test results, together with automatically measured P2 and P3 timings, are available early on day 3 (Figure 4.12A). The Eeva Test results are also highlighted as colored squares around

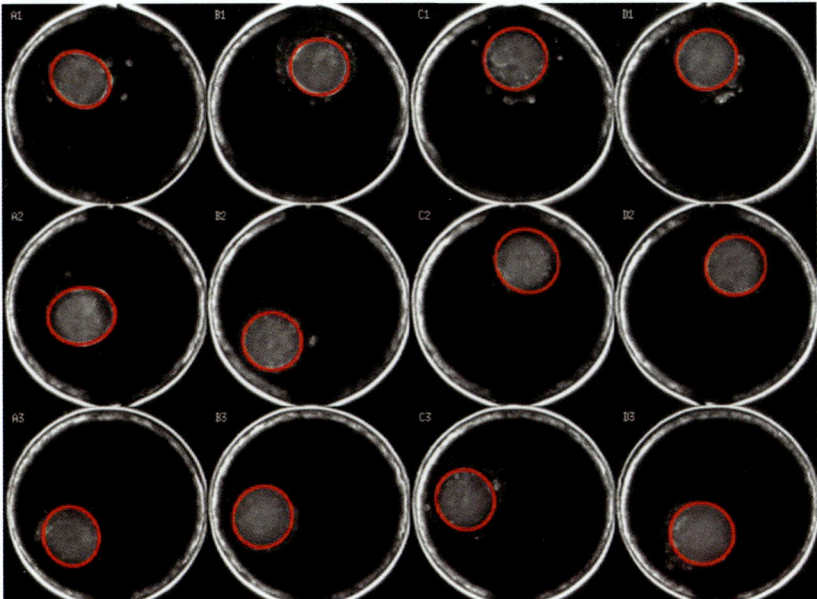

Video 4.7 Automated image analysis that tracks the number of cells for all the embryos on the Eeva Dish. Reproduced with permission from Auxogyn Inc.

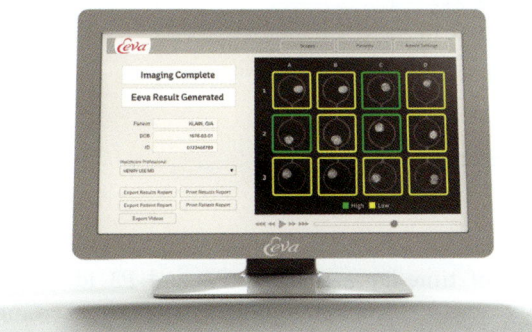

Figure 4.10 Image of the Eeva Station. Reproduced with permission from Auxogyn Inc.

microwells on the dish view, providing a quick glance at the results for the entire dish (Figure 4.12B).

Through the Eeva Station, three types of reports serving potentially different purposes in a clinic are available: 1. The "Results Report" includes the Eeva Test results together with P2 and P3 timings for embryologists to combine the information with morphology for embryo selection (Figure 4.13); 2. The "Image Report" shows images of all embryos in the dish labeled with the test results to provide embryologists snapshots of embryos (Figure 4.14); 3. The "Patient Report" includes snapshots of images

selected by embryologists for communication with patients (Figure 4.15). In addition to the patient report, the Eeva Station also allows the user to export videos of individual embryos to provide directly to patients (Video 4.8).

4.3. The Eeva Test: clinically validated aiding in embryo selection

Often the most elusive requirement for introducing a new technology is that it is validated to improve the standard of care. When performing a retrospective study, it is relatively straightforward to determine a statistically significant difference in timing parameters when comparing videos from two embryo groups (e.g., implanted vs. non-implanted embryos). However, it is challenging and significantly less common to demonstrate comparable and consistent performance of the timing parameter on an independent set of prospectively collected test data from multiple centers. It is rarer still that new potential selection schemes are compared to traditional embryo selection methods (i.e., morphology). In this section, we describe results from the first prospective, multi-center study that validated the use of time-lapse for improving embryo selection, and from a follow-up

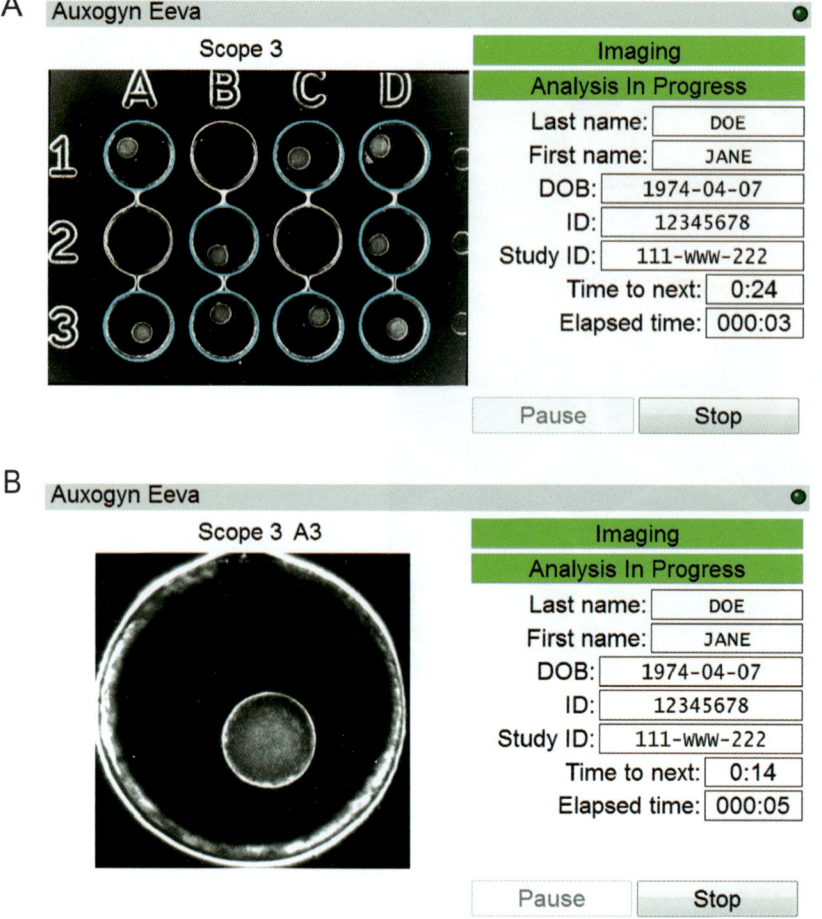

Figure 4.11 Screenshots from the Eeva Station for reviewing images. (A) Image review for the entire dish. (B) Image review for an individual well. Reproduced with permission from Auxogyn Inc.

study demonstrating that embryologists with diverse backgrounds benefit from adding Eeva Test results to traditional morphology. We also review several clinical cases exemplifying how Eeva Test results have been used to provide actionable information for different types of IVF patients.

4.3.1. Clinical validation of the Eeva Test

A prospective, five-center clinical trial was conducted from June 2011 to February 2012 to 1. assess the predictive power of P2 and P3 parameters in a diverse and clinical data set independent of Wong *et al.*; 2. develop a novel image analysis technology to automatically measure P2 and P3; and 3. test embryo selection using morphology vs. morphology plus Eeva Test results (ClinicalTrials.gov #NCT01369446) [24]. The study results first demonstrated that specific ranges of time-lapse markers P2 and P3 identified embryos with high developmental competence, and the ranges were consistent with Wong *et al.* and other successive studies [13]. The study further tested the performance and utility of software that automatically measures P2 and P3 and generates an automatic classification of high or low developmental potential. Using timing intervals that were refined for a clinical dataset pooled from five IVF centers (P2: 9.33–11.45 hours and P3: 0–1.73 hours), the Eeva Test results automatically differentiated a usable blastocyst from an arrested embryo with significantly improved diagnostic specificity (85%) compared to traditional morphology (57%, p < 0.0001). When Eeva Test results were provided to experienced embryologists in addition to morphology, specificity of these embryologists' assessment of embryos was improved. Specificity, or the ability to correctly predict which

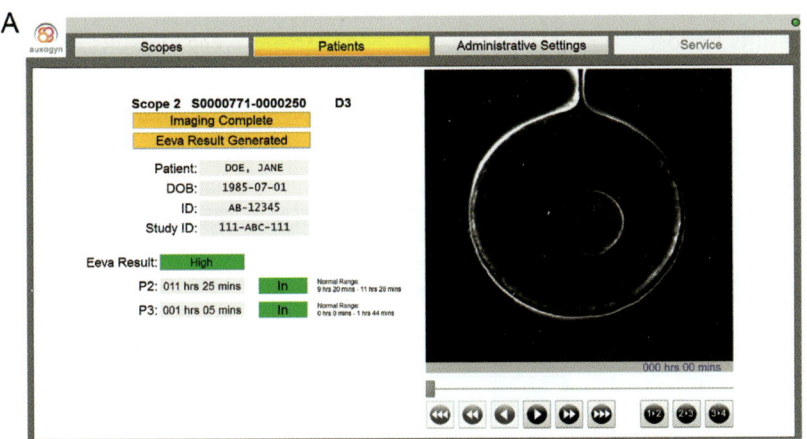

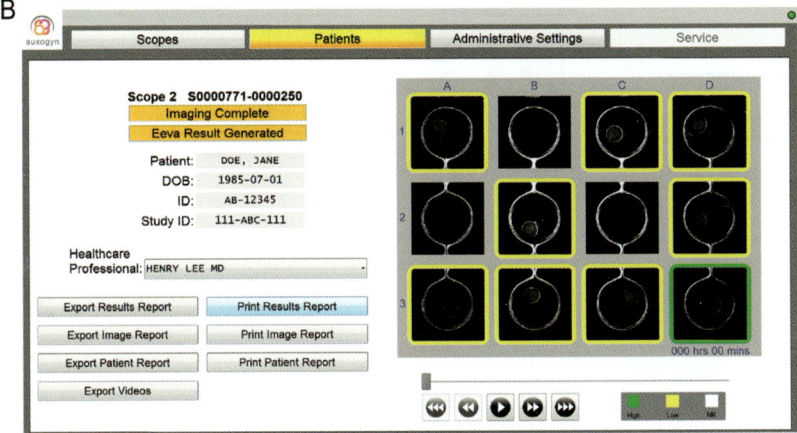

Figure 4.12 Screenshots from the Eeva Station for reviewing Eeva Test results. (A) View of Eeva Test results for an individual microwell. (B) View of Eeva Test results for the entire dish. Reproduced with permission from Auxogyn Inc.

embryos will arrest, is highly relevant to embryo selection since traditional morphology is most limited in selecting among "good morphology" embryos. It should be noted that the five IVF clinics that contributed clinical data to the development and validation of the Eeva Test each underwent their own standard procedures for stimulation, egg retrieval, embryo culture, and insemination. Incorporation of such diverse data allows for broad applicability of the Eeva Test in diverse clinical embryology laboratories, without requiring each laboratory to undergo a burdensome data collection and model-building effort.

In a separate report, utility of the Eeva Test adjunctive to traditional morphology was further studied by Diamond *et al.* to characterize its impact on a new set of embryologists' embryo selection decisions [26]. Five embryologists of diverse clinical practices, laboratory training, and geographical areas predicted embryo developmental potential using day 3 morphology alone and using morphology followed by Eeva Test results. To assess embryologists' performance, odds ratios (OR) and other diagnostic measures were calculated by comparing prediction results to true blastocyst outcomes. When Eeva Test results were used adjunctively with traditional morphology, the odds of an embryo forming a blastocyst was 3.51-fold (95% CI = 2.62–4.69) higher in the group predicted to develop into blastocysts than in other embryos. In contrast, the OR using morphology alone was 2.69 (95% CI = 2.06–3.50). This improvement in OR with Eeva Test results was also assessed in the subset of morphologically good and fair embryos. By traditional morphology, the OR for this subset dropped to 1.68, slightly better than random prediction (p < 0.0001). Adding Eeva Test results improved OR to 2.57, a 53% increase over traditional morphology and significantly better than random

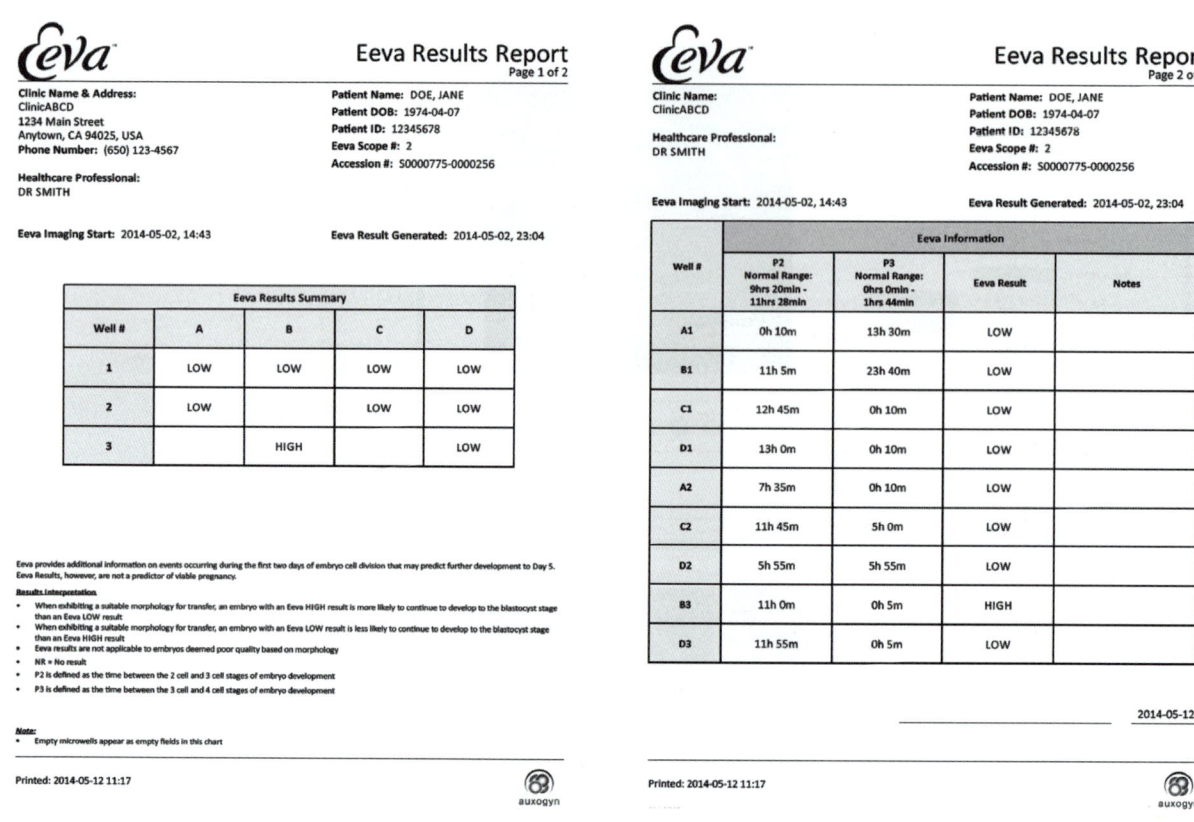

Figure 4.13 A sample Eeva Test Results Report. Reproduced with permission from Auxogyn Inc.

prediction (p < 0.0001, Figure 4.16). In addition to OR, Eeva Test results also helped improve the positive predictive value (PPV) over morphology alone (54% vs. 43%, p = 0.02), while maintaining the same level of negative predictive value (NPV, 68% vs. 68%, Figure 4.17). These results indicate that the Eeva Test aids embryologists of diverse backgrounds by distinguishing among similar-looking embryos that are evaluated first by morphological criteria. Using Eeva Test results as an adjunct to morphology, every individual embryologist's prediction performance was improved (Figure 4.18).

4.3.2. Reduced variability using the Eeva Test

The Diamond *et al.* study also assessed variability of prediction performance among embryologists [26]. Using the Eeva Test adjunctively to morphology, the variability in performance across all five embryologists was reduced from a range of 1.06 (OR = 1.14 to 2.20) to a range of 0.45 (OR = 2.33 to 2.78, Figure 4.18).

Notably, the five embryologists in this second study represent a separate and diverse range of clinical practices, laboratory training (<3 to >10 years), and geographical areas. Furthermore, the embryologist with the greatest improvement in odds ratio was one of the senior embryologists with more than 10 years of training in morphology grading. Since intra- and inter-operator variability in morphological grading has been shown to negatively impact IVF success rates [2, 27], adjunctive use of the Eeva Test may improve the standardization, reproducibility, and ultimate success of day 3 embryo selection.

4.3.3. Patient cases exemplifying embryo selection using Eeva Test results

Here we review three IVF patient cases from clinical studies where embryologists used Eeva Test results to aid their embryo selection decisions. These cases were chosen to exemplify the use of the Eeva Test in combination with traditional morphology, for a variety of patient case scenarios.

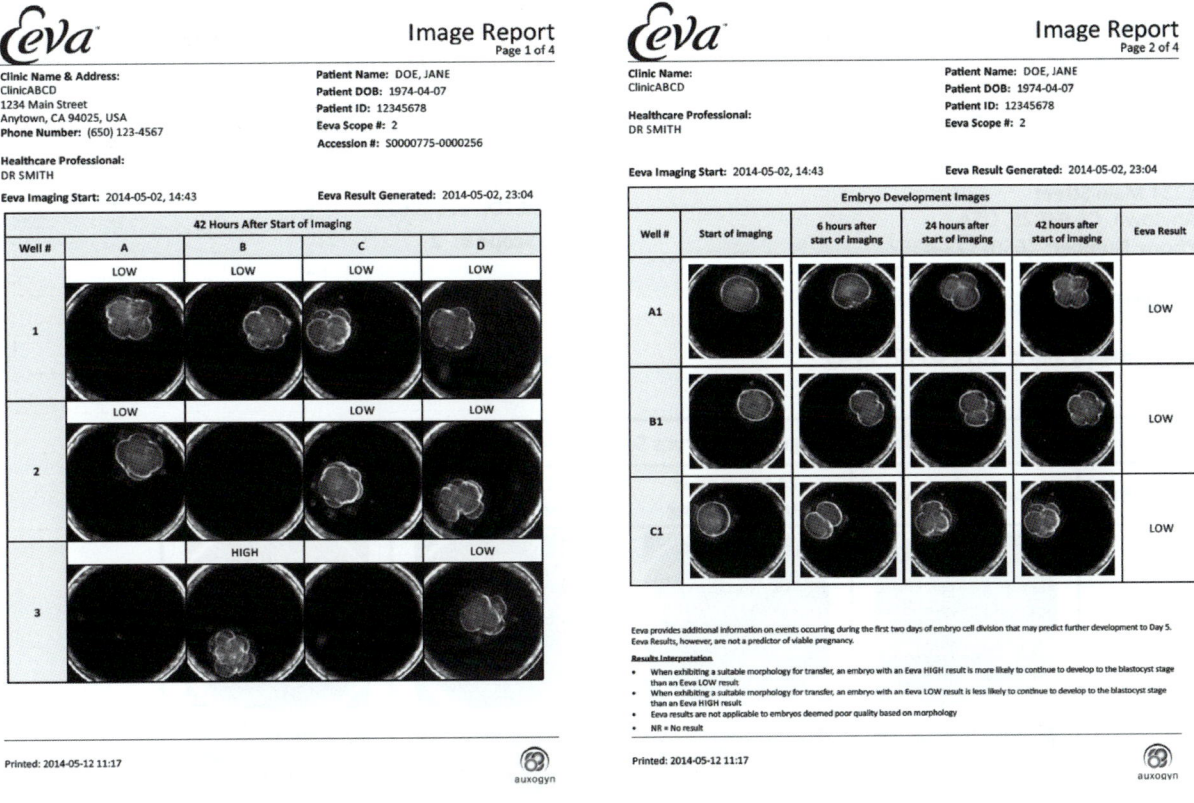

Figure 4.14 A sample Image Report. Reproduced with permission from Auxogyn Inc.

4.3.3.1. Case I: day 3 transfer

A 38-year-old patient with unexplained reason for infertility underwent programmed ovarian stimulation in 2013. After follicle stimulation, nine oocytes were retrieved, and traditional IVF insemination was performed for all oocytes. Six oocytes were fertilized and achieved 2PN (pronuclei) stage on day 1. All embryos were monitored by the Eeva System. On day 3, three embryos (A2, C1, C2) had good morphology: eight cells with even blastomeres and minimal degree of fragmentation (Table 4.1). By morphology alone, embryologists seeking an elective single embryo transfer on day 3 would have chosen any one of these three top candidate embryos, with expected variation among different embryologists in which top embryo was selected for transfer. However, the Eeva Test revealed that only top candidate C1 was an Eeva High embryo, thereby providing additional information to assist the embryologist in their decision. Using morphology followed by Eeva Test results, the clinic transferred embryo C1 (Video 4.8),

and fetal heartbeat was detected 4 weeks later. The patient gave birth to a healthy girl (2830 g) at 39 weeks 5 days.

4.3.3.2. Case II: day 3 transfer

A 32-year-old patient underwent programmed ovarian stimulation in 2013. Her partner was previously diagnosed with male infertility. After follicle stimulation, 13 oocytes were retrieved, and ICSI was performed for all oocytes. Ten oocytes were properly fertilized and achieved 2PN stage on day 1. All embryos were monitored by the Eeva System. On day 3, unfortunately there were no top (grade A, eight cells, less than 10% fragmentation, and even symmetry) embryos for selection. Eight embryos (A1, B1, D1, B2, C2, D2, A3, B3) had fair (grade B) morphology (Table 4.2). By morphology alone, embryologists seeking an elective single embryo transfer on day 3 would need to consider and select among all eight grade B embryos, with expected variation among different embryologists in which embryo would be selected for

63

Patient Report
Page 1 of 2

Clinic Name & Address:
ClinicABCD
1234 Main Street
Anytown, CA 94025, USA
Phone Number: (650) 123-4567

DR SMITH

Eeva Imaging Start: 2014-05-02, 14:43

Patient Name: DOE, JANE
Patient DOB: 1974-04-07
Patient ID: 12345678
Eeva Scope #: 2

Eeva Result Generated: 2014-05-02, 23:04

		Sample Images of Embryos Selected For Transfer		
Well #	**Start of imaging**	**24 hours after start of imaging**	**42 hours after start of imaging**	**Last Captured Image: 2012-07-16 13:53**
A1				
B1				
A2				

Eeva provides additional information on events occurring during the first two days of embryo cell division that may predict further development to Day 5. Eeva Results, however, are not a predictor of viable pregnancy.

This report contains images of your embryo(s) that were selected by your embryologist for transfer. These embryos were selected using the following sequential approach: First your embryologist performed standard embryo grading techniques to identify and select the best quality embryos deemed suitable for transfer. Second, your embryologist utilized Eeva results to aid in further selecting the embryo(s) to be transferred.

Printed: 2014-05-12 11:17

auxogyn

Figure 4.15 A sample Patient Report. Reproduced with permission from Auxogyn Inc.

transfer. The Eeva Test revealed that among these candidates, only two embryos (A1 and D2) were Eeva High, thereby providing additional information to assist the embryologist in their decision. Using morphology followed by Eeva Test results, the clinic transferred embryo D2 (Video 4.9), and fetal heartbeat was detected 4 weeks later. The patient gave birth to a healthy girl (3490 g) at 39 weeks 4 days.

4.3.3.3. Case III: day 5 transfer

A 29-year-old patient with unexplained reason for IVF underwent programmed ovarian stimulation in 2013. After stimulation, 26 oocytes were retrieved and

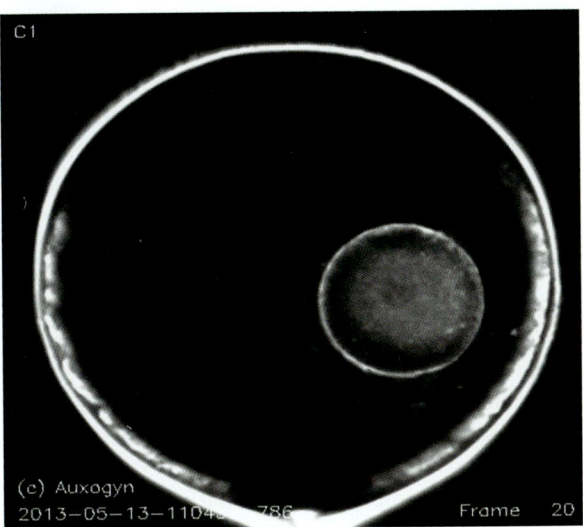

Video 4.8 Video for transferred embryo (C1) in patient case I. Reproduced with permission from Auxogyn Inc.

inseminated by ICSI. A total of 18 fertilized embryos were imaged by the Eeva System. On day 5, five embryos developed into expanded blastocysts with good inner cell mass morphology and good trophectoderm morphology (Gardner Grading Scale [28]). By morphology alone, these blastocysts had similar chances of being selected for transfer on day 5. However, the Eeva Test revealed that one of these embryos had a very short P2 value, which suggested the embryo had undergone an abnormal cleavage (one cell dividing to more than two daughter cells) during the first mitosis. Further video review confirmed that an early abnormal cleavage event had occurred for this embryo, even though the embryo developed into a morphologically good, expanded blastocyst (Video 4.10). It has been shown that abnormal cleavage events in early cell cycles are associated with aneuploidy and extremely low implantation rates [29, 30]. Another good morphology blastocyst without abnormal cleavage was transferred (Video 4.11), and the patient had ongoing pregnancy before she was released back to her obstetrician/gynecologist for prenatal care. This case demonstrates how application of Eeva Test results may be used to help avoid the transfer of a morphologically good blastocyst that had experienced an abnormal cleavage event. Such automatically available insight into what embryos have experienced during development has the potential to assist in both day 3 and day 5 transfer cases.

4.4. Ongoing research using the Eeva Test

Since the introduction of the Eeva System, the community of Eeva System users and developers has been

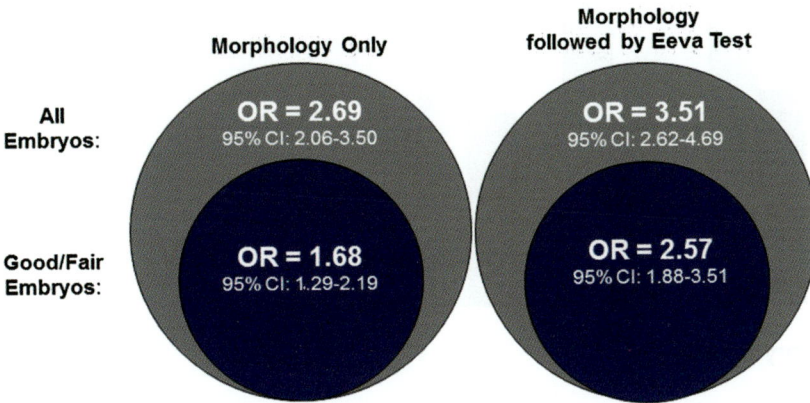

Figure 4.16 Odds ratio for predicting embryo development using Morphology Only (left) and Morphology followed by Eeva Test (right). Odds ratios and 95% confidence intervals were calculated for all embryos (represented in gray) and for the subset of embryos graded as good/fair (represented in dark blue). Reprinted from the *Journal of Assisted Reproduction and Genetics*, 2014:1–8, Diamond, *et al*, Using the Eeva Test adjunctively to traditional day 3 morphology is informative for consistent embryo assessment within a panel of embryologists with diverse experience (with permission from Elsevier).

proactively engaged in scientific and clinical research to advance the field. In this section, we review several examples of research using the Eeva System in both clinical and basic scientific settings.

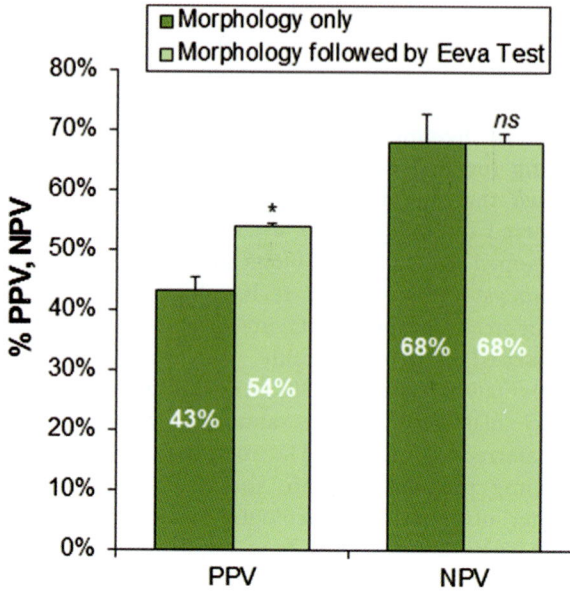

Figure 4.17 Mean positive predictive value (PPV) and mean negative predictive value (NPV) across all embryologists predicting blastocyst formation using Morphology Only and Morphology followed by the Eeva Test, among good/fair embryos. *p = 0.02, ns = not significant (error bars represent upper 95% confidence intervals). Reprinted from the *Journal of Assisted Reproduction and Genetics*, 2014:1–8, Diamond, *et al*, Using the Eeva Test adjunctively to traditional day 3 morphology is informative for consistent embryo assessment within a panel of embryologists with diverse experience (with permission from Elsevier).

4.4.1. Eeva Test results correlate to implantation and pregnancy

With the initial critical steps in developing and validating the Eeva Test as a predictive and automated clinical assay completed, further work has been taken to assess the impact of Eeva Test results on implantation and pregnancy outcomes. Towards this goal, VerMilyea *et al.* published results from a multicenter blinded study that was performed to examine the correlation between Eeva Test results and implantation/pregnancy rates [31]. A total of 331 transferred embryos with known implantation from 205 patients enrolled at six IVF clinics were analyzed. The study concluded that Eeva High embryos had a significantly higher probability of successful implantation (37%, 41/111) than Eeva Low embryos (23%, 50/220, p = 0.003, Figure 4.19).

Eeva Test results were also found to be correlated with clinical pregnancy rates. Patients were divided into two groups: patients with at least one Eeva High embryo transferred and those with no Eeva High embryo transferred. The patients' clinical characteristics for the two groups were compared, including egg age, number of eggs retrieved, number of 2PNs on day 1, and number of embryos transferred. There was no statistically significant difference found for any of the clinical characteristics assessed. However, patients with at least one Eeva High embryo transferred had significantly higher clinical pregnancy rates than those with no Eeva High embryos transferred (51% vs. 39%, p = 0.04, Table 4.3). Additional analysis of embryo implantation revealed a

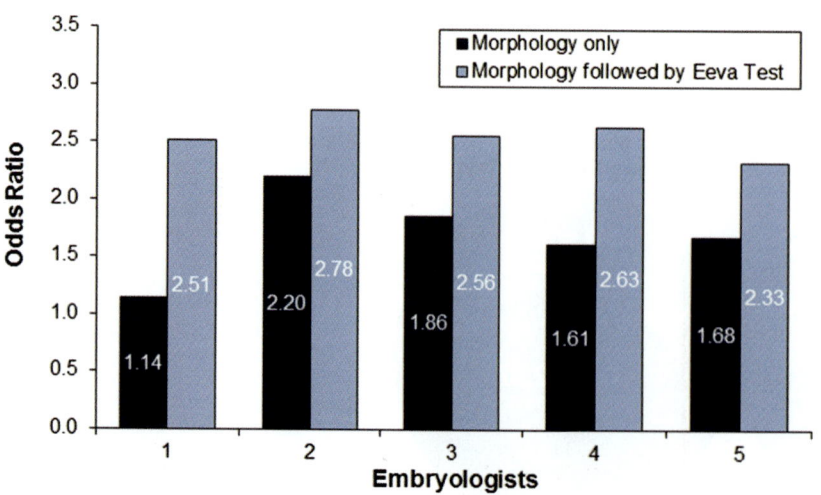

Figure 4.18 Consistent improvement in odds ratios for individual embryologists who predicted blastocyst formation using Morphology Only and Morphology followed by Eeva Test, among good/fair embryos. Reprinted from the *Journal of Assisted Reproduction and Genetics*, 2014:1–8, Diamond, *et al*, Using the Eeva Test adjunctively to traditional day 3 morphology is informative for consistent embryo assessment within a panel of embryologists with diverse experience (with permission from Elsevier).

Table 4.1 Embryo morphology on day 3, Eeva Test results and clinical decision of the embryos for Patient Case I

Embryo	Cell number	Symmetry	Fragmentation	Overall grade	Eeva Test results	P2 (hrs)	P3 (hrs)	Clinical decision
A1	6	Uneven	No Frag	B	High	10.92	0.25	Frozen
A2	8	Even	No Frag	A	Low	12.25	0.42	Frozen
B1	6	Uneven	No Frag	B	Low	6.83	5.50	Frozen
B2	6	Uneven	No Frag	B	Low	9.58	2.92	Frozen
C1	8	Even	No Frag	A	High	11.25	0.42	Transfer
C2	8	Even	No Frag	A	Low	0.08	0.42	Frozen

Table 4.2 Embryo morphology on day 3, Eeva Test results and clinical decision of the embryos for Patient Case 2

Embryo	Cell number	Symmetry	Fragmen-tation	Overall grade	Eeva Test results	P2 (hrs)	P3 (hrs)	Clinical decision
A1	7	Even	1–10%	B	High	9.50	0.17	Frozen
B1	8	Even	1–10%	B	Low	0.58	0.17	Frozen
C1	Completed fragmented		>25%	D	No result			Discard
D1	9	Even	No Frag	B	Low	2.67	0.17	Frozen
A2	2	Even	1–10%	D	Low	19.25	0.17	Discard
B2	7	Uneven	1–10%	B	Low	3.75	0.17	Discard
C2	8	Even	1–10%	B	Low	13.08	0.17	Frozen
D2	9	Even	1–10%	B	High	10.83	1.08	Transfer
A3	7	Uneven	1–10%	B	Low	0.08	0.17	Frozen
B3	6	Even	1–10%	B	Low	12.75	2.08	Frozen

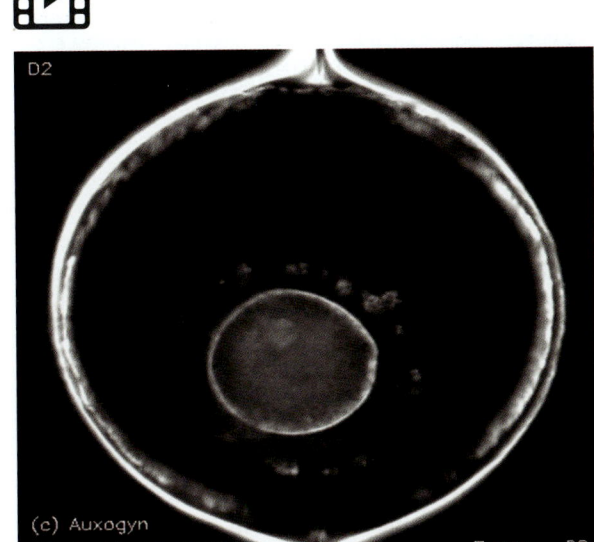

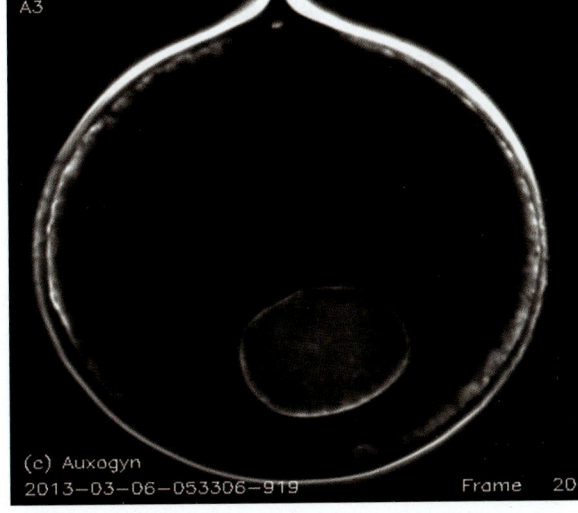

Video 4.9 Video for transferred embryo (D2) in patient Case II. Reproduced with permission from Auxogyn Inc.

Video 4.10 Video for an embryo that underwent abnormal cleavage during the first cell division, and eventually formed an expanded blastocyst (Patient Case III). Reproduced with permission from Auxogyn Inc.

Table 4.3 Clinical pregnancy rates and clinical characteristics for patients whose embryos were assessed by the Eeva System and who had at least one Eeva High embryo transferred or no Eeva High embryos transferred

Patient group	No. of patients	Egg age	No. of eggs retrieved	No. of 2PN	No. of embryos transferred	Pregnancy rate	Implantation rate
At least one Eeva High transferred	105	33.0±4.9	17.7±8.4	11.0±5.5	1.8±0.8	51% (54/105)[a]	34% (65/192)[b]
No Eeva High embryos transferred	100	33.2±5.3	17.0±8.7	10.1±5.5	1.8±0.7	39% (39/100)[a]	25% (45/182)[b]

[a] p = 0.04;
[b] p = 0.03. Adapted from VerMilyea *et al.*, Computer-automated time-lapse analysis results correlate with embryo implantation and clinical pregnancy: A blinded, multi-centre study, *Reproductive Biomedicine Online*, 2014 Dec; 29(6): 729–36.

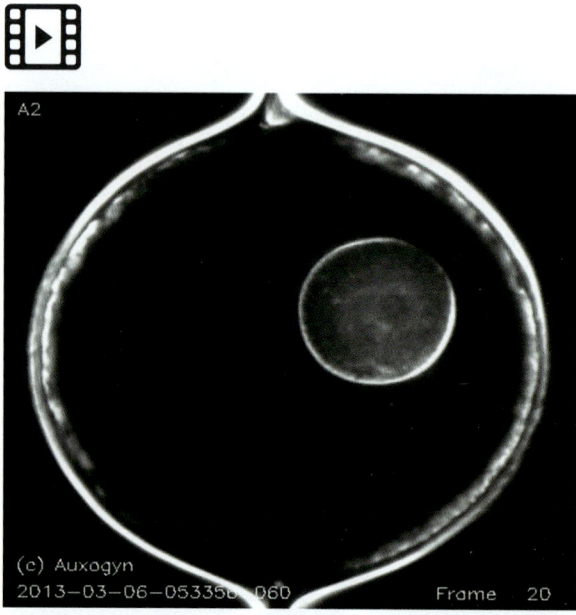

Video 4.11 Video for an embryo without abnormal cleavage that formed an expanded blastocyst, and was transferred in Patient Case III. Reproduced with permission from Auxogyn Inc.

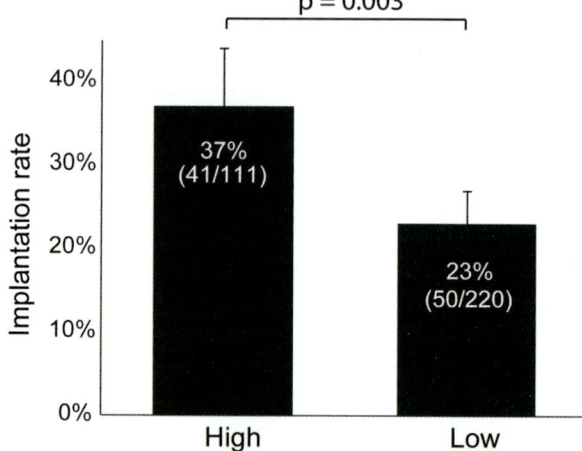

Figure 4.19 Implantation rates for Eeva two-category High vs. Low scored embryos. Error bars represent 95% upper confidence limit. The difference in implantation rates between High vs. Low embryos is statistically significant: p = 0.003. Reprinted from VerMilyea *et al.*, Computer-automated time-lapse analysis results correlate with embryo implantation and clinical pregnancy: A blinded, multi-centre study, *Reproductive Biomedicine Online*, 2014 Dec; 29(6): 729–36, with permission from Elsevier.

statistically significant difference between the two groups of patients (34% vs. 25%, p = 0.03). The results from VerMilyea *et al.* therefore add evidence to a growing body of literature surrounding cell division timings P2 and P3, and their relationship to implantation [14] and pregnancy [23, 32].

4.4.2. Consistency of Eeva Test results across clinics

A novel and key aspect of the study by VerMilyea *et al.* was that it examined the correlation between

Eeva Test results and embryo implantation across different IVF centers, which each followed their own standard procedures for embryo culture and embryo selection [31].

Consistency was assessed using a potential next generation, three-category version of the Eeva Test results (High/Medium/Low). Overall, Eeva High embryos had the highest likelihood of implantation (37%), followed by Eeva Medium (35%), and Eeva Low (15%); and the difference in implantation rates between Eeva High vs. Eeva Low embryos and Eeva Medium vs. Eeva Low embryos was statistically

Table 4.4 Clinical pregnancy rates and clinical characteristics for patients whose embryos were assessed by the Eeva System and who had at least one Eeva High embryo transferred; at least one Eeva Medium embryo transferred; or no Eeva High/Medium transferred

Patient group	No. of patients	Egg age	No. of eggs retrieved	No. of 2PN	No. of embryos transferred	Pregnancy rate	Implantation rate
At least one Eeva High transferred	105	33.0±4.9	17.7±8.4	11.0±5.5	1.8±0.8	51% (54/105)[a]	34% (65/192)[b]
No Eeva High and at least one Eeva Medium transferred	53	33.4±5.3	16.0±7.4	10.5±5.0	1.8±0.7	43% (23/53)	29% (28/97)
No Eeva High or Medium transferred	47	32.9±5.3	18.2±10	9.6±6.2	1.8±0.7	34% (16/47)[a]	20% (17/85)[b]

[a] p = 0.02;
[b] p = 0.0009. Adapted from VerMilyea *et al.*, Computer-automated time-lapse analysis results correlate with embryo implantation and clinical pregnancy: A blinded, multi-centre study, *Reproductive Biomedicine Online*, 2014 Dec; 29(6): 729–36.

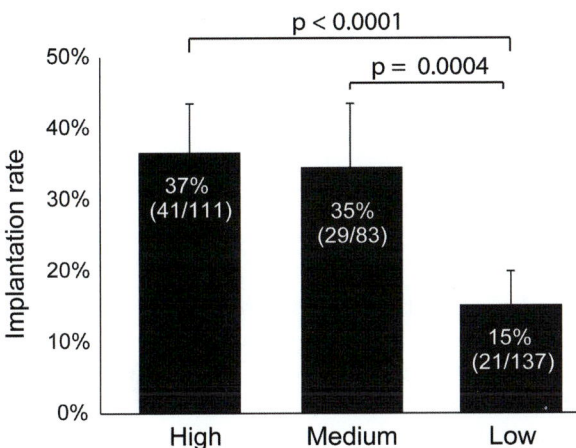

Figure 4.20 Implantation rates for Eeva three-category High, Medium and Low scored embryos. Error bars represent 95% upper confidence limit. P values comparing implantation rates between High vs. Low, High vs. Medium and Medium vs. Low are p < 0.0001, p = 0.4, and p = 0.0004, respectively. Reprinted from VerMilyea *et al.*, Computer-automated time-lapse analysis results correlate with embryo implantation and clinical pregnancy: A blinded, multi-centre study, *Reproductive Biomedicine Online*, 2014 Dec; 29(6): 729–36.

significant (p < 0.0001, p = 0.0004, respectively, Figure 4.20). Patients with one or more Eeva High embryos transferred had significantly higher clinical pregnancy rates and embryo implantation rates than those with no High or Medium embryos transferred (clinical pregnancy rates 51% vs. 34%, p = 0.02; implantation rates 34% vs. 20%, p = 0.0009, Table 4.4). Importantly, Eeva Test results correlated to embryo implantation consistently among these different IVF centers (Figure 4.21), despite the variations in protocols among these centers, including insemination technique (IVF vs. ICSI), oxygen concentration at 5% vs. 20%, VitroLife vs. Sage vs. Irvine Scientific culture media, sequential vs. single step culture protocols, etc. These results suggest that the difference in implantation and pregnancy rates between Eeva High and Eeva Low embryo transfers can be augmented when a Medium category is introduced and that the Eeva Test has broad applicability in diverse clinical IVF laboratories.

4.4.3. Advancing scientific knowledge of embryo development using the Eeva System

Although the Eeva Test is designed specifically to aid clinical decision-making and reduce time and labor requirements in the IVF laboratory, inherently, the time-lapse capabilities of the Eeva System are also useful for basic embryology research. Here we describe several research studies using the Eeva System to advance scientific knowledge of embryo development.

Insight into the mechanisms of human embryo aneuploidy. More than half of human embryos at the cleavage stage are believed to be affected by aneuploidy, which leads to failed implantation, miscarriage, or the birth of a baby with a related disorder. Using the Eeva System to study dynamic behavior of human embryos, Chavez *et al.* discovered a novel mechanism that may contribute to human embryo aneuploidy [15]. By combining non-invasive time-lapse imaging with karyotypic reconstruction of all blastomeres in four-cell human embryos, the

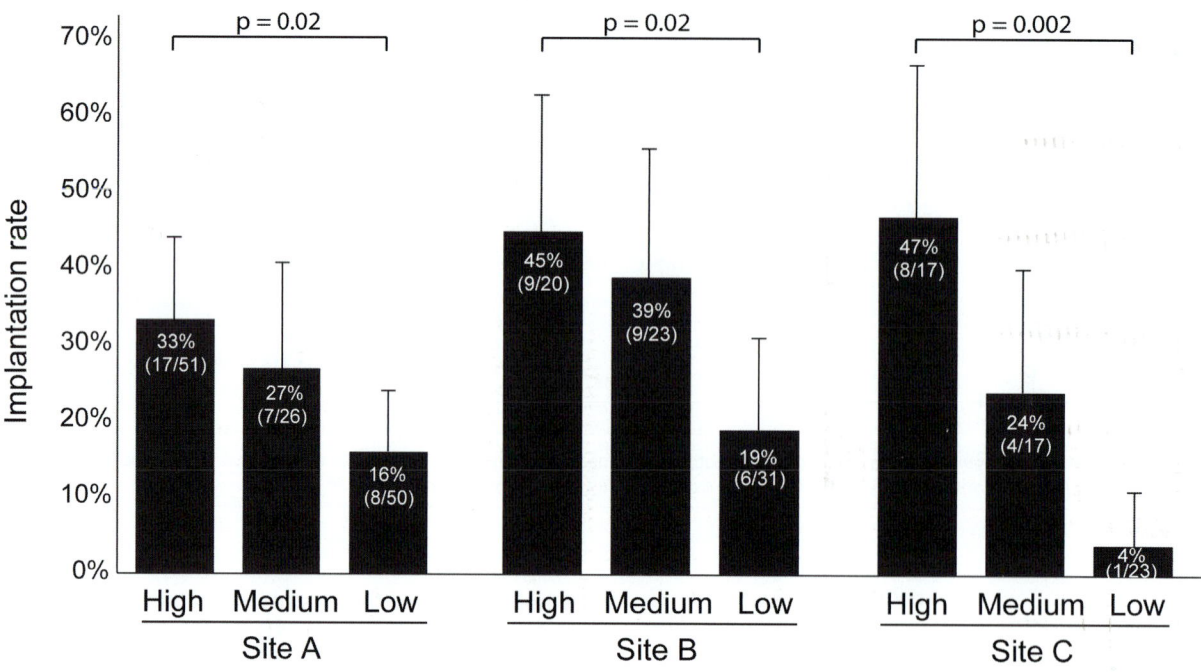

Figure 4.21 Implantation rates for embryos with High, Medium and Low scores from three clinical sites with at least 50 embryos of known implantation data per site. Error bars represent 95% upper confidence limit. For all three sites, the difference in implantation rates between High vs. Low embryos is statistically significant: (Chi-square test): p = 0.02 (site A); p = 0.02 (site B); p = 0.002 (site C). Reprinted from VerMilyea *et al.*, Computer-automated time-lapse analysis results correlate with embryo implantation and clinical pregnancy: A blinded, multi-centre study, *Reproductive Biomedicine Online*, in press, with permission from Elsevier.

researchers found that precise cell cycle parameter timings were observed in all euploid embryos to the four-cell stage, whereas 30% of aneuploid embryos exhibited parameter values within normal timing windows. Furthermore, they discovered that chromosome-containing fragments/micronuclei frequently emerge during embryo development and may persist or become reabsorbed. The authors therefore hypothesized that dynamic fragmentation could be a mechanism leading towards various states of human embryonic aneuploidy. As part of this study, a preliminary automation tool compatible with the framework of the Eeva System was developed to assist in the detection of fragmentation dynamics (Video 4.12). Although further work is needed, this study demonstrates potential for the application of time-lapse and automated image analysis tools to aneuploidy risk assessment.

Insight into atypical embryo dynamics. In addition to fragmentation, human embryos undergo a wide spectrum of dynamic behaviors during preimplantation development. Limited understanding of these dynamic behaviors and their association

to embryo quality existed before commercially available time-lapse technology was introduced in IVF clinics. Using the Eeva System, Athayde Wirka *et al.* published a comprehensive characterization of the prevalence and consequence of various embryo dynamic behaviors, and identified novel dynamic embryo phenotypes in human embryos from IVF clinics [33]. Three phenotypes were highly prevalent (found in > 70% of the patients) and could be selected for transfer based on traditional morphology: 1. Abnormal syngamy phenotype was defined by the presence of disordered pronuclear movement, accompanied by delayed dispersion of the pronuclear envelopes (Video 4.13); 2. abnormal first cytokinesis phenotype was defined by the presence of oolema ruffling before completion of the first cytokinesis (Video 4.14); 3. Abnormal cleavage phenotype was defined as a single cell division event producing more than two cells (Video 4.15). Embryos with any of these phenotypes were found to have poorer developmental potential (lower blastocyst formation rate and implantation rate). Interestingly, the abnormal cleavage phenotype was

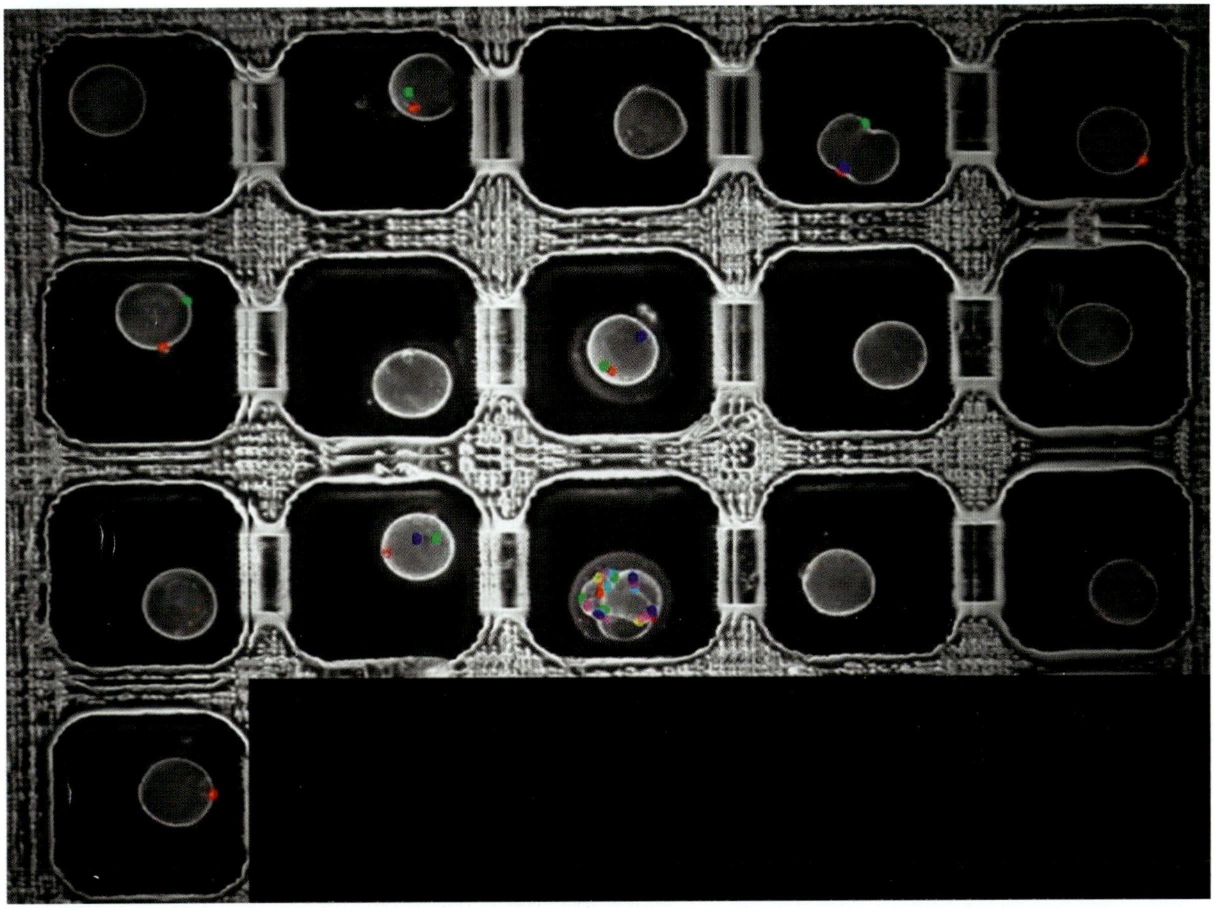

Video 4.12 Automated segment detection of fragments for human embryos. Reprinted by permission from Macmillan Publishers Ltd: Nature Communications. Chavez *et al.*, Dynamic blastomere behaviour reflects human embryo ploidy by the four-cell stage, *Nature Communications* 3:1251, copyright 2012.

highly enriched in embryos selected for day 3 transfer by morphology only (29%). These study results, in combination with other studies reporting poorer developmental potential for abnormally cleaving embryos [29, 30], suggest that the Eeva System and potential automation capabilities may be useful for excluding low potential embryos for transfer to improve IVF success rates.

Insight into the contribution of sperm to embryo development. While studies examining IVF human embryos provide the most relevant information for clinical application, it is often difficult to perform mechanistic studies with human embryos due to ethical and logistical limitations. For this reason,

the Eeva System has been used in animal model studies that allow for deeper investigation into the underlying mechanism of embryo development and demise. Among the animal models, Rhesus macaques (*Macaca mulatta*) and humans share a high degree of similarity in terms of female reproductive physiology, fundamental aspects of pre-implantation development, and frequencies of aneuploidy at the cleavage stage as well as successful progression to the blastocyst stage [34]. Using Rhesus macaques as the model system, Burruel *et al.* used the Eeva System to study the effect of oxidative damage to sperm on embryo development [35]. Burruel *et al.* discovered that oxidative stress exposed to sperm by reactive

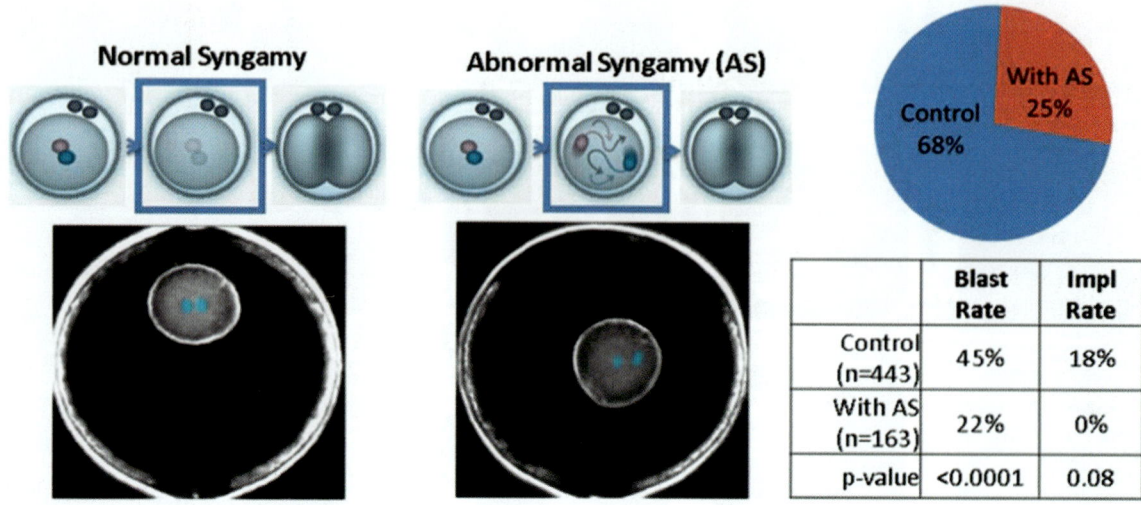

Abnormal Syngamy

Normal Syngamy

Abnormal Syngamy (AS)

Control 68%

With AS 25%

	Blast Rate	Impl Rate
Control (n=443)	45%	18%
With AS (n=163)	22%	0%
p-value	<0.0001	0.08

Abnormal syngamy is associated with poorer developmental potential

Athayde Wirka et al. *Fertil & Steril,* 2014

Video 4.13 Video of embryos with and without abnormal syngamy phenotype, and data showing association between the phenotype and poor developmental outcome. Data are from Athayde Wirka K *et al.* Atypical embryo phenotypes identified by time-lapse microscopy: high prevalence and association with embryo development. *Fertil Steril* 2014; 101(6): 1637–48.e5. Adapted with permission from Elsevier.

oxygen species can result in abnormal cleavage events that lead to embryonic and fetal demise. The results of the study, although performed in an animal model, have implications for human conditions impacted by oxidative stress including inherited disease, diet, environmental factors, and cancer.

Research studies using the Eeva System have improved our understanding of human embryo development. It can be expected that time-lapse technology will continue to add value in a research setting and help answer fundamental questions about underlying mechanisms for dynamic embryo behavior. Further technological improvements in automation and optics promise to increase the efficiency of research, and will help translate such research findings into new solutions for reproductive medicine.

4.5. Conclusion

The Eeva Test is a first-of-its-kind prognostic test that has been validated in a multicenter study to add value to embryologists' morphological evaluations by improving embryo assessment and reducing inter-observer variability. It combines both highly robust predictive markers of embryo development and a novel automated assay to rapidly obtain quantitative measurements of key time-lapse markers, which have been shown to be valuable across IVF laboratories under different protocols. Components of the Eeva Test fit into the current busy workflow in IVF clinics, which minimizes the extra time and labor associated with other time-lapse systems. With these advantages, the Eeva Test provides uniquely valuable information

Abnormal First Cytokinesis (A1^cyt)

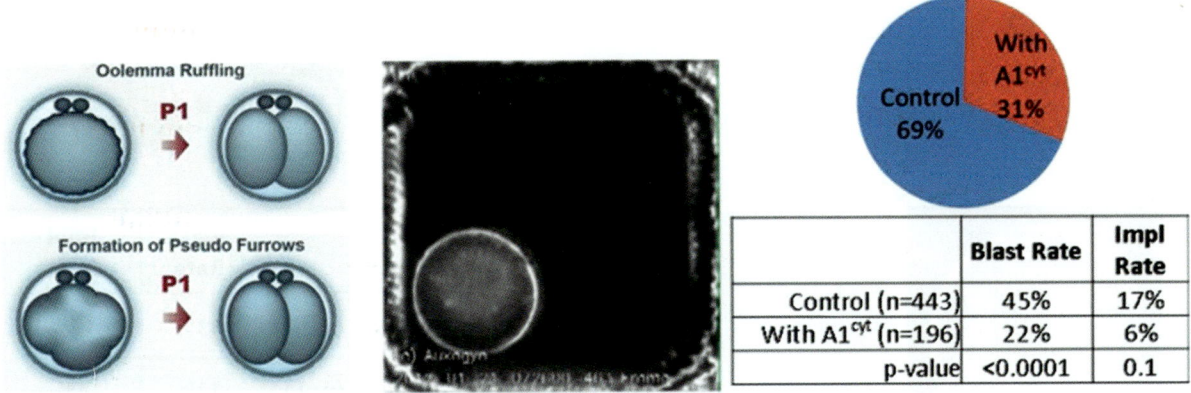

	Blast Rate	Impl Rate
Control (n=443)	45%	17%
With A1^cyt (n=196)	22%	6%
p-value	<0.0001	0.1

A1^cyt phenotype is associated with poorer developmental potential

Athayde Wirka et al. *Fertil & Steril,* 2014

Video 4.14 Video of an embryo with abnormal first cytokinesis phenotype, and data showing association between the phenotype and poor developmental outcome. Data are from Athayde Wirka K *et al*. Atypical embryo phenotypes identified by time-lapse microscopy: high prevalence, and association with embryo development. *Fertil Steril* 2014; 101(6): 1637–48.e5. Adapted with permission from Elsevier.

to help embryologists and clinicians make decisions with ease and confidence, and thus allows for wide adoption of time-lapse technology into IVF clinics.

There is a growing body of evidence that demonstrates the benefits of time-lapse technology for IVF patients and clinics. In particular, the Eeva Test is a pioneering, fully automated prognostic test that has been validated to aid diverse embryologists in improving embryo selection. Cleared by the FDA through the de novo pathway for innovative, first-of-its-kind devices, the Eeva System will serve as a benchmark for evaluating the safety and efficacy of future devices in its category. Future generations of the Eeva System are being developed to further improve its predictive power and ease of

use; such enhancements will be clinically validated and, ideally, integrated with minimal disruption to the lab.

Acknowledgments

We acknowledge the physicians, embryologists, and patients who participated in the development and validation of the Eeva Test. We are grateful to the IVF center at VU University Medical Center Amsterdam for kindly sharing data, and IKS International for providing images. We also thank the clinical, scientific, and computer vision groups at Auxogyn Inc. for their technical contributions and insightful discussion.

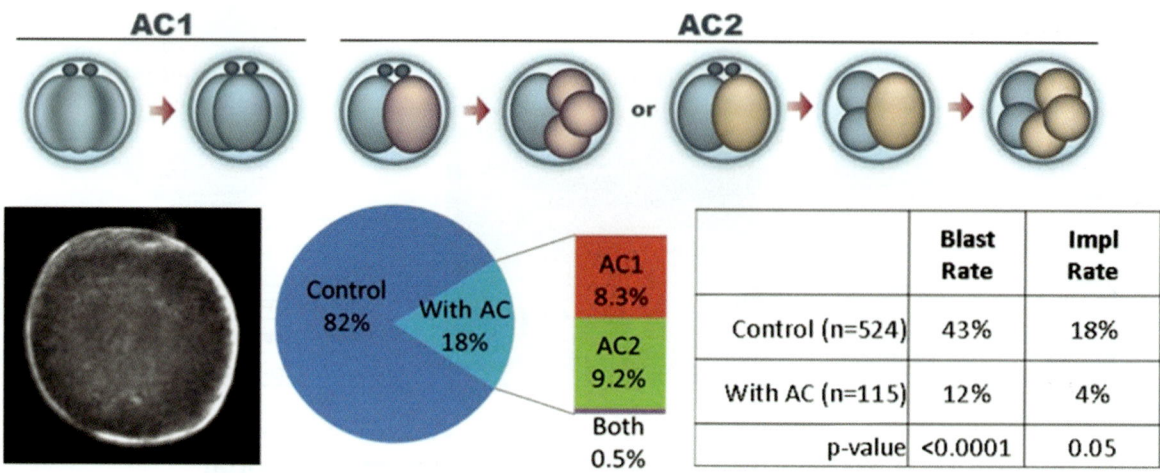

- **AC1 and AC2 embryos are often selected for Day 3 transfer** (28.6%)
- **AC embryos are often good quality** (46.9% 6–10 cells, ≤10% frag)
- **Implantation Rate: 3.7%**

Athayde Wirka et al. *Fertil & Steril,* 2014

Video 4.15 Video of an embryo with abnormal cleavage phenotype, and data showing association between the phenotype and poor developmental outcome. Data are from Athayde Wirka K *et al.* Atypical embryo phenotypes identified by time-lapse microscopy: high prevalence and association with embryo development. *Fertil Steril* 2014; 101(6): 1637–48.e5. Adapted with permission from Elsevier.

References

1. Abeyta M, Behr B. Morphological assessment of embryo viability. *Semin Reprod Med* 2014; 32(02): 114–26.

2. Paternot G, Devroe J, Debrock S, D'Hooghe TM, Spiessens C. Intra- and inter-observer analysis in the morphological assessment of early-stage embryos. *RBQE* 2009; 7: 105.

3. Kirkegaard K, Agerholm IE, Ingerslev HJ. Time-lapse monitoring as a tool for clinical embryo assessment. *Hum Reprod* 2012; 27(5): 1277–85.

4. Wong C, Chen AA, Behr B, Shen S. Time-lapse microscopy and image analysis in basic and clinical embryo development research. *Reprod Biomed Online* 2013; 26(2): 120–9.

5. Chen AA, Tan L, Suraj V, Reijo Pera R, Shen S. Biomarkers identified with time-lapse imaging: discovery, validation, and practical application. *Fertil Steril* 2013; 99 (4): 1035–43.

6. Aparicio B, Cruz M, Meseguer M. Is morphokinetic analysis the answer? *Reprod Biomed Online* 2013; 27(6): 654–63.

7. Harper J, Cristina Magli M, Lundin K, Barratt CLR, Brison D. When and how should new technology be introduced into the IVF laboratory? *Hum Reprod* 2012; 27(2): 303–13.

8. Bodurtha J, Strauss JF, 3rd. Genomics and perinatal care. *N Engl J Med* 2012; 366(1): 64–73.

9. Martini E, Flaherty SP, Swann NJ, Payne D, Matthews CD. Analysis

of unfertilized oocytes subjected to intracytoplasmic sperm injection using two rounds of fluorescence in-situ hybridization and probes to five chromosomes. *Hum Reprod* 1997; 12(9): 2011–18.

10. Hardarson T, Lofman C, Coull G, Sjogren A, Hamberger L, Edwards RG. Internalization of cellular fragments in a human embryo: time-lapse recordings. *Reprod Biomed Online* 2002; 5(1): 36–8.

11. Mio Y, Maeda K. Time-lapse cinematography of dynamic changes occurring during in vitro development of human embryos. *Am J Obstet Gynecol* 2008; 199(6): 660 e1–5.

12. Pribenszky C, Matyas S, Kovacs P, Losonczi E, Zadori J, Vajta G. Pregnancy achieved by transfer of a single blastocyst selected by time-lapse monitoring. *Reprod Biomed Online* 2010; 21(4): 533–6.

13. Wong C, Loewke K, Bossert N, Behr B, De Jonge C, Baer T, et al. Non-invasive imaging of human embryos before embryonic genome activation predicts development to the blastocyst stage. *Nat Biotechnol* 2010; 28(10): 1115–21.

14. Meseguer M, Herrero J, Tejera A, Hilligsoe KM, Ramsing NB, Remohi J. The use of morphokinetics as a predictor of embryo implantation. *Hum Reprod* 2011; 26(10): 2658–71.

15. Chavez SL, Loewke KE, Han J, Moussavi F, Colls P, Munne S, et al. Dynamic blastomere behaviour reflects human embryo ploidy by the four-cell stage. *Nat Commun* 2012; 3: 1251.

16. Wong C, Chen A, Behr B, Shen S. Time-lapse microscopy and image analysis in basic and clinical embryo development research. *Reprod Biomed Online* 2013 Feb; 26(2): 120–9.

17. Kaser DJ, Racowsky C. Clinical outcomes following selection of human preimplantation embryos with time-lapse monitoring: a systematic review. *Hum Reproduction Update* 2014; 20(5): 617–31.

18. Munoz M, Cruz M, Humaidan P, Garrido N, Perez-Cano I, Meseguer M. Dose of recombinant FSH and oestradiol concentration on day of HCG affect embryo development kinetics. *Reprod Biomed Online* 2012; 25(4): 382–9.

19. Dal Canto M, Coticchio G, Mignini Renzini M, De Ponti E, Novara PV, Brambillasca F, et al. Cleavage kinetics analysis of human embryos predicts development to blastocyst and implantation. *Reprod Biomed Online* 2012; 25(5): 474–80.

20. Ciray HN, Aksoy T, Goktas C, Ozturk B, Bahceci M. Time-lapse evaluation of human embryo development in single versus sequential culture media – a sibling oocyte study. *J Assist Reprod Genet* 2012; 29(9): 891–900.

21. Basile N, Morbeck D, Garcia-Velasco J, Bronet F, Meseguer M. Type of culture media does not affect embryo kinetics: a time-lapse analysis of sibling oocytes. *Hum Reprod* 2013; 28 (3): 634–41.

22. Kirkegaard K, Hindkjaer JJ, Ingerslev HJ. Effect of oxygen concentration on human embryo development evaluated by time-lapse monitoring. *Fertil Steril* 2013; 99(3): 738–44 e4.

23. Rubio I, Galán A, Larreategui Z, Ayerdi F, Bellver J, Herrero J, et al. Clinical validation of embryo culture and selection by morphokinetic analysis: a randomized, controlled trial of the EmbryoScope. *Fertil Steril* 2014; 102(5): 1287–94.

24. Conaghan J, Chen AA, Willman SP, Ivani K, Chenette PE, Boostanfar R, et al. Improving embryo selection using a computer-automated time-lapse image analysis test plus day 3 morphology: results from a prospective multicenter trial. *Fertil Steril* 2013; 100(2): 412–9 e5.

25. Ebner T, Shebl O, Moser M, Mayer RB, Arzt W, Tews G. Group culture of human zygotes is superior to individual culture in terms of blastulation, implantation and life birth. *Reprod Biomed Online* 2010; 21 (6): 762–8.

26. Diamond M, Suraj V, Behnke E, Yang X, Angle M, Lambe-Steinmiller J, et al. Using the Eeva Test$^{™}$ adjunctively to traditional day 3 morphology is informative for consistent embryo assessment within a panel of embryologists with diverse experience. *J Assist Reprod Genet* 2014: 1–8.

27. Baxter Bendus AE, Mayer JF, Shipley SK, Catherino WH. Interobserver and intraobserver variation in day 3 embryo grading. *Fertil Steril* 2006; 86(6): 1608–15.

28. Gardner DK, Schoolcraft WB. In vitro culture of human blastocysts. In: R Jansen, D Mortimer, editors. *Toward Reproductive Certainty: Fertility and Genetics Beyond*. Carnforth, UK: Parthenon Publishing; 1999. pp. 378–88.

29. Zaninovic N, Ye Z, Zhan Q, Clarke R, Rosenwaks Z. Cell stage onsets, embryo developmental potential and chromosomal abnormalities in embryos exhibiting direct unequal cleavages (DUCs). *Fertil Steril* 2013; 100(3): S242.

30. Rubio I, Kuhlmann R, Agerholm I, Kirk J, Herrero J, Escribá M-J, et al. Limited implantation success of direct-cleaved human zygotes: a time-lapse study. *Fertil Steril* 2012; 98(6): 1458–63.

31. VerMilyea MD, Tan L, Anthony JT, Conaghan J, Ivani K, Gvakharia M, et al. Computer-automated time-lapse analysis results correlate with embryo implantation and clinical

pregnancy: A blinded, multi-centre study. *Reprod Biomed Online*, 2014; 29(6): 729–36.

32. Meseguer M, Rubio I, Cruz M, Basile N, Marcos J, Requena A. Embryo incubation and selection in a time-lapse monitoring system improves pregnancy outcome compared with a standard incubator: a retrospective cohort study.

Fertil Steril 2012; 98(6): 1481–9.e10.

33. Athayde Wirka K, Chen AA, Conaghan J, Ivani K, Gvakharia M, Behr B, et al. Atypical embryo phenotypes identified by time-lapse microscopy: high prevalence and association with embryo development. *Fertil Steril* 2014; 101(6): 1637–48.e5.

34. Phillips KA, Bales KL, Capitanio JP, Conley A, Czoty PW, 't Hart BA, et al. Why primate models matter. *Am J Primatology* 2014; 76 (9): 801–27.

35. Burruel V, Klooster K, Barker C, Reijo Pera R, Meyers S. Abnormal early cleavage events predict early embryo demise: sperm oxidative stress and early abnormal cleavage. *Sci Rep* 2014; 4: 6598.

Morphological cytoplasmic oocyte alterations: embryo kinetics of dysmorphic oocytes

Belén Aparicio-Ruiz and Maria Jose de los Santos

5.1. Introduction

Before mature oocytes arrive at IVF laboratories, oocytes need to initiate a rather long trip from the arrested phase to the diplotene stage of prophase I, which occurs in the fetal ovary [1, 2]. Oocytes inside endocrinologically inactive primordial follicles are able to develop to mature oocytes at the time of ovulation. However, before this, they have to go through a series of follicular growing phases, the first is independent of gonadotrophins and the second is dependent on them [3]. This trip takes around 180 days and is critical for oocytes to progressively acquire oocyte competence [1].

During the journey, the expression of a new set of genes, cytoplasmic organization, reorganization, and formation of new organelles and generation of the glucoprotein layer of the zona pellucida take place. It has been stated that the presence of endocytic vacuoles, reorganization of organelles, and appearance of necrotic areas arise in the final stage of maturation, during the time of the first polar body extrusion [4].

For example, the Golgi apparatus and the endoplasmic reticulum undergo reconfiguration and go to the cortex of the cell [5], and ooplasma may also accumulate lipid droplets, multivesicular and crystalline bodies, which may remain until ovulation and may play a role during fertilization and further embryo development.

From a consensus point of view, although a morphologically normal metaphase II [6] oocyte should be spherical in shape, and have no large perivitelline space, a homogeneous cytoplasm with no inclusions, and zona pellucida of around 15–20 μm thick, physiological variation throughout oocyte growth may lead to differences in oocyte phenotypes. For example, the hormonal follicular environment appears to be related to the incidence of certain oocyte morphological aspects, such as polar body fragmentation, a large perivitelline space, and presence of cytoplasmic inclusions.

Actually it is interesting to note that for some authors, the so-called "normal oocyte phenotype" is actually the least normal in terms of frequency. As a matter of fact, between 60 and 70% of the oocytes from stimulated cycles present one phenotypic variation or more [7, 8].

Experts from the Istanbul consensus paper have expressed that, given the unknown biological significance of the majority of oocyte dysmorphisms, their main interest and concern lie in very few oocyte phenotypes; e.g., oocytes with severe central granularity, also called cluster oocytes; oocytes with a smooth endoplasmic reticulum; oocytes with vacuoles > 14 μm; oocytes with big polar bodies [9].

Many reviews and scientific articles on oocyte dysmorphisms and their association with adverse IVF outcomes have been published. Yet despite extensive reviews [10, 11], a consensus about the predictive value to consider an egg to be bad or good based merely on its morphology is lacking.

This chapter offers the opportunity to evaluate oocyte morphology from another perspective. To date, no data are available on the kinetics of the embryos generated from such oocytes immediately after fertilization.

Therefore the main objective of the current chapter was to look at oocytes from a kinetic point of view to describe the cell cycle events that take place during embryo development. This retrospective study included 5,252 denuded oocytes from 450 egg donation cycles which underwent ICSI between 2012 and 2014.

Time-Lapse Microscopy in In-Vitro Fertilization, ed. Marcos Meseguer. Published by Cambridge University Press.
© Cambridge University Press 2016.

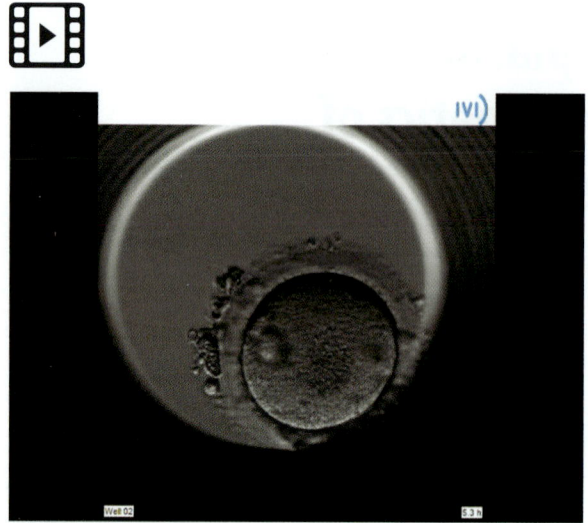

Video 5.1 Even though it is not an especially excessive granulated oocyte, central granularity in the cytoplasm can be observed. This video shows an example of direct cleavage, from one to three cells in less than 5 hours.

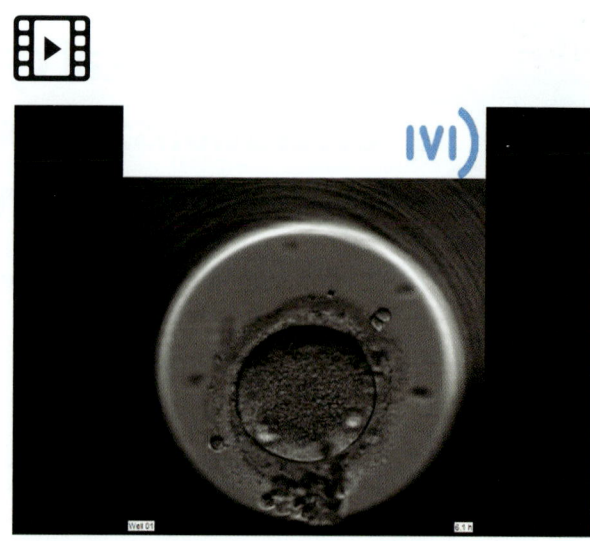

Video 5.2 Oocyte with pronounced central granularity. It is a good example of an embryo with cleavage rates within the ranges compatible with implantation.

5.1.1. Assessment of oocyte phenotypic characteristics

Oocyte assessment was recorded during microinjection and was confirmed under an Embryosocope by a senior embryologist.

Oocyte phenotypes were grouped into 13 morphological categories based on the cytoplasmic and extracytoplasmic features observed in oocytes. Oocytes were classified as follows: normal oocytes (N) (with homogeneous or very slightly granulated cytoplasm, a non-fragmented, normal size polar body, and without an overlarge perivitelline space, and a normal zona pellucida); oocytes with altered necrotic bodies in the cytoplasm (NB); oocytes with refractile bodies (RB); oocytes with homogeneous very granular cytoplasm (HGC); oocytes with severe centrally located granular cytoplasm (CLGC); smooth endoplasmic reticulum (SER); oocytes with vacuoles (VAC); oocytes with multiple abnormalities (M).

Morphological extra-cytoplasmic alterations were also recorded as follows: oocytes with a large perivitelline space (LPS); oocytes with abnormal polar bodies (PB); elongated oocytes (E) (Video 5.21); oocytes with debris in the perivitelline space (Db).

Oocytes were grouped as follows according to the appearance of the zona pellucida: oocytes with normal zona (nZP); oocytes with an elongated zona (eZP); oocytes with a pigmented or dark zona (pZP);

oocytes with a thick zona (thckZP); oocytes with a thin zona (tZP); oocytes with an evident bilayered zona (bZP); oocytes with a zona of irregular thickness (irZP); oocytes with a rough inner zona (izrZP).

Provided below is a brief description of each of these phenotypic characteristics assessed in the oocytes.

5.1.2. Homogeneous granular and located granular cytoplasm

In some cases, it was possible to observe granulation in the cytoplasm, which was homogeneously or centrally localized. A centrally located granular cytoplasm (CLCG) is diagnosed as being a larger, dark, granular area. This phenotype may obey the central localization of organelles in the cytoplasm. Severity of granulation is based on the diameter of the granular area and depth (Video 5.1 and Video 5.2).

5.2. Smooth endoplasmic reticulum (SER)

Transmission electron microscopy has confirmed that this dimorphism is the result of a massive accumulation of tubular-type smooth endoplasmic reticulum clusters [13]. Even though there is currently no clear explanation for the mechanisms that lead to the appearance of SER aggregates, different groups have observed a positive correlation between presence of

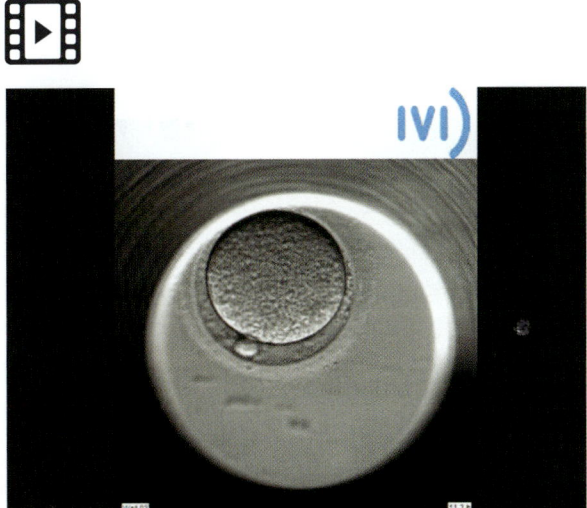

Video 5.3 Very granular oocyte with abnormal fertilization and cleavage pattern. This embryo shows an abnormal development. Observe how the granular aspect of cytoplasm changes over time.

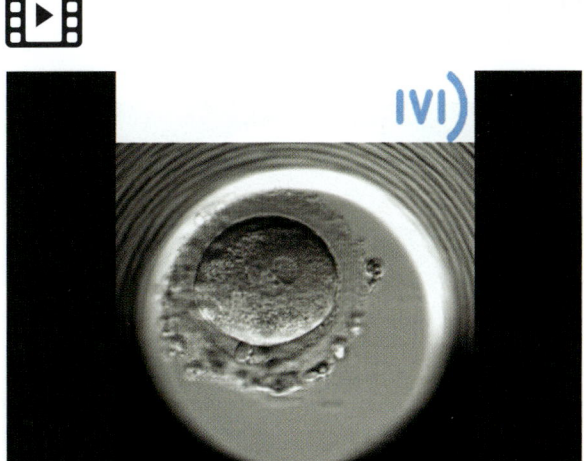

Video 5.4 Very granular oocyte with normal fertilization and cleavage pattern. As the previous oocyte the granular aspect of cytoplasm changes over time which may support the idea that granularity may be a dynamic feature and therefore the phenotype can be difficult to evaluate within a single observation.

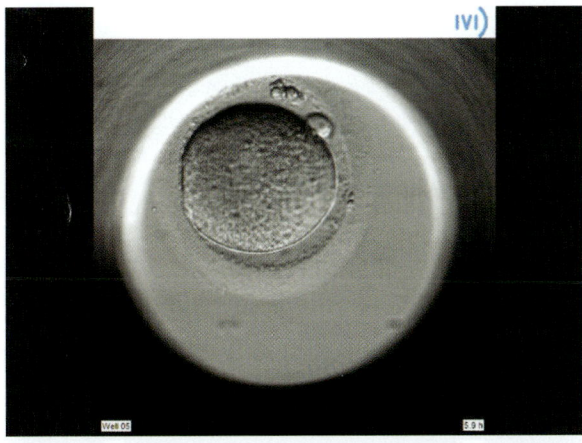

Video 5.5 In this video we can observe the presence of smooth endoplasmic reticulum as well as some vacuoles. Interestingly, we can point out that vacuoles are protuberant when pronuclei appear but disappear in the first cleavage. This oocyte is slightly uneven at two-cell stage and shows slow cleavage pattern at the four-cell stage.

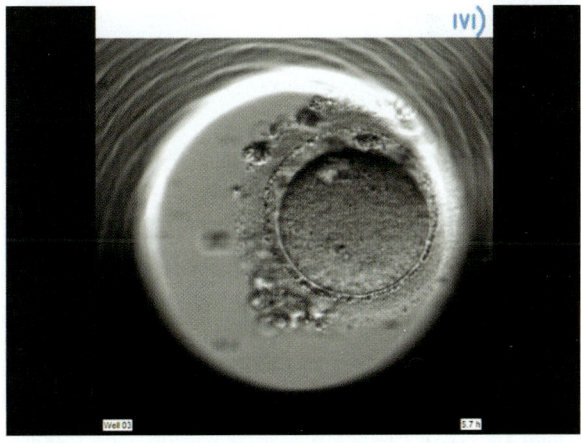

Video 5.6 In this case smooth endoplasmic reticulum can also be observed, this time combined with refractile bodies. According to the embryo development this SER positive oocyte has a good prognosis.

SER and serum estradiol concentrations on the day of ovulation induction, as well as concentrations of anti-Mullerian hormone [14, 15, 16] (Video 5.5 and 5.6).

5.3. Vacuoles

One of the commonest oocyte dysmorphisms is cytoplasmic vacuolization. Size might vary, as may number, and this can be observed in 5–12% of oocytes. Appearance of vacuoles can be spontaneous or through the fusion of pre-existing vesicles which derive from the Golgi apparatus or the smooth endoplasmic reticulum [17] (Video 5.7 and Video 5.8).

5.4. Refractile and necrotic bodies

Both these phenotypes include a variety of cytoplasmic features that are considered small necrotic areas of around 10 μm in size, composed of lipid droplets

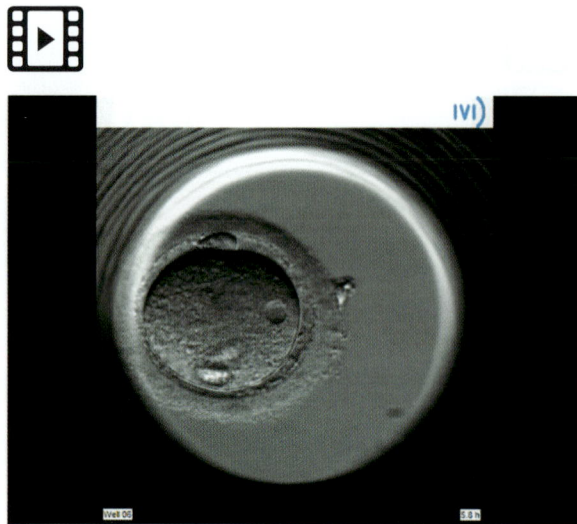

Video 5.7 Oocyte with a small vacuole. If we go further on embryo development we can clearly see how this vacuole stays in one of the blastomeres throughout the development. And, as expected, there is no influence on embryo cleavage timings.

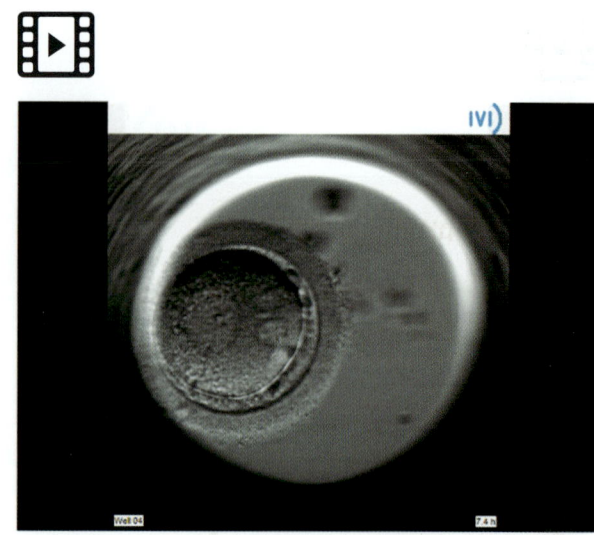

Video 5.8 Oocyte with a big vacuole. The blastomere with the vacuole however does not present any kind of delay in cleavage, which suggests that eventually even vacuoles with important dimensions do not necessarily affect embryo morphokinetics.

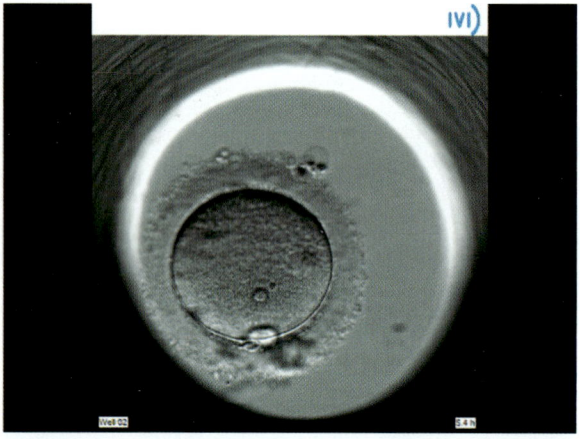

Video 5.9 Oocyte having a refractile body, in this case it is located near the polar body. It remains in one of the blastomeres without fragmenting, and with no apparent influence on embryo development.

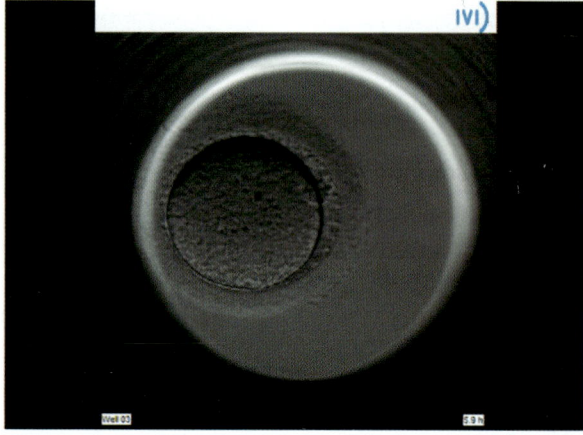

Video 5.10 Another example of refractile body. In this case vacuoles, cytoplasm polarization, and severe membrane irregularities appear throughout embryo development.

that can be found isolated or in groups [18, 19] (Video 5.9, Video 5.10, Video 5.11, and Video 5.12).

5.5. Large perivitelline space (LPS)

This dysmorphism can be clearly visualized under an inverted microscope due to an increase in the perivitelline space, and there is no contact, so the oocyte seems to float within the interior of the zona pellucida. Some authors have reported that an increase in the perivitelline space can be related with extreme oocyte maturation [20, 21] (Video 5.13 and Video 5.14).

5.6. Cellular debris in the perivitelline space

The presence of cellular debris in the perivitelline space is considered an extra-cytoplasmic anomaly. Even though there is currently no evidence for the influence

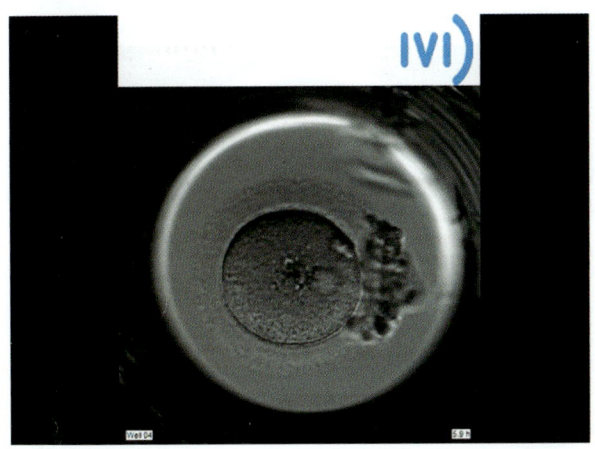

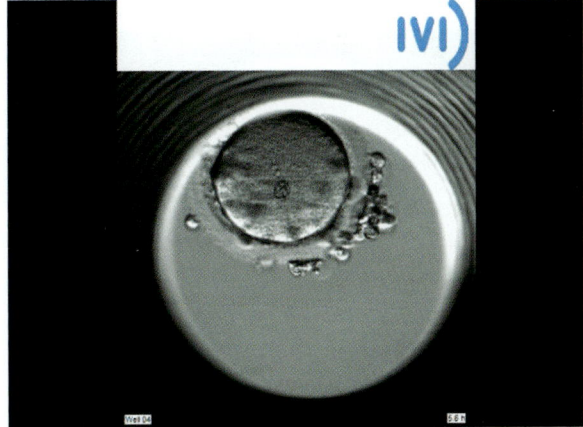

Video 5.11 and 5.12 In these videos we can see two oocytes with a necrotic body. One of them is relatively short, where we cannot see the embryo development but we found it interesting because the dismorphism studied is very clear.

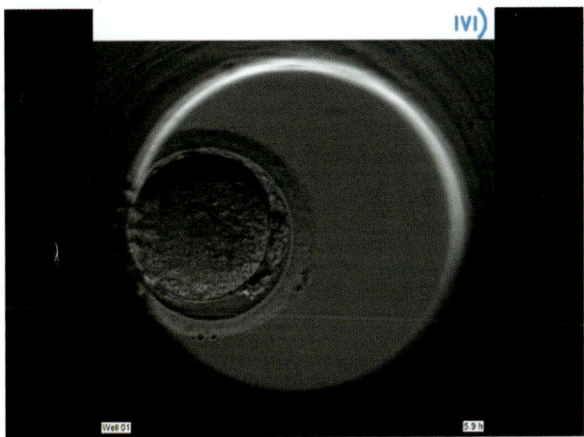

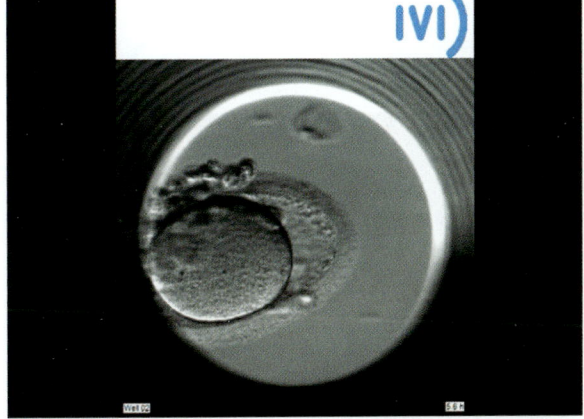

Video 5.13 In this video we can see a larger perivitelline space than in a normal oocyte however it is not exaggeratedly pronounced. A delayed t5 timing can be observed although blastulation finally occurs within normal ranges.

Video 5.14 Large perivitelline space generated by an elongated zona pellucida. During first cleavage one of the blastomeres rapidly occupies the extra space, affecting the organization of blastomeres but not the cleavage timings.

of this dysmorphism, and no quantitative measurement to determine it, some authors have related it to internal zona pellucida deterioration, while others have observed that the presence of detritus may be a sign of an "overdose" of gonadotrophins used in stimulation protocols [22] (Video 5.15).

5.7. Dysmorphisms in the zona pellucida

The zona pellucida undergoes changes throughout embryo development not only in its thickness but also

in its general appearance. Most of these dysmorphisms are clearly visible when ICSI is performed because, presently, the oocyte needs to be turned to establish the best position to perform injection in order to avoid damage to the metaphase spindle (Video 5.16, Video 5.17, Video 5.18, Video 5.19, and Video 5.20).

5.8. First polar body (PB)

It seems that the morphology of the first PB changes within hours of in vitro culture and, as such, may vary depending on the observation time [23].

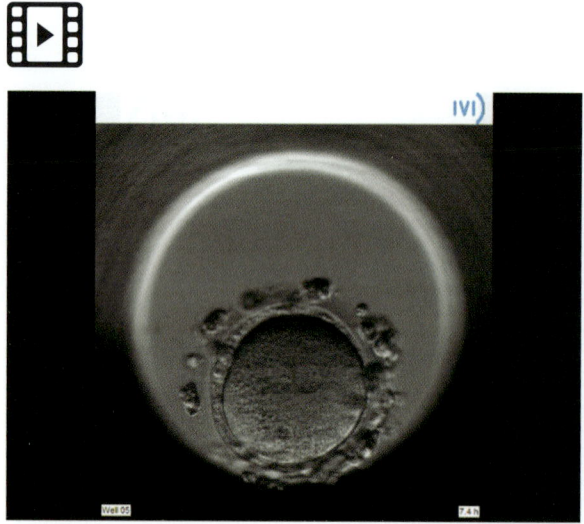

Video 5.15 Small amounts of cellular debris are seen between the zona pellucida and the cytoplasm that tend to fade away with the embryo development.

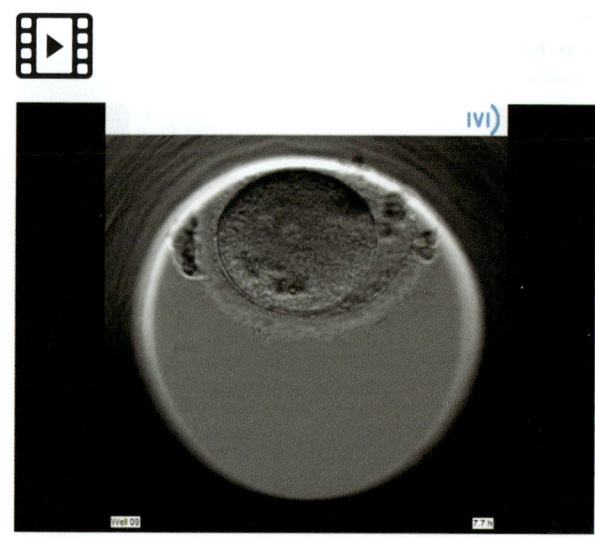

Video 5.16 This video shows an oocyte which combines a large perivitelline space with an elongated zona pellucida.

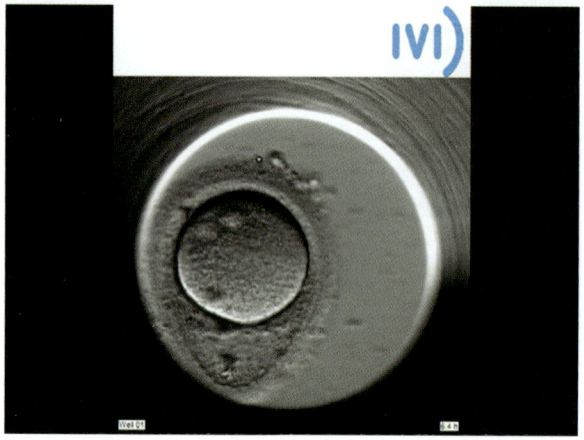

Video 5.17 Oocyte tabicated zona pellucida having normal embryo development.

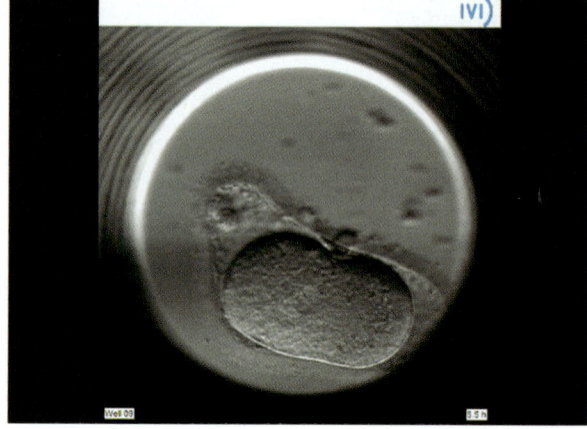

Video 5.18 In this video the pronucleus behavior is anomalous. We can observe how they separate before fading. It is interesting to see the abnormal zona pellucida. We can see how blastomeres adopt the left amorphous space after different rounds of cleavage.

Different dysmorphisms are related to the first polar body (Video 5.21), such as number or fragmentation. However, with the exception of the very large first polar body, the correlation of these dymorphisms with the potential of this oocyte is not clear.

5.8.1. Assessment of embryo characteristics

Embryo morphology was assessed on days 2 and 3 by considering the number of blastomeres, symmetry, and granularity of blastomeres, type and percentage of fragmentation, presence of multinucleated blastomeres, and degree of compaction. In accordance with these phenotype features, cleavage stage embryos were scored according to four categories, A, B, C, and D, which were partially described during the Istanbul consensus workshop on embryo assessment [9]. Of these four categories, A contained the top quality embryos and D comprised the lowest embryo quality category, implying a poorer ability to implant.

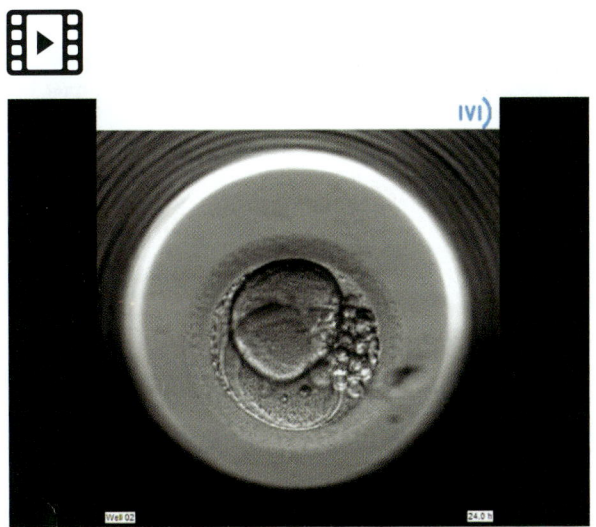

Video 5.19 Cleavage events in an oocyte with no dismorphisms.

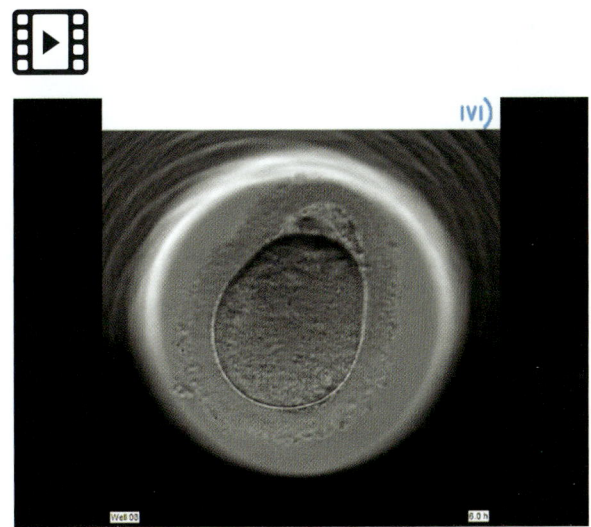

Video 5.20 In this case the zona pellucida of the oocyte is not only elongated but also thicker than a normal one.

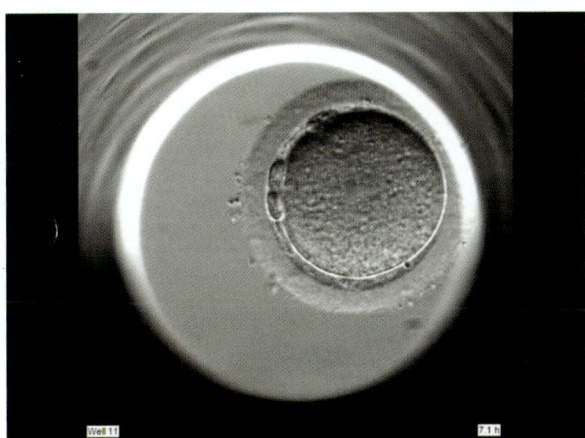

Video 5.21 Polar bodies often present some dismorphisms. In this case we can see an oocyte with a fragmented polar body, which is frequently seen in mature oocytes. As can be observed, the second polar body was extruded shortly after sperm injection.

5.9. Embryo score and culture conditions

Following injection, oocytes were placed in the Embryoscope. Fertilization was assessed 16–19 h post ICSI, based on the images acquired with the time-lapse monitoring system. Successful fertilization was assessed at 16–19 h post ICSI based on the digital images acquired with the time-lapse monitoring system. Embryo morphology was evaluated on days 2 (44–48 h post ICSI) and 3 (64–72 h post ICSI), based on the acquired digital images, and taking into account the cell number, symmetry and granularity, type and percentage of fragmentation, presence of multinucleated blastomeres, and compaction. Embryo selection for transfer was based on the algorithm developed by Meseguer [12], which is based on both morphology and a hierarchal model, which classifies embryos according to the timings of the most relevant kinetic parameters.

5.10. Time-lapse evaluation of morphokinetic parameters

A retrospective analysis of the acquired images of each embryo was done with an external computer, Embryo Viewer workstation (EV) (Unisense FertiliTech, Aarhus, Denmark), using an image analysis software in which all the considered embryo developmental events were annotated together with the corresponding timing of the events in hours after ICSI microinjection. In this study, the EV was used to identify the precise timing of the first divisions: division to two cells (t2), three cells (t3), four cells (t4), five cells (t5), and timing when blastulation started (tb).

Time of cleavage was defined as the first observed time-point when the newly formed blastomeres were completely separated by confluent cell membranes. Timings were expressed as hours post ICSI in all cases.

83

Table 5.1 Distribution of oocyte dysmorphisms, fertilization ability, and embryo quality

Oocyte phenotype	Phenotype distribution	Fertilization n (5%)	OR IC 95%	GQ embryos n (%)	OR CI 95%
Normal	3456 (66.9%)	2715 (78.6%)*		1368 (50.4%)	
NB	91 (1.8%)	66 (72.5%)	0.72 (0.41–1.27)	39 (59.1%)	0.43–1.26
RB	249 (4.8%)	198 (79.5%)	1.06 (0.77–1.46)	100 (50.5%)	1.47–1.52
HGC	285 (5.5%)	194 (68.1%)*	0.58* (0.45–0.76)	95 (48.7%)	0.37–1.17
CLGC	266 (5.2%)	202 (75.9%)	0.86 (0.63–1.15)	108 (53.5%)	0.51–2.15
SER	76 (1.5%)	57 (75.0%)	0.82 (0.48–1.39)	33 (57.9%)	0.33–1.80
VAC	42 (0.8%)	27 (64.3%)*	0.49* (0.26–0.93)	12 (44.4%)	0.38–1.39
M	53 (1.0%)	35 (66.0%)*	0.53* (0.30–0.94)	13 (36.4%)	0.41–1.35
LPS	434 (6.6%)	297 (74.6%)	0.82 (0.63–1.06)	141 (47.5%)	0.23–1.47
PB	136 (2.6 %)	104 (76.5%)	0.89 (0.59–1.33)	52 (50%)	0.38–1.26
E	48 (0.9%)	35 (72.9%)	0.74 (0.39–1.40)	18 (51.4%)	0.17–1.1
A	54 (1.0%)	41 (77.4%)	0.93 (0.49–1.78)	12 (29.3%)	0.13–071
Db	62 (1.2%)	45 (72.6%)	0.72 (0.45–1.15)	20 (44.4%)	0.26–128

Only fertilization rates were significantly different among some types of oocyte dysmorphisms, * $p < 0.05$.

Of the 5,252 oocytes included in the study, we found that the proportion considered to be "normal" was, as expected, the highest (38.3%), followed by OR (25%), AC (12.1%), fPB (10%), LPS (6.6%), M (5.9%), and PB (1.6%). The zona pellucida phenotypes of the oocytes arising from our study populations were distributed as follows: 3% eZP, 4.5% tZP, 3.2% thZP, 2.1% irZP, 0.2% mZP, 72.7% nZP, 0.6% pZP, 0.9% bZP. In order to simplify the data, we divided the zona pellucida phenotypes into two main groups (nZP and abZP).

5.11. Fertilization and oocyte morphology

Only three types of oocyte morphologies showed statistically poorer fertilization ability if compared to the oocytes considered normal. These oocytes were those with multiple abnormalites (66.0%), oocytes with vacuoles (64.3%), and the heavily granulated ones (68.15), whereas the normal oocytes had a 78.6% fertilization rate (Table 5.1). Fertilization was also similar among the oocytes with normal and abnormal zones.

Table 5.2 Distribution of zona pellucida morphology, fertilization ability, and embryo quality

Zona pellucida phenotypes	Zona pellucida distribution	Fertilization n (%)	OR CI 95%	GQ embryos n (%)	OR CI 95%
nZP	4312 (81.6%)	3329 (77.4%)		1678 (50.4%)	
eZP	152 (3.0%)	116 (76.3%)		52 (44.8%)	0.48–6.65
tZP	172 (3.3%)	135 (78.5%)	0.74–5.65	74 (54.5%)	0.46–8.69
thZP	276 (5.3%)	221 (80.1%)	0.40–4.57	101 (45.7%)	0.42–5.41
IrZP	42 (0.8%)	29 (69.0%)	0.84–6.91	16 (57.1%)	0.34–4.59
Dark ZP	16 (0.3%)	10 (62.5%)	0.61–5.36	4(40.0%)	0.54–7.63
bZP	52 (1.0%)	36 (69.2%)	0.74–6.42	19 (52.8%)	0.48–6.65
izrZP	141 (2.7%)	106 (75.2%)	0.65–5.68	62 (57.5%)	0.46–7.70

Regarding fertilization and embryo quality, no significant differences were observed among the groups.

5.12. Morphokinetics and oocyte morphology

With regards to embryo morphology, the average percentage of good quality embryos found in the present study was 50%. Similar percentages of good quality embryos were found among the oocytes carrying different phenotypes (Table 5.1), and also between the oocytes with normal or abnormal zona pellucidas (Table 5.2).

Likewise when the kinetics parameters were evaluated, no discrepancies were seen among all the types of oocyte phenotypes analyzed (Table 5.3), or between the embryos originating from oocytes with normal or abnormal zona pellucidas (Table 5.4).

5.13. Discussion

The present chapter shows that certain oocyte phenotypes, such as presence of multiple abnormalities, vacuoles, and heavily granulated cytoplasm, were associated with a significant 10% decrease in fertilization ability. No effect was seen when normal zona pellucida morphology was observed. Alteration of the zona pellucida can result in fertilization failure following conventional insemination [24]. However, due to the performance of ICSI in all the analyzed cases, its influence on fertilization failure vanished.

Correct completion of fertilization and embryo development is associated with proper oocyte maturation, and our findings reflect that some of the studied oocyte dysmorphisms may reflect a defect in protein synthesis signaling pathways.

However, it is interesting to note that oocytes with a similar phenotype that overcame those deficiencies responsible for fertilization failure, may actually have a normal cytoplasmic environment like the oocytes with normal phenotypes, which are able to provide the essential proteins for subsequent PN migration. Indeed, no differences were observed in the precise timings related to fertilization events, such as second polar body extrusion (tPB), pronuclear appearance (tPNa), and pronuclear fading (tPNf), which actually fell within the ranges previously established by our team [12] and were comparable with the timing displayed by the normal oocytes. All these are indirect signs of the right processing of sperm components inside the cytoplasm.

We observed that once fertilization had been overcome, embryo quality seemed to be very hard to predict by only simple oocyte morphological assessment or by the appearance of the zona pellucida. These results are also supported by the fact that no differences were found in the kinetic evaluation data of the embryos generated from both the oocyte and zona pellucida morphology.

Endocytic vacuoles in oocytes have been reported to appear in late oocyte maturation stages [4]. Our findings demonstrate that although they have more limitations for fertilization, fertilization rates are still

Table 5.3 Kinetic timings (hours) in each type of oocyte dysmosphism

Oocyte phenotype	tPB2	TPNa	tPNf	t2	cc2	s2	t5
Normal	3.5 (3.4–3.6)	7.9 (7.7–8.0)	23.3 (22.8–23.6)	26.0 (25.6–26.4)	12 (11.6–12.5)	3.9 (3.6–4.2)	47.7 (46.8–48.5)
NB	3.4 (2.8–3.9)	8.2 (7.3–9.2)	24.4 (22.2–26.7)	24.4 (22.2–26.7)	11.7 (8.7–14.8)	4.2 (1.6–6.7)	43.6 (38.6–48.7)
RB	3.5 (3.2–3.7)	7.9 (7.2–8.6)	23.0 (21.5–24.5)	25.2 (23.8–26.5)	12.0 (10.3–13.7)	3.5 (2.2–4.8)	46.5 (42.9–50.0)
HGC	3.9 (3.3–4.6)	8.5 (7.8–9.2)	24.9 (22.8–27.2)	25.8 (24.3–27.3)	11.8 (9.8–13.6)	4.8 (3.4–6.2)	46.8 (44.1–49.6)
CLGC	3.8 (3.5–4.0)	8.3 (7.7–8.9)	22.9 (21.4–24.3)	27.1 (25.2–28.9)	12.6 (11.1–14.1)	3.7 (2.5–4.9)	50.1 (46.7–53.3)
SER	3.9 (3.2–4.7)	7.2 (6.4–7.9)	22.5 (19.9–25.1)	25.3 (21.8–29.9)	12.3 (9.7–14.9)	6.8 (2.2–11.4)	42.1 (37.0–47.0)
VAC	3.4 (2.4–4.3)	9.3 (6.9–11.8)	24.5 (19.2–29.7)	27.1 (24.1–30.1)	10.8 (5.4–16.2)	4.6 (0.5–8.6)	48.0 (41.6–64.3)
M	3.1 (2.6–3.6)	8.1 (6.1–10.1)	25.1 (21.3–28.8)	28.3 (26.7–30.1)	9.4 (4.2–14.6)	8.7 (0.5–16.9)	39.1 (24.1–64.1)
LPS	3.3 (3.1–3.5)	7.8 (7.4–8.4)	22.2 (21.1–23.9)	25.9 (24.9–26.9)	11.9 (10.9–12.9)	3.7 (2.6–4.7)	47.5 (45.1–50.0)
PB	3.3 (3.0–3.7)	8.5 (7.4–9.7)	25.9 (24.2–27.7)	26.3 (24.6–28.0)	11.9 (9.6–14.2)	3.8 (2.2–5.3)	45.5 (41.4–49.6)
E	3.8 (3.1–4.4)	7.5 (6.3–8.7)	25.0 (23.7–26.3)	25.7 (22.6–28.7)	13.4 (9.9–17.0)	1.2 (0.5–3.4)	52.5 (45.4–69.4)
A	3.6 (3.0–4.3)	7.3 (5.6–9.0)	25.2 (21.4–29.1)	25.6 (22.5–28.9)	11.6 (7.9–15.2)	2.8 (1.3–4.3)	44.9 (38.8–50.9)
Db	4.4 (3.6–5.2)	6.9 (5.4–8.5)	19.7 (19.9–25.1)	25.8 (23.3–28.2)	12.4 (8.5–16.2)	3.8 (1.4–6.0)	48.4 (42.9–63.8)

Values between brackets are 95% confidence intervals.

Table 5.4 Kinetic timings (hours) in embryos with normal and abnormal zona pellucidas.

Zona pellucida phenotype	tPB2	TPNa	tPNf	t2	cc2	s2	t5
Normal ZP	3.5 (3.4–3.6)	7.9 (7.7–8.0)	23.3 (22.8–23.6)	26.2 (25.6–26.4)	11.9 (11.6–12.5)	4.0 (3.6–4.2)	47.3 (46.8–48.5)
Abnormal ZP	3.5 (2.8–4.3)	7.9 (7.3–8.5)	22.9 (20.9–24.9)	25.4 (23.9–26.8)	12.7 (8.7–14.8)	3.1 (2.3–4.1)	47.3 (44.1–50.4)

Values between brackets are 95% confidence intervals.

high. Perhaps the degree of vacuolization is responsible for the stronger or weaker impact on fertilization and embryo development, which is in line with previous publications [17, 25, 26].

Ebner [25] conducted a study which focused on vacuoles and drew new conclusions about this oocyte dysmorphism. Their results showed that the effect of vacuoles on embryo development was size-

and time-dependent. They also observed that, for example, vacuolization on day 5 had a negative impact on blastocyst formation, and that day 4 was the most critical day in terms of spontaneous vacuolization.

The other two types of morphology to present worse fertilization ability were oocytes which may combine abnormalities and very granulated ones. We do not know whether extremely granulated oocytes are signs of aging oocytes. Kahraman's group [27] evaluated the effect of granulation and observed no differences in fertilization rates, embryo quality, and pregnancy rates between oocytes with or without granulation. However, implantation and ongoing pregnancy rates seemed low in those cases with CLCG oocytes.

Multiple abnormalities are often seen in oocytes with combined oocyte dysmorphisms, which are severe in some cases, and are responsible for the lower fertilization results.

Special attention has been paid to SER-positive oocytes, and some groups have suggested that the presence of SER in the cytoplasm of an oocyte may interfere with normal calcium stores and calcium oscillations during fertilization and may, therefore, have detrimental effects on embryo development and implantation [16, 28]. However, no differences were found in terms of fertilization-associated events (PB extrusion, PN appearance, PN fading) and of embryo development until day 3. Therefore we question the real impact of Ca^{++} oscillations and signaling on such types of events.

Akarsu [29] presented the first case report with SER aggregations in all retrieved oocytes. This study was done in three consecutive ICSI cycles, two of which ended in clinical ongoing pregnancies, but with multiple fetal anomalies. In Mateizel's study [30], one patient, with all the oocytes presenting SER, delivered a healthy newborn with no major malformations.

The influence of other types of phenotypes studied, such as necrotic and refractile bodies, is a matter of controversy. We found that they had the same cleavage pattern as the embryos originating from normal oocytes, which is in agreement with some authors [23, 31], but disagrees with the results published by others, who have suggested that the individual identification of oocyte dysmorphisms may be a prognostic tool for blastocyst development [32].

Similar findings were observed with oocytes with a minor variation in the PB that differed from major alterations, such as an immature polar body, two clearly separated polar bodies, oversized or undersized polar bodies which, once again, showed similar timings and cleavage pattern to normal oocytes.

In conclusion, the new approach to measure embryo cleavage timings of the embryos originating from a variety of oocyte phenotypes shows that the vast majority exhibit a similar pattern. Some categories of dysmorphisms, such as presence of vacuoles and multiple abnormalities, have been significantly associated with a lower fertilization rate, but once fertilization occurred, embryo development measurements were similar among groups. Similarly, zona pellucida phenotypes were independent of the aspect of the oocyte morphology, and did not affect either fertilization events or embryo kinetics after ICSI.

In conclusion, several intra-cytoplasmic and extracytoplasmic abnormalities have been described, and the kinetic timing of both fertilization and embryo development has been measured. Our results confirm that variations in neither oocyte morphology nor the zona pellucida are predictive of oocyte competence after ICSI. Only some oocyte phenotypes may have reduced oocyte competence, which can be detected at the time of fertilization.

References

1. Human ovarian follicular development: from activation of resting follicles to preovulatory maturation. *Annales d'endocrinologie*: Elsevier; 2010.

2. Sathananthan AH. Ultrastructure of the human egg. *Hum Cell* 1997 Mar;10(1):21–38.

3. Hutt KJ, Albertini DF. An oocentric view of folliculogenesis and embryogenesis. *Reprod Biomed Online* 2007;14(6): 758–764.

4. Van Blerkom J. Occurrence and developmental consequences of aberrant cellular organization in meiotically mature human oocytes after exogenous ovarian hyperstimulation. *J Electron Microsc Tech* 1990;16(4): 324–346.

5. Sathananthan AH. Ultrastructural changes during meiotic maturation in mammalian oocytes: unique aspects of the human oocyte. *Microsc Res Tech* 1994;27(2): 145–164.

6. Rienzi L, Vajta G, Ubaldi F. Predictive value of oocyte morphology in human IVF: a systematic review of the literature. *Hum Reprod Update* 2011 Jan-Feb;17(1): 34–45.

7. Rienzi L, Balaban B, Ebner T, Mandelbaum J. The oocyte. *Hum Reprod* 2012 Aug;27 Suppl 1: i2–21.

8. Tejera A, Herrero J, de Los Santos M, Garrido N, Ramsing N, Meseguer M. Oxygen consumption is a quality marker for human oocyte competence conditioned by ovarian stimulation regimens. *Fertil Steril* 2011;96(3):618–623. e2.

9. Balaban B, Brison D, Calderón G, Catt J, Conaghan J, Cowan L, et al. The Istanbul consensus workshop on embryo assessment: proceedings of an expert meeting. *Hum Reprod* 2011; 26(6):1270–1283.

10. Rienzi L, Romano S, Albricci L, Maggiulli R, Capalbo A, Baroni E, et al. Embryo development of fresh 'versus' vitrified metaphase II oocytes after ICSI: a prospective randomized sibling-oocyte study. *Hum Reprod* 2010 Jan;25(1): 66–73.

11. Swain JE, Pool TB. ART failure: oocyte contributions to unsuccessful fertilization. *Hum Reprod Update* 2008 Sep-Oct; 14(5):431–446.

12. Meseguer M, Herrero J, Tejera A, Hilligsoe KM, Ramsing NB, Remohi J. The use of morphokinetics as a predictor of embryo implantation. *Hum Reprod* 2011 Oct;26(10): 2658–2671.

13. Otsuki J, Okada A, Morimoto K, Nagai Y, Kubo H. The relationship between pregnancy outcome and smooth endoplasmic reticulum clusters in MII human oocytes. *Hum Reprod* 2004 Jul; 19(7):1591–1597.

14. Otsuki J, Okada A, Morimoto K, Nagai Y, Kubo H. The relationship between pregnancy outcome and smooth endoplasmic reticulum clusters in MII human oocytes. *Hum Reprod* 2004 Jul;19 (7):1591–1597.

15. Sá R, Cunha M, Silva J, Luís A, Oliveira C, Teixeira da Silva J, et al. Ultrastructure of tubular smooth endoplasmic reticulum aggregates in human metaphase II oocytes and clinical implications. *Fertil Steril* 2011;96(1):143–149. e7.

16. Ebner T, Moser M, Shebl O, Sommerguber M, Tews G. Prognosis of oocytes showing aggregation of smooth endoplasmic reticulum. *Reprod Biomed Online* 2008;16(1): 113–118.

17. Fancsovits P, Murber Á, Gilán ZT, Rigó Jr J, Urbancsek J. Human oocytes containing large cytoplasmic vacuoles can result in pregnancy and viable offspring. *Reprod Biomed Online* 2011;23 (4):513–516.

18. Meriano JS, Alexis J, Visram-Zaver S, Cruz M, Casper RF. Tracking of oocyte dysmorphisms for ICSI patients may prove relevant to the outcome in subsequent patient cycles. *Hum Reprod* 2001 Oct;16 (10):2118–2123.

19. Suzuki K, Yoshimoto N, Shimoda K, Sakamoto W, Ide Y, Kaneko T, et al. Cytoplasmic dysmorphisms in metaphase II chimpanzee oocytes. *Reprod Biomed Online* 2004;9(1):54–58.

20. Miao YL, Kikuchi K, Sun QY, Schatten H. Oocyte aging: cellular and molecular changes, developmental potential and reversal possibility. *Hum Reprod Update* 2009 Sep-Oct;15(5): 573–585.

21. Mikkelsen AL, Lindenberg S. Morphology of in-vitro matured oocytes: impact on fertility potential and embryo quality.

Hum Reprod 2001 Aug;16 (8):1714–1718.

22. Hassan-Ali H, Hisham-Saleh A, El-Gezeiry D, Baghdady I, Ismaeil I, Mandelbaum J. Perivitelline space granularity: a sign of human menopausal gonadotrophin overdose in intracytoplasmic sperm injection. *Hum Reprod* 1998 Dec;13(12): 3425–3430.

23. Balaban B, Urman B, Sertac A, Alatas C, Aksoy S, Mercan R. Oocyte morphology does not affect fertilization rate, embryo quality and implantation rate after intracytoplasmic sperm injection. *Hum Reprod* 1998 Dec;13 (12):3431–3433.

24. Henkel R, Franken D, Habenicht U. Zona pellucida as physiological trigger for the induction of acrosome reaction. *Andrologia* 1998;30(4–5):275–280.

25. Ebner T, Moser M, Sommergruber M, Gaiswinkler U, Shebl O, Jesacher K, et al. Occurrence and developmental consequences of vacuoles throughout preimplantation development. *Fertil Steril* 2005;83 (6):1635–1640.

26. Van Blerkom J, Henry G. Oocyte dysmorphism and aneuploidy in meiotically mature human oocytes after ovarian stimulation. *Hum Reprod* 1992 Mar;7(3): 379–390.

27. Kahraman S, Yakin K, Donmez E, Samli H, Bahce M, Cengiz G, et al. Relationship between granular cytoplasm of oocytes and pregnancy outcome following intracytoplasmic sperm injection. *Hum Reprod* 2000 Nov;15 (11):2390–2393.

28. Otsuki J, Okada A, Morimoto K, Nagai Y, Kubo H. The relationship between pregnancy outcome and smooth endoplasmic reticulum clusters in MII human oocytes. *Hum Reprod* 2004;19 (7):1591–1597.

29. Akarsu C, Çağlar G, Vicdan K, Sözen E, Biberoğlu K. Smooth endoplasmic reticulum aggregations in all retrieved oocytes causing recurrent multiple anomalies: case report. *Fertil Steril* 2009;92(4):1496. e1–1496. e3.

30. Mateizel I, Van Landuyt L, Tournaye H, Verheyen G. Deliveries of normal healthy babies from embryos originating from oocytes showing the presence of smooth endoplasmic reticulum aggregates. *Hum Reprod* 2013 Aug;28(8):2111–2117.

31. De Sutter P, Dozortsev D, Qian C, Dhont M. Oocyte morphology does not correlate with fertilization rate and embryo quality after intracytoplasmic sperm injection. *Hum Reprod* 1996 Mar;11(3):595–597.

32. Braga, Daniela Paes Almeida Ferreira, Setti AS, de Cássia S Figueira R, Machado RB, Iaconelli Jr A, Borges Jr E. Influence of oocyte dysmorphisms on blastocyst formation and quality. *Fertil Steril* 2013;100(3): 748–754.

Evaluating embryo development stages using time-lapse microscopy (TLM)

Thomas Freour

6.1. Early division

6.1.1. Description

The early embryo division, also named as early cleavage, first cleavage, or first embryo division, has been defined as that which generally occurs at 25–27 hours post-ICSI, giving rise to a two-cell embryo. This cellular phenomenon and its impact on pregnancy rate in humans were first published in 1984 by the Edwards group [1]. The transition from the zygote to the two-cell embryo depends on a large number of highly regulated cellular events, which are initiated by intracytoplasmic calcium waves induced by fertilization. Asynchrony in the first division between two simultaneously fertilized zygotes and their different developmental competence and capacity to reach the blastocyst stage could be due to differences in the ability of each oocyte to respond to this calcium-based stimulus, skill that is progressively acquired during oogenesis/folliculogenesis, suggesting that oocyte maturity could pre-determine the timing of first embryo division.

Apart from oocyte maturity, several other factors have been evocated as regulators of first division. While sperm characteristics have not been shown to impact early embryo division [2], meiotic spindle parameters were correlated with the timing of first cleavage [24]. Some authors have reported that delayed first division could be caused by a longer DNA replication phase (S) in the cellular cycle, due to the existence of chromosomal abnormalities at the zygote stage.

Therefore, in summary, the molecular and genetic conditions of gametes, and particularly the oocyte, affect the start of the first cell cycle, which is of crucial importance for zygote to embryo transition and embryo genome activation. Besides the obvious role of intrinsic oocyte factors, such as genes, proteins, organelles, mitochondria, on first division, the mechanisms explaining the huge variations in the time of first embryonic division are not clear, suggesting a possible role for culture conditions. By the way, early cleavage can be considered as a relevant marker of correct fertilization with appropriate gametes.

6.1.2. Impact on embryonic development

Several authors have reported that use of early cleaved embryos for transfer results in higher implantation and pregnancy rates [3, 4, 5, 6, 7]. However, some of these studies should be interpreted with care, as double embryo transfers associating early cleaved and non-early cleaved embryos were performed. In studies including only "pure" early cleaved embryo transfer, significantly higher blastulation rate, lower abortion rate, and higher evolutive pregnancy rate were found in early cleavage group [7]. It should be noted that most authors reported better morphology on the day of transfer in early cleaved embryos than in non-early cleaved [5, 7].

Although most studies are consistent in identifying early cleavage as a good predictor of embryo competence and implantation capability, it should be noted that consensus classifications, such as those of the Istanbul Consensus Group, do not recommend systematically early cleavage in embryo selection parameters [8].

6.2. Parameters evaluated on day 2 (D2) and day 3 (D3)

6.2.1. Introduction

Not all embryos generated in an IVF laboratory will have the morphological requirements compatible

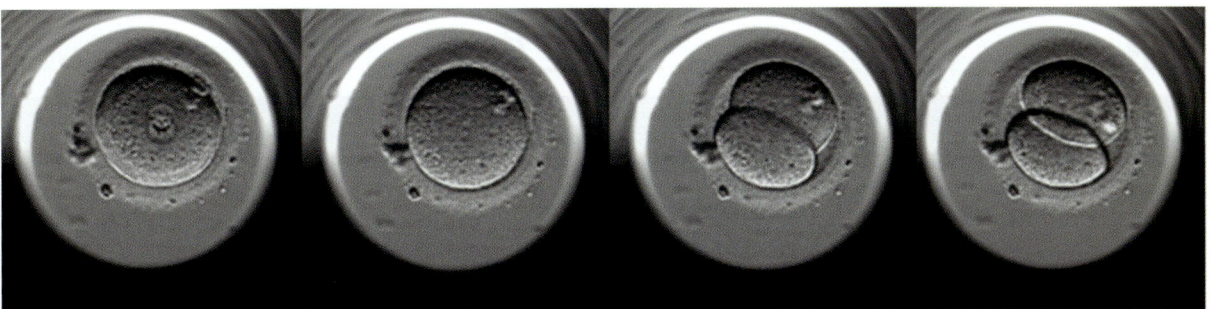

Figure 6.1 The first division results in an embryo of two cells called blastomeres, generally of the same size. Cells of different sizes can be obtained mainly by the loss of cytoplasm that will result in cell fragments. This figure represents a sequence of images from the stage of zygote to obtain a two-cell embryo after passing through the pronuclei fading stage and start of mitotic division.

with transfer or cryopreservation according to state-of-the art classifications [6, 8].

The choice of the embryo with the highest implantation potential is the main purpose of clinical embryology, but to date this selection is mainly based on morphological criteria whose predictive performances are modest. The two main reasons for this are the following: firstly, successful implantation does not only depend on the embryo, but also requires a receptive endometrium; secondly, morphological parameters routinely used for embryo quality assessment are only punctual and do not reflect the dynamic and complex process of early embryo development in vitro, thus yielding biased and partial information.

This is now changing thanks to new methods of embryo quality assessment, including time-lapse devices such as the EmbryoScope, which give access to the whole embryo development process and lead to the identification of cellular events that could have an impact on embryo implantation potential [9, 10].

The zygote obtained after the completion of fertilization will enter into a phase of successive mitotic divisions which will ideally end in the blastocyst stage after 5 to 6 days of in vitro culture. Pre-implantation embryo development is a complex and dynamic process, during which the embryo will go through a series of various morphological features that are used by embryologists as a decision-making tool for deciding which embryos have the highest implantation potential and select them for transfer and/or freezing.

Morphology-based classification of early-stage embryos relies on the following parameters: number of blastomeres, developmental speed, blastomere size, and symmetry, presence and abundance of fragments, presence of multinucleated blastomeres, cytoplasm aspect (vacuoles, vesicles, granularity), polarization and contour of the blastomeres, degree of compaction, and zona pellucida aspect. Thus, an embryo with few cells, asymmetric, with many fragments, or multinucleated blastomeres, will have a worse prognosis than an embryo with an appropriate number of symmetric cells and not fragmented.

In general, it can be considered that embryos can be classified according to their potential implantation in:

-**Optimal (or good-quality, top-quality) embryos:** embryos that have no characteristic of poor prognosis and proper development. These embryos will be transferred or frozen.

-**Suboptimal (or fair-quality) embryos:** embryos presenting with one or more characteristics associated with lower quality, but still eventually compatible with transfer if no other embryo with better morphology is available. An alternative strategy for these suboptimal embryos can be prolonged culture up to the blastocyst stage (D5–6) in order to get more information on their developmental competence before re-evaluating the possibility of transfer or freezing. These embryos tend to have significantly lower implantation rates than optimal ones, but this obviously depends on the type and extent of morphological defects that have been observed.

-**Non-viable (or atypical, poor-quality) embryos:** embryos presenting with several characteristics associated with a lack or extremely low implantation potential, or embryos whose development is blocked. These embryos will generally be discarded, except in cases of regulatory constraints.

91

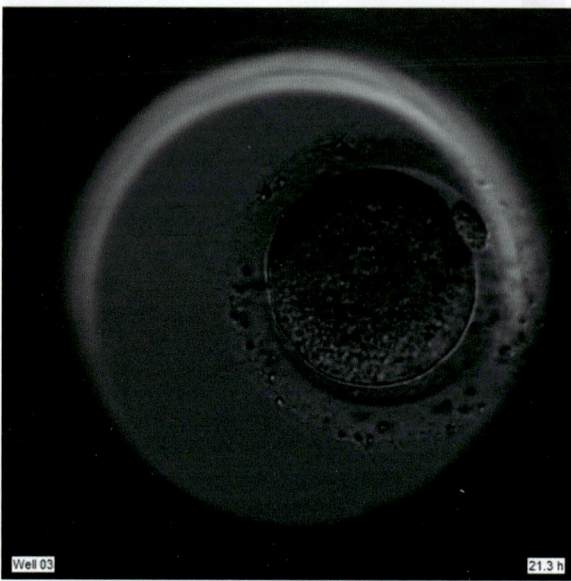

Well 03   21.3 h

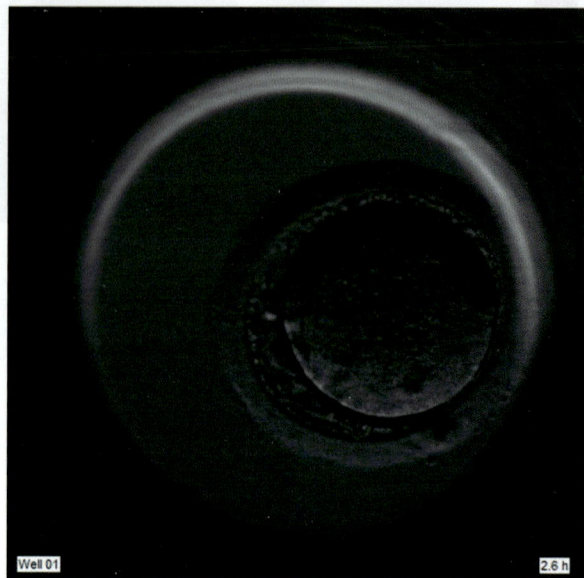

Well 01   2.6 h

Video 6.1 Embryo with proper early development. Regular and almost synchronous division rate, cell regularity, and absence of debris are in favor of high implantation potential.

Video 6.2 Embryo presenting with one or more characteristics associated with lower quality, but still eventually compatible with transfer if no other embryo with better morphology is available. In this case, the embryo presents early fragmentation, uneven blastomeres at the two-cell stage and at further developmental stages, and asynchronous divisions.

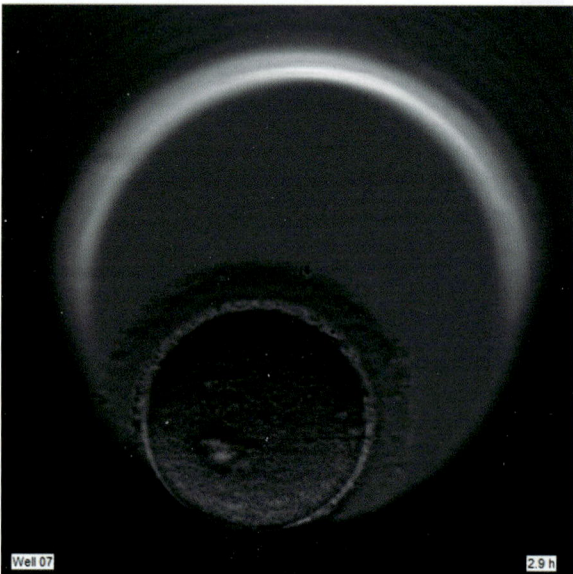

Well 07   2.9 h

Video 6.3 Embryo presenting with several characteristics associated with a lack or extremely low implantation potential, i.e., early fragmentation, irregular cells, direct cleavage from zygote to more than two blastomeres, and slow division rate.

Despite the limits of embryo quality assessment based on morphology, the morphological appearance of an embryo is the most widely used parameter by embryologists in order to evaluate the implantation potential of embryos obtained in IVF cycles. Features mentioned earlier, such as the number and symmetry of blastomeres, degree of fragmentation, etc. are correlated with a successful clinical outcome.

6.2.2. Cell number and division rate

One of the morphological indicators of embryo quality which can be easily determined is the number of cells on D2 and/or D3. Several studies have shown that too fast (> 9 cells on D3) or too slow (< 7 cells on D3) embryos are significantly less likely to reach the blastocyst stage, and therefore implant, probably due to a greater number of chromosomal abnormalities [6, 11].

Although there are numerous studies that correlate the number of cells on D2 and D3 and embryo implantation potential, they were always associated with other parameters, such as symmetry, fragmentation or multinucleation. As these morphological

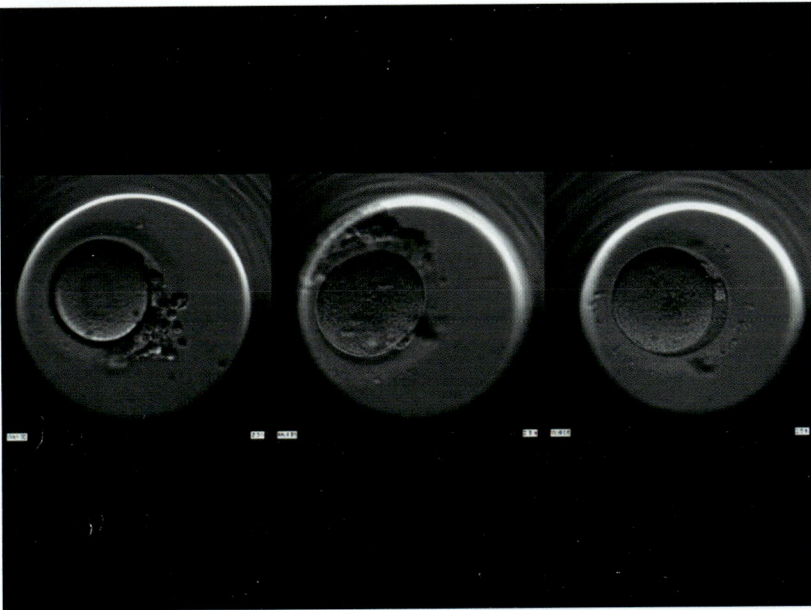

Video 6.4 Three embryos with various cell numbers and division rates on day 2.

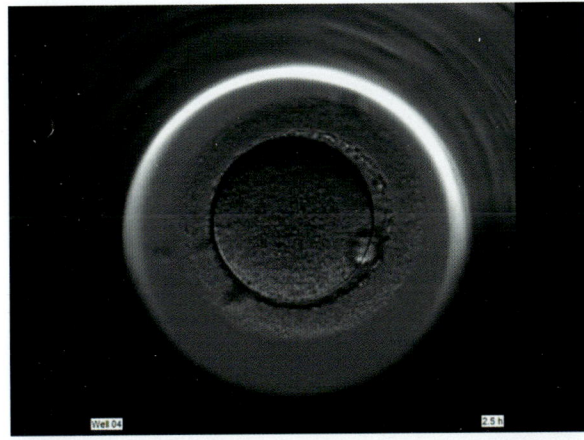

Video 6.5 Embryo with proper early development, with four regular cells and no fragment on day 2.

factors are correlated to each other, it is therefore difficult to determine their respective predictive interest. However, cell number has been often presented as the most significant morphological parameter [6].

Cell number should be assessed at specific time points [12]:

-day 1 (26 ± 1 h post-ICSI, 28 ± 1 h post-IVF): two-cell stage
-day 2 (44 ± 1 h): four-cell stage
-day 3 (68 ± 1 h): eight-cell stage

Embryos with a higher or lower number of cells, i.e., faster or slower development, than the numbers presented here have been shown to have lower implantation capacity by several authors reporting that the bigger the difference in cell number, the lower the implantation rate [12, 13].

For example, on day 2 (at 44 ± 1 h post-insemination):

-good prognosis: four-cell
-low prognosis: two, three or five cells
-very low prognosis: ≥ five cells

Concerning evaluation on D3, embryo quality assessment also depends on the number of cells, the highest quality being observed for embryos with seven to eight cells at 68 ± 1 h post-insemination. However, it also depends on the rate of cellular division between D2 and D3. Therefore, cellular division rate between D2 and D3 should be taken into account when evaluating embryos on D3.

6.2.3. Percentage and type of cell fragmentation

Embryo fragmentation is one of the most common phemonena in IVF cycles. A fragment is defined as an anuclear extra-cellular cytoplasmic structure

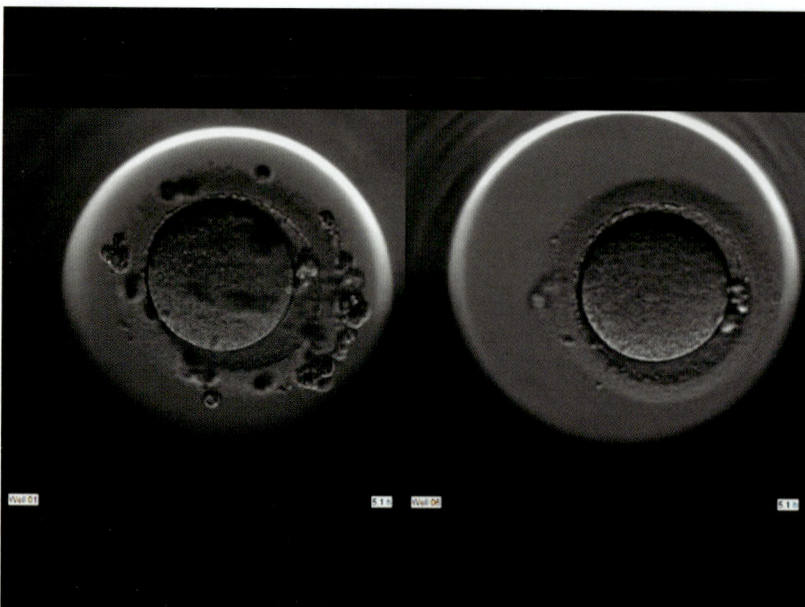

Video 6.6 Two embryos with low prognosis and imperfect development development up to day 2. The first demonstrates a relatively slow development, presenting as a two regular cells embryo on day 2. The second embryo demonstrates a higher cleavage rate, reaching a five irregular cells stage on day 2, which is considered faster than the ideal division rate.

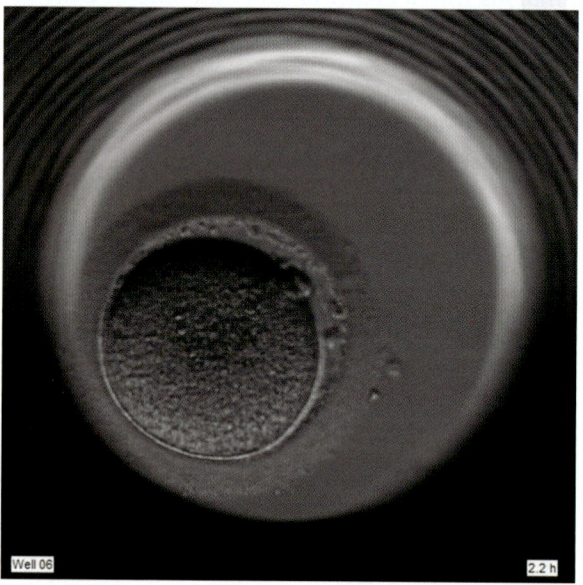

Video 6.7 Embryo with more than five cells on day 2. This embryo has a significantly too high division rate, leading to a very low implantation potential.

Video 6.8 Embryo moving from two cells at D2 to <= 8 cells at D3.

surrounded by cytoplasmic membrane. Most of the fragments are formed during the pronuclear stage, the first and/or second mitotic division. It should be noted that it can be challenging to make the distinction between a large fragment and a small blastomere. Their presence in high quantity has been linked to abnormal cell division, apoptosis, chromosomal segregation errors or alteration of embryo

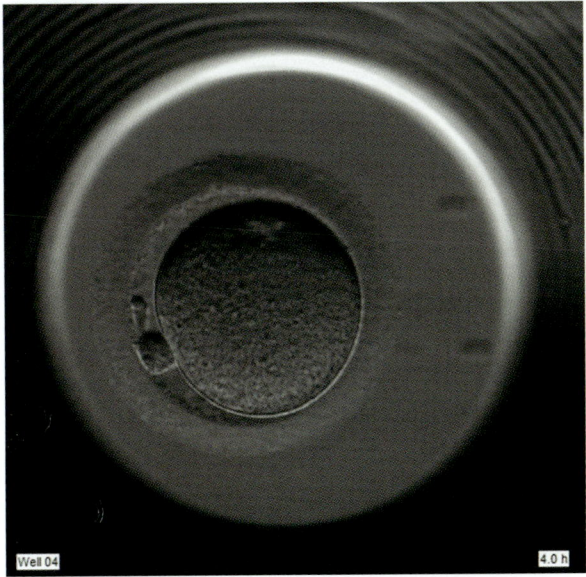

Video 6.9 Embryo with focal fragmentation.

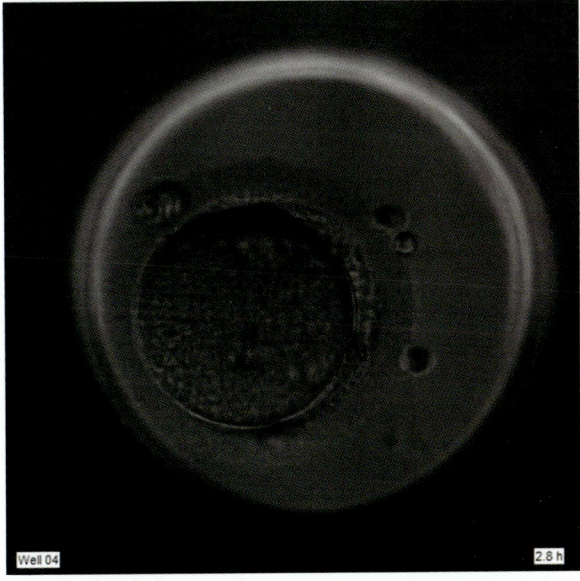

Video 6.10 Embryo with scattered fragmentation.

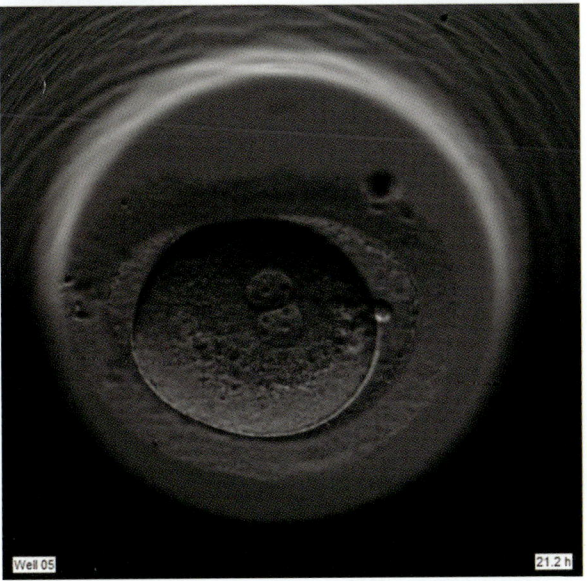

Video 6.11 Embryo with early fragmentation, occurring as soon as the first cleavage is completed.

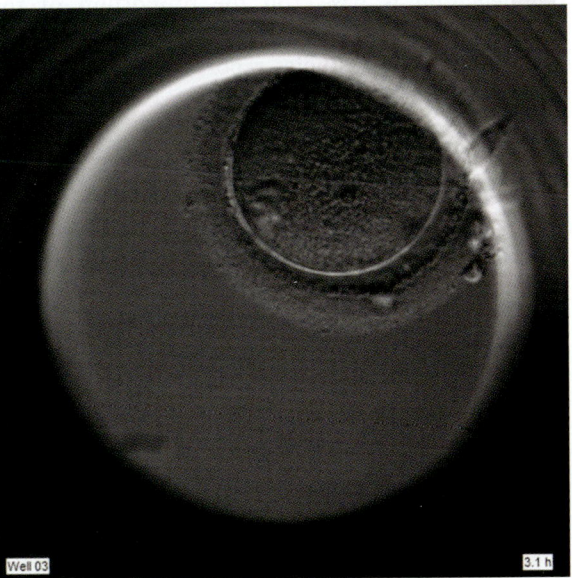

Video 6.12 Embryo with relatively little fragmentation at the two-cell stage, but increasing fragmentation throughout cell divisions up to the eight-cell stage.

membrane, all leading to reduced implantation rate [6, 11]. The mechanisms underlying this loss of implantation capacity have not been clearly identified, but could be the ones leading to fragmentation itself (such as apoptosis) and/or the modification of blastomere cellular junctions and further compaction processes. Molecular differences have also been described between fragmented and non-fragmented

95

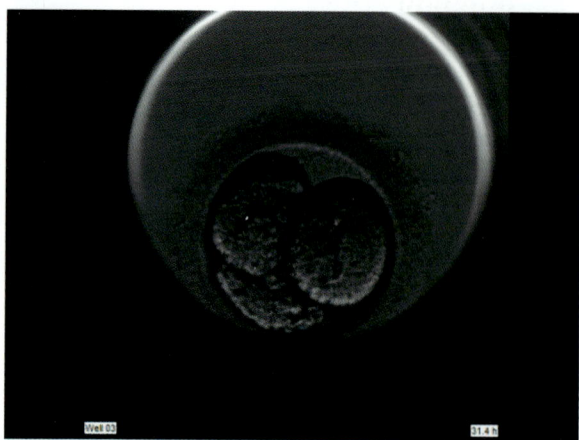

Video 6.13 Embryo with poor early development, but finally reaching the blastocyst stage on day 5 with relatively acceptable morphology.

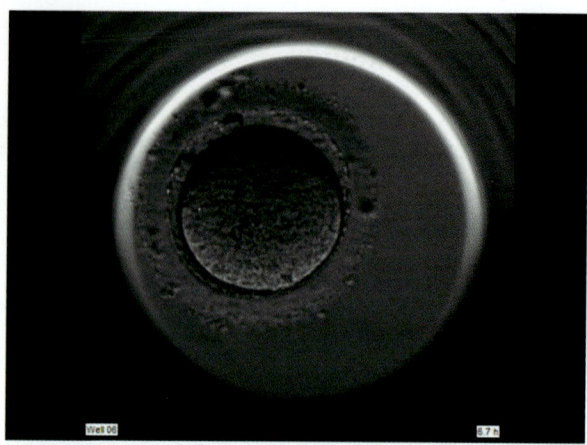

Video 6.14 Embryo with proper development, blastomeres being regular and of the same size at all developmental stages.

embryos, at the genetic and cytoskeletal points of view.

The most frequently used scoring system for fragmentation is based on the proportion of the volume of the embryo covered (< 10%, 10–25%, > 25%) [12]. Most studies report lower blastulation and implantation rates with increasing fragment score [12].

Besides fragmentation itself, the spatial repartition of fragments and the timing of fragmentation occurence can also be of interest. Concerning fragment spatial localization, scattered and concentrated distribution can eventually be distinguished, but their respective impact on implantation is unclear.

The kinetics of fragmentation occurence is also of importance, as it significantly impacts embryo quality: the faster they appear, the worse the embryo development is.

It should be noted that the predictive interest of fragmentation alone is hard to evaluate. Indeed, fragmentation and cell number are two closely related parameters, as a high fragmentation score can disturb cellular division and lead to lower cell number.

However, even if highly fragmented embryos should not be chosen first for transfer or freezing, some of them might be able to develop up to the blastocyst stage, suggesting that they should not be systematically discarded at cleavage stage but rather be kept in prolonged culture [14].

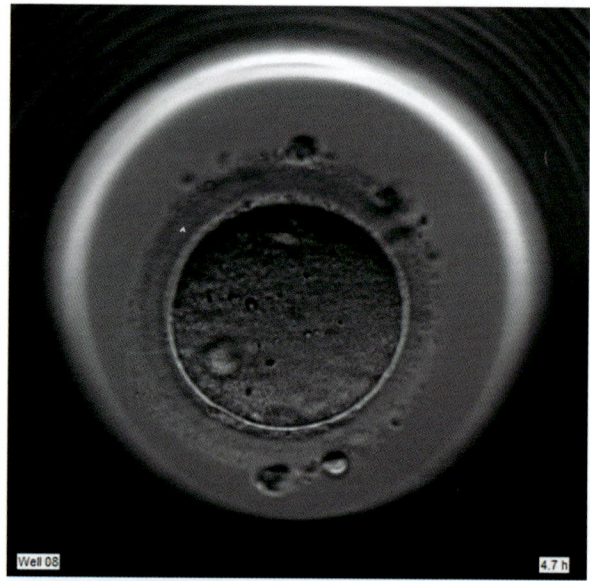

Video 6.15 Embryo with direct cleavage from the zygote stage to a three-cell stage and subsequent direct cleavage from the three-cell stage to a six-cell stage.

6.2.4. Blastomere size and regularity

Embryo development is based on successive mitoses, each blastomere division theoretically leading to two even daughter cells. Theoretically, embryos will

(A)

(B)

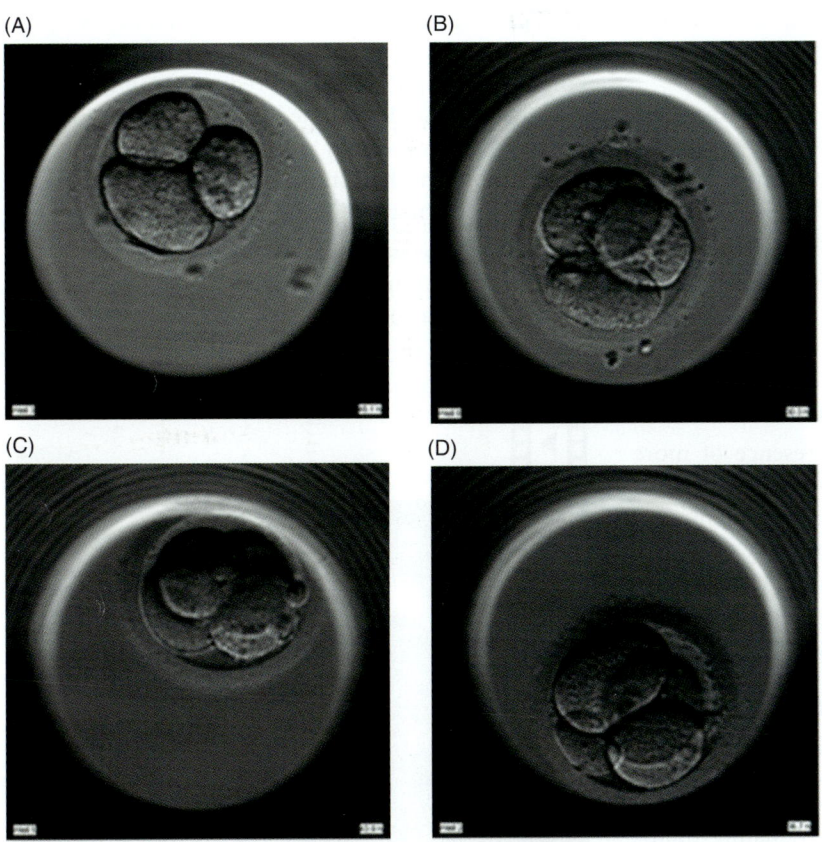

Figure 6.2 (A) Stage specific asymmetrical three-cell embryo, (B) non-stage specific symmetrical three-cell embryo, (C) Stage specific symmetrical four-cell embryo, (D) non-stage specific asymmetrical four-cell embryo.

(C)

(D)

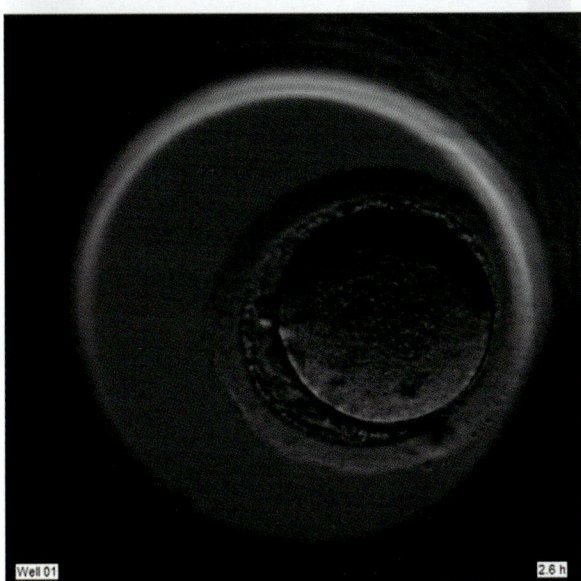

Video 6.16 Embryo with irregular cells from the first cleavage to day 3, and asynchronous divisions.

undergo synchronous and symmetrical divisions along the first three days of development, featuring two, four, eight cells and so on, all of them similar in size.

Blastomere size depends on embryo cleavage stage and cleavage regularity. It is generally considered that blastomeres with > 25% difference in diameter can be scored as uneven or irregular [12]. Most studies evaluating the asymmetry of the blastomeres as a marker of embryo quality have concluded its deleterious impact on implantation rate [6, 11]. However, as asynchronous cell divisions can be seen physiologically in humans, all embryos will not always follow this "ideal" program and will display for a short period a different number of blastomeres, presenting with asymmetry. Therefore, cell number and blastomere size should be analyzed together, and unequal cell size should not be systematically be considered as a signal of poor prognosis.

However, conventional punctual morphology assessment does not usually allow embryologists to distinguish between a three-cell unequal embryo that will rapidly undergo the next cell division and lead to a symmetric four-cell embryo, and a three-cell embryo staying at an unequal stage for a long period, i.e., more than 5 hours.

One of the major advantages of time-lapse is that it allows the dynamic analysis of blastomere divisions and the monitoring of cell cycle length. Thus, unequal stages can be easily observed, and their length measured in order to distinguish between a stage-specific unequal blastomere size and a non-stage-specific size of poorer prognosis.

6.2.5. Other parameters

6.2.5.1. Visualization of nuclei and multinucleation

Multinucleation is defined as the presence of more than one nucleus within at least one blastomere. Its frequency ranges from 20 to 40% of human embryos, but it is considerably higher when other morphological abnormalities are also present, such as non-stage-specific cell number [12, 15]. These embryos have been shown to present with a higher incidence of chromosomal abnormalities than non-multinucleated ones [6, 16], and most studies confirmed that these embryos had significantly lower implantation potential than non-multinucleated ones [6, 12, 15]. Both the number of multinucleated cells and the number of nuclei per cell are of interest in evaluating multinucleation and its deleterious effect on implantation.

As this phenomenon occurs transiently and can be sometimes hidden by fragments or cytoplasmic granularity, its observation is difficult and not very frequent when conventional morphology assessment is made once or twice a day at specific time-points. Moreover, multinucleated embryos at one stage, for example two-cell, can be exempt from multinucleation at the following stage [15].

Time-lapse systems, such as the Embryoscope®, allow improved evaluation of multinucleation, as embryo development is observed and monitored continuously. It should be noted that the nuclei cannot be observed when the cells are not in interphase.

Various mechanisms have been proposed to explain the formation of multinucleation:

1. Nuclear replication without cytokinesis [17].
2. Fragmentation of the nucleus or defective migration at mitotic anaphase.
3. Expression of apoptosis.

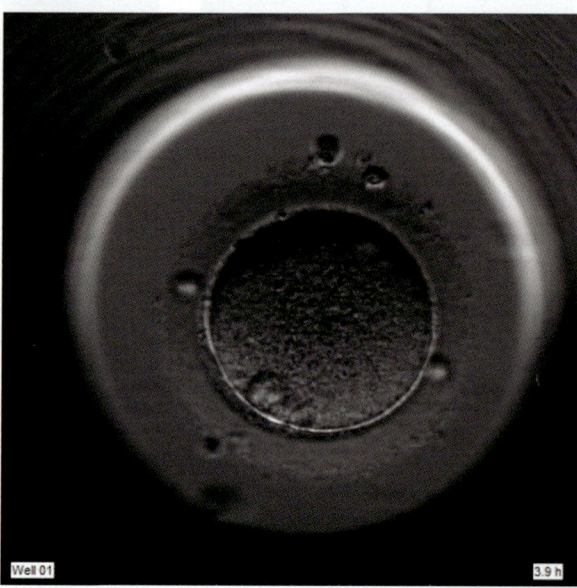

Video 6.17 Embryo multinucleated at the two-cell stage, but with no sign of multinucleation at the four-cell stage.

(A)

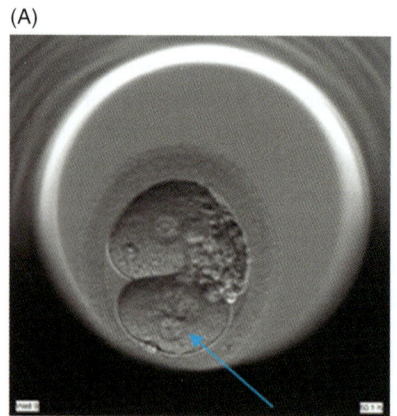

(B)

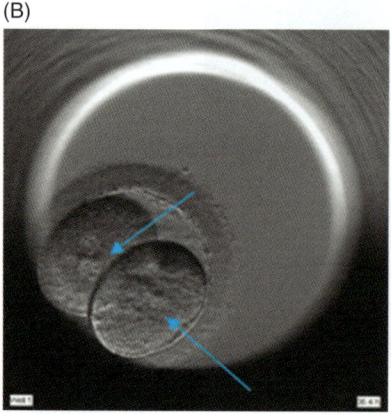

Figure 6.3 (A) image of a two-cell embryo showing one mononucleated cell, (B) image of a two-cell embryo showing two multinucleated cells (two nuclei).

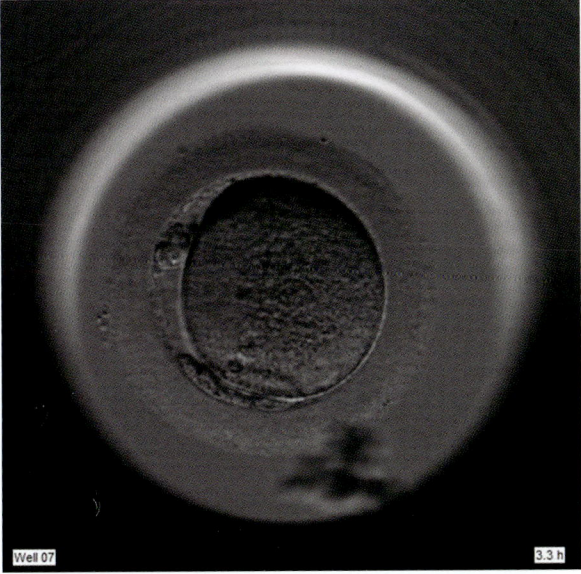

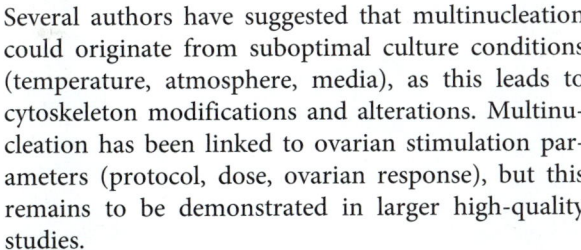

Video 6.18 Embryo trying to cleave at the two-cell stage with one of the two blastomeres failing, resulting in one multinucleadted blastomere at the three-cell stage.

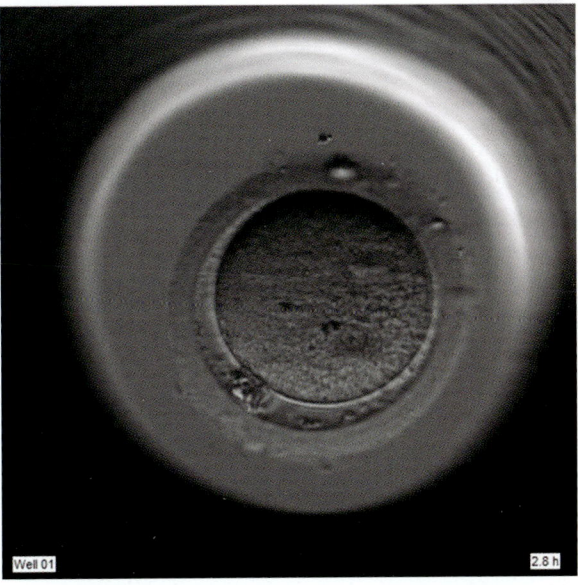

Video 6.19 Embryo trying to cleave at the two-cell stage and failing, leading to the formation of two binucleated blastomeres.

Several authors have suggested that multinucleation could originate from suboptimal culture conditions (temperature, atmosphere, media), as this leads to cytoskeleton modifications and alterations. Multinucleation has been linked to ovarian stimulation parameters (protocol, dose, ovarian response), but this remains to be demonstrated in larger high-quality studies.

The vast majority of the studies published on multinucleation as a tool for embryo quality assessment report lower blastocyst formation rate, lower implantation rate and finally dramatically lower live birth rate with multinucleated embryos than with non-multinucleated ones [12, 15]. Therefore, it is generally considered that multinucleation assessment should be a part of embryo scoring, and that multinucleated embryos should be excluded from transfer or freezing, even if few authors do not confirm this strategy. Time-lapse systems will allow more relevant studies on multinucleation frequency and consequences on embryo development. Whether multinucleated embryos have specific morphokinetic patterns, i.e., out-of-range cleavage timings and/or abnormal cellular divisions, remains to be demonstrated.

6.2.5.2. Polarization of blastomeres

Blastomere polarization is characterized by an acytoplasmic translucent ring around blastomeres, corresponding to zones without organelles. This is generally associated with darkened cytoplasm with centralized granularity, due to the retraction of the organelles towards the center of the cell [12]. This has been correlated with cell lysis and, therefore, with poor implantation prognosis [18].

6.2.5.3. Presence of vacuoles

Presence of vacuoles is a frequent observation in IVF, especially in oocytes, but sometimes also in embryos. These vacuoles vary in size, shape, and number, and can be grouped in three main categories [19]:

-those present in oocyte from the beginning (ovum pickup)
-those artificially created by ICSI procedure
-those arising later during embryo development

Even if the presence of vacuolated blastomeres has been associated with poor prognosis in some studies, there is no evidence in the literature that the quantity, size or distribution of these vacuoles in cells are correlated with implantation rate [8].

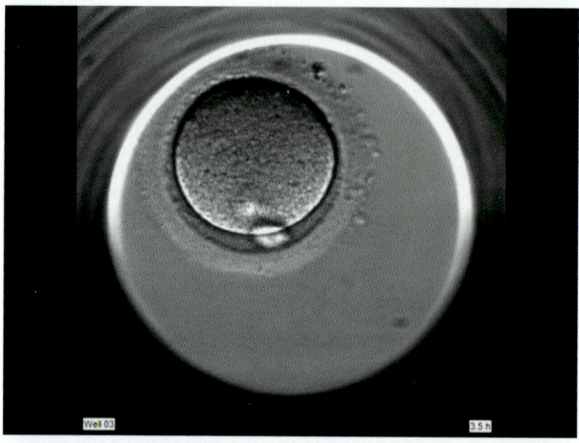

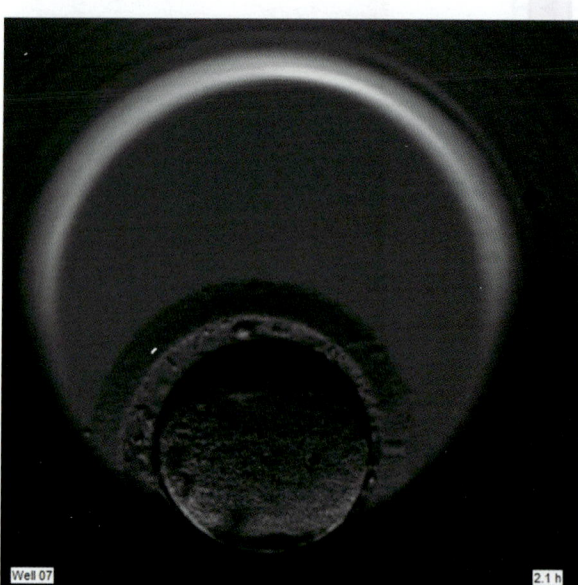

Video 6.20 and Video 6.21 Two videos of embryos multinucleated at the two-cell stage directly cleaving to a five-cell stage.

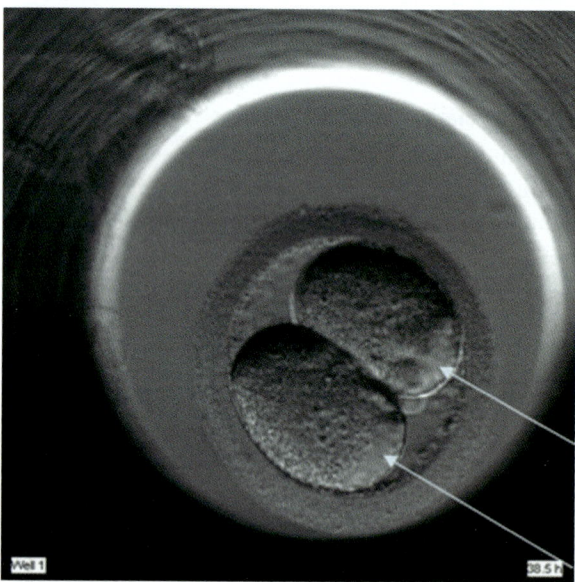

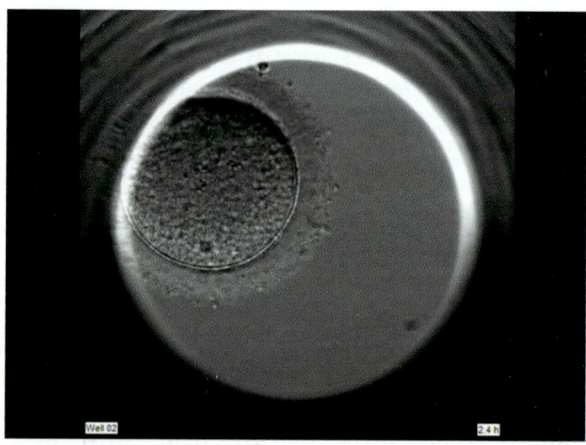

Video 6.22 Embryo with some cytoplasmic vacuoles.

Figure 6.4 Two-cell embryo with polarized blastomeres (marked with arrows).

6.2.5.4. Zona pellucida

The characteristics of the zona pellucida (ZP) are important when evaluating oocyte and embryo quality. Indeed, its appearance, shape, thickness, and birefringence have been reported to be predictive of the fertilization ability of the oocyte, and abnormalities in ZP structure have been evocated to be responsible for hatching failure, leading to a low implantation rate [20, 21]. Normal zona pellucida generally has an approximate thickness of 17 μm, with a spherical or round shape, no double layer caused by partitioning of the internal area, absence of bumps or irregularities, and a translucent appearance [20].

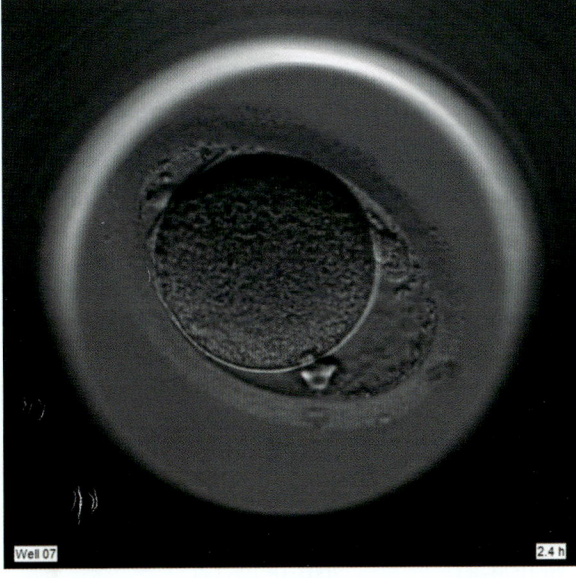

Well 07   2.4 h

Video 6.23 Embryo showing a partitioned zona pellucida.

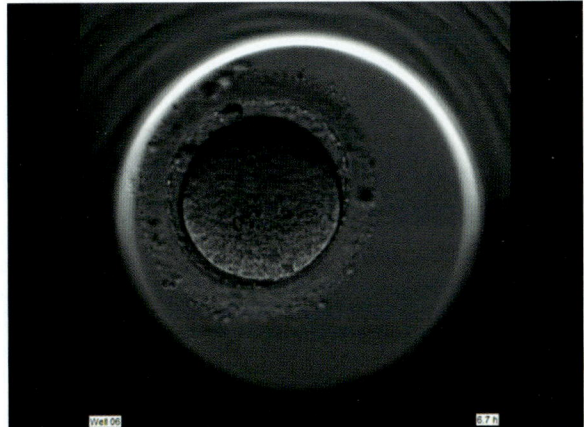

Well 00   3.7 h

Video 6.24 Embryo presenting with early compaction on day 3. Cell membranes start to oppose to each other as the junctions are setting up as soon as the eight-cell stage.

6.2.5.5. Early adhesion/compaction

Embryo compaction, consisting in the formation of intercellular junctions leading to cell membrane rapprochement and progressive disappearance of cell boundaries, usually appears on day 4 of embryo development, at the morula stage. Early compaction is defined as onset of compaction on day 3. Although it is rather uncommon [22], it seems to be a good prognosis factor, although the data reported in the literature are not sufficient to include this parameter in a classification scheme, probably partly due to the subjectivity of this observation [12, 22, 23].

6.2.5.6. Cytoplasmic granularity

This characteristic is common in D3 embryos, and might reflect embryonic activation with both nuclear and cytoplasmic activation. It is not known whether this morphological feature really affects any aspects of embryo development and implantation, but most studies do not report any association between blastomere granularity and developmental potential in D3 human embryos [23].

6.2.5.7. Scores

In order to standardize practices in ART centers, professional organizations such as ASRM, ESHRE or ALPHA have proposed scoring systems, including most relevant morphological variables according to the literature [6]. Although these classification schemes are obviously useful for training and standardization, their clinical importance in improving pregnancy rates remain to be demonstrated.

6.2.5.8. Revisiting embryo morphology assessment with time-lapse

Time-lapse systems allow objective and accurate evaluation of embryo development, not only quantitatively, but also qualitatively. Indeed, the continuous acquisition of high-resolution images allows the observation of dynamic events, such as fragment resorption or abnormal cleavage, which cannot be seen with conventional punctual morphological assessment. As abnormal cleavage, either due to abnormal cytokinesis duration, unusual morphological behavior before completing cleavage or presence of multiple cleavage furrows, is of critical importance for the evaluation of implantation potential of the embryos and can hardly be observed with conventional morphology, time-lapse systems pave the way for improved morphological embryo evaluation.

Thus, the onset of time-lapse technology should not lead one to forsake conventional morphology assessment, but these new insights into early embryo development should rather lead to re-evaluation of the existing classifications in order to refine them.

6.2.6. Conclusion

Although the morphological parameters that should be observed under the microscope in order to evaluate embryo quality are well known and described, embryo morphology-based scoring remains insufficient to guarantee the selection of the best embryo(s) for transfer. Time-lapse allows the monitoring of kinetics and chronology of embryonic development, thus providing much more information on embryo quality. Embryo classification systems should be revised according to morphology evolution and dynamics rather than static evaluation.

Bibliography

1. Edwards RG, Fishel SB, Fehilly CB, Purdy JM, Slater JM, Cohen J, et al. Factors influencing the success of in-vitro fertilization for alleviating human infertility. *J In Vitro Fert Embryo Transf.* 1984;1:3–23.

2. Berger DS, Abdelhafez F, Russell H, Goldfarb J, Desai N. Severe teratozoospermia and its influence on pronuclear morphology, embryonic cleavage and compaction. *Reprod Biol Endocrinol.* 2011;22;9:37.

3. Giorgetti C, Hans E, Terriou P, Salzmann J, Barry B, Chabert-Orsini V, et al. Early cleavage: an additional predictor of high implantation rate following elective single embryo transfer. *Reprod Biomed Online.* 2007;14:85–91.

4. Hesters L, Prisant N, Fanchin R, Méndez Lozano DH, Feyereisen E, Frydman R, et al. Impact of early cleaved zygote morphology on embryo development and in vitro fertilization-embryo transfer outcome: a prospective study. *Fertil Steril.* 2008;89:1677–84.

5. Lian WL, Xin ZM, Jin HX, Song WY, Peng ZF, Sun YP. Effects of early-cleavage embryo transfer on in vitro fertilization-embryo transfer pregnancy outcomes. *Clin Exp Obstet Gynecol.* 2013;40:319–22.

6. Machtinger R, Racowsky C. Morphological systems of human embryo assessment and clinical evidence. *Reprod Biomed Online.* 2013;26:210–21.

7. Terriou P, Giorgetti C, Hans E, Salzmann J, Charles O, Cignetti L, et al. Relationship between even early cleavage and day 2 embryo score and assessment of their predictive value for pregnancy. *Reprod Biomed Online.* 2007;14:294–9.

8. Alpha Scientists in Reproductive Medicine and ESHRE Special Interest Group of Embryology. The Istanbul consensus workshop on embryo assessment: proceedings of an expert meeting. *Hum Reprod.* 2011;26:1270–83.

9. Herrero J and Meseguer M. Selection of high potential embryos using time-lapse imaging: the era of morphokinetics. *Fertil Steril.* 2013;99:1030–4.

10. Kirkegaard K, Agerholm IE, Ingerslev HJ. Time-lapse monitoring as a tool for clinical embryo assessment. *Hum Reprod.* 2012;27:1277–85.

11. Holte J, Berglund L, Milton K, Garello C, Gennarelli G, Revelli A, Bergh T. Construction of an evidence-based integrated morphology cleavage embryo score for implantation potential of embryos scored and transferred on day 2 after oocyte retrieval. *Hum Reprod.* 2007;22:548–57.

12. Prados FJ, Debrock S, Lemmen JG, Agerholm I. The cleavage stage embryo. *Hum Reprod.* 2012;27 Suppl 1:i50–71.

13. Racowsky C, Vernon M, Mayer J, Ball GD, Behr B, Pomeroy KO, et al. Standardization of grading embryo morphology. *Fertil Steril.* 2010;94:1152–3.

14. Guerif F, Frapsauce C, Chavez C, Cadoret V, Royere D. Treating women under 36 years old without top-quality embryos on day 2: a prospective study comparing double embryo transfer with single blastocyst transfer. *Hum Reprod.* 2011;26:775–81.

15. Van Royen E, Mangelschots K, Vercruyssen M, De Neubourg D, Valkenburg M, Ryckaert G, et al. Multinucleation in cleavage stage embryos. *Hum Reprod.* 2003;18:1062–9.

16. Ambroggio J, Gindoff PR, Dayal MB, Khaldi R, Peak D, Frankfurter D, et al. Multinucleation of a sibling blastomere on day 2 suggests unsuitability for embryo transfer in IVF-preimplantation genetic screening cycles. *Fertil Steril.* 2011;96:856–9.

17. Pickering SJ, Taylor A, Johnson MH, Braude PR. An analysis of multinucleated blastomere formation in human embryos. *Hum Reprod.* 1995;10:1912–22.

18. Veek LL. Preembryo grading and degree of cytoplasmic fragmentation. In: *An Atlas of Human Gametes and Conceptuses: An Illustrated Reference for Assisted Reproductive Technology.* New York, USA: Parthenon Publishing, 1999: pp. 46–51.

19. Ebner T, Moser M, Sommergruber M, Gaiswinkler U, Shebl O, Jesacher K, et al. Occurrence and developmental consequences of vacuoles throughout preimplantation

development. *Fertil Steril.* 2005;83:1635–40.

20. Gabrielsen A, Lindenberg S, Petersen K. The impact of the zona pellucida thickness variation of human embryos on pregnancy outcome in relation to suboptimal embryo development. A prospective randomized controlled study. *Hum Reprod.* 2001;16:2166–70.

21. Sun YP, Xu Y, Cao T, Su YC, Guo YH. Zona pellucida thickness and clinical pregnancy outcome following in vitro fertilization. *Int J Gynaecol Obstet.* 2005;89:258–62.

22. Le Cruguel S, Ferré-L'Hôtellier V, Morinière C, Lemerle S, Reynier P, Descamps P, et al. Early compaction at day 3 may be a useful additional criterion for embryo transfer. *J Assist Reprod Genet.* 2013;30:683–90.

23. Skiadas CC, Jackson KV, Racowsky C. Early compaction on day 3 may be associated with increased implantation potential. *Fertil Steril.* 2006;86:1386–91.

24. Tomari H, Honjou K, Nagata Y, Horiuchi T. Relationship between meiotic spindle characteristics in human oocytes and the timing of the first zygotic cleavage after intracytoplasmic sperm injection. *J Assist Reprod Genet.* 2011;28:1099–104.

Embryo quality (classification and selection)

Alberto Tejera, Natalia Basile, Maura Caiazzo, and Marcos Meseguer

7.1. Embryo morphology

At the moment, standard morphological assessment remains the most common method to evaluate embryo quality. Even though other non-invasive evaluation methods are being developed and constantly improved, morphology continues to be the most widespread and effective system in use. However, concern about the effects of handling the embryos outside stable culture conditions has limited the frequency of microscopic observations.

One of the key disadvantages of evaluating embryos according only to morphological parameters is that this assessment is not always done at the same time-points and these variations can be relevant. For instance, the same day 2 embryo may be at the two-cell stage early on day 2 or already at the four-cell stage if assessed later.

Furthermore, the morphological appearance of an embryo can change rapidly, which may be relevant for day 2 and day 3 evaluations [1] and can result in a classification that is not objectively correct. In addition, morphologically variable embryos may nonetheless go on to implant successfully [2].

The evaluation of different morphological criteria at different embryo stages has been studied in order to correlate them with the probability of successful embryo implantation.

Evaluation at these stages has traditionally formed the basis for determining embryo quality. However, while there is a certain amount of agreement on what constitutes a good or poor quality embryo, it is very difficult to find a consensus between laboratories when attempting to evaluate intermediate quality embryos.

Moreover, the subjectivity in embryo quality can result in highly variable embryo scores depending on the embryologist.

7.2. Cleavage stage classification

Established embryo scoring systems are relatively crude, being based on the number of blastomeres and embryo quality [3]. Embryologists from ALPHA and the European Society of Human Reproduction and Embryology (ESHRE) have recently elaborated some of the pitfalls of morphology to grade embryos at a consensus meeting in Istanbul [4].

An exhaustive review of all the parameters related to embryo assessment was included in the previous chapter, therefore in this chapter we will focus on embryo quality classification based on two different points of view: standard morphological assessment (according to the consensus) and morphokinetics.

7.3. Standard morphological assessment

7.3.1. Assessment of cell number and rate of division

There are very few studies that provide a clear system to correlate different implantation rates with the number of cells or rate of division.

Time-Lapse Microscopy in In-Vitro Fertilization, ed. Marcos Meseguer. Published by Cambridge University Press.

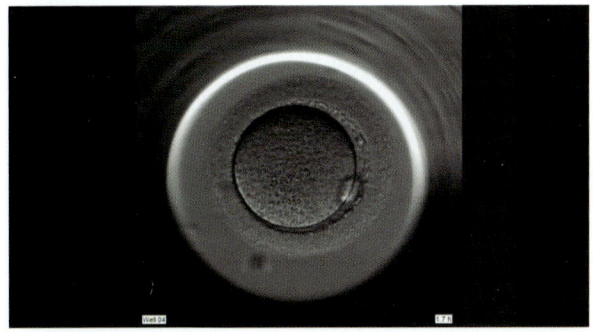

Video 7.1 Day 2 embryo assessment.

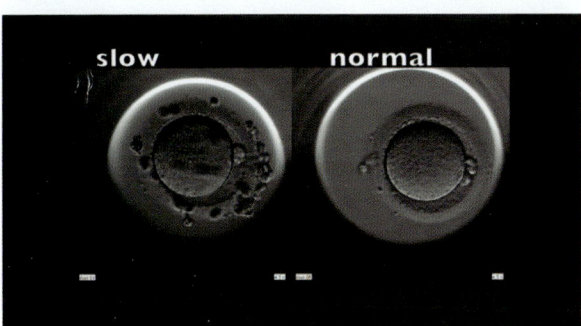

Video 7.3 Slow vs. normal embryo development.

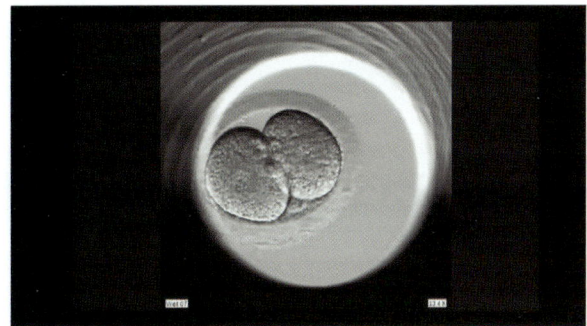

Video 7.2 Day 3 embryo assessment.

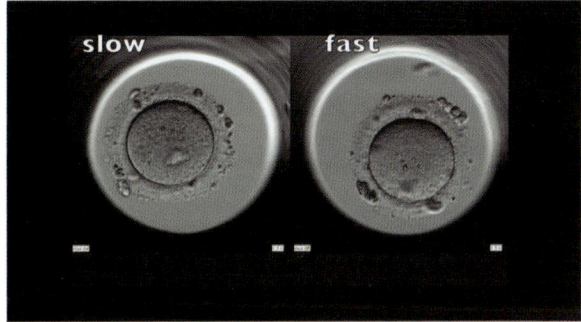

Video 7.4 Slow vs. fast embryo development.

According to the consensus, optimal embryos are defined as those with four cells on day 2, and eight cells on day 3 at each of the nominated time-points post-insemination [5]:

Day 2 embryo assessment 44 ± 1 h four-cell stage (Video 7.1)

Day 3 embryo assessment 68 ± 1 h eight-cell stage (Video 7.2)

On the other hand, embryos that cleave slower or faster than the expected rates have a reduced implantation potential (Video 7.3 and Video 7.4)

It was noted, however, that this may change in the future, depending on the culture media being used.

7.3.2. Fragmentation

The presence of cell fragmentation is common in human embryos and does not always correlate with a low implantation rate. As long as the degree of fragmentation does not exceed 20% [6] or even 25% [7], the implantation of the embryo will not be compromised. In addition to the percentage of fragments, their size and distribution may also have a correlation with implantation. Considering that the majority of large fragments may be confused with cells the following description has been established: a fragment is an extra-cellular membrane-bound cytoplasmic structure that is 45 mm diameter in a day 2 embryo and 40 mm diameter in a day 3 embryo.

A predominance of large fragments is known as a "Type IV" fragmentation pattern [8] and is associated with a very low probability of implantation.

According to this consensus the relative degrees of fragmentation were defined as: mild (< 10%) (Video 7.5); moderate (10–25%) (Video 7.6) and severe (> 25%) (Video 7.7). The percent values are based on the cell equivalents, therefore for a four-cell embryo, 25% fragmentation would be equivalent to one cell in volume.

105

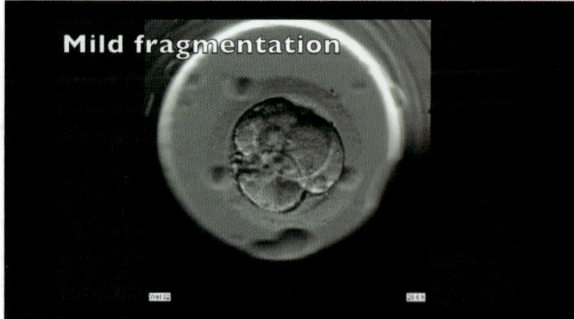

Video 7.5 Embryo fragmentation (mild fragmentation).

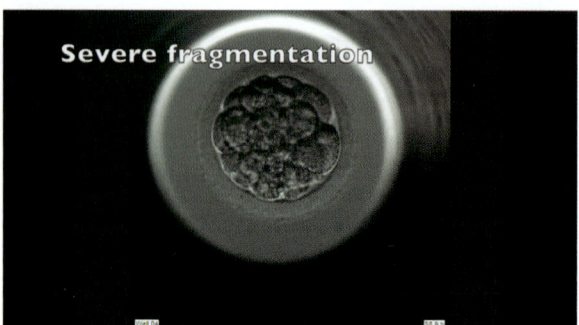

Video 7.7 Embryo fragmentation (severe fragmentation).

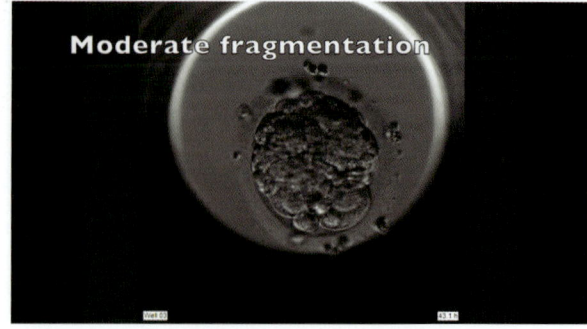

Video 7.6 Embryo fragmentation (moderate fragmentation).

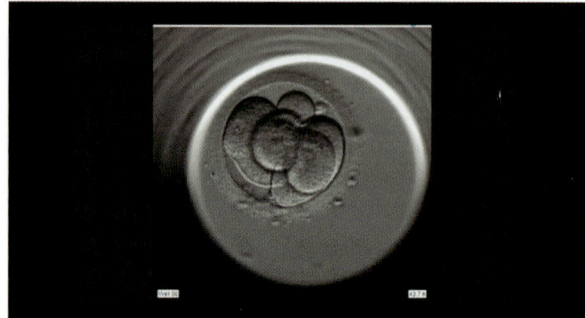

Video 7.8 Fragment reabsorption.

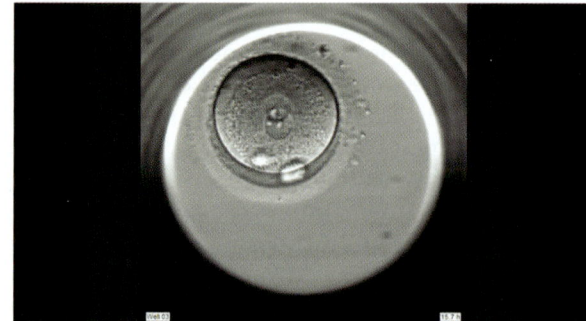

Video 7.9 Multinucleation at two-cell stage.

Continuous monitoring of embryo development has demonstrated that fragmentation can be a dynamic phenomenon and that the fragments can move within the embryo or undergo reabsorption (Video 7.8). This consensus doesn´t include the impact of fragment localization on implantation rates.

7.3.3. Multinucleation

Multinucleation was defined as the presence of more than one nucleus in a blastomere, and includes micronuclei (Video 7.9). The consensus shows a correlation between the presence of two or more nuclei in a cell with a decreased implantation potential, probably because the presence of multinucleated blastomeres implies an increase in the rate of chromosomal anomalies [9], and as a consequence, it increases the rate of abortion, although children have been born from multinucleated blastomere embryos [10–12].

According to this consensus the multinucleation assessment should be performed on day 2 (at a discrete time-point of 44 ± 1 h post-insemination) (Video 7.10), and it has been established that the observation of multinucleation in one cell is sufficient for the embryo to be considered multinucleated. It was further agreed that multinucleation assessment on day 3 would be complicated due to

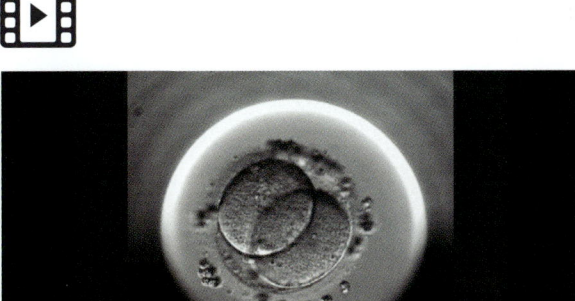

Video 7.10 Multinucleation on day 2.

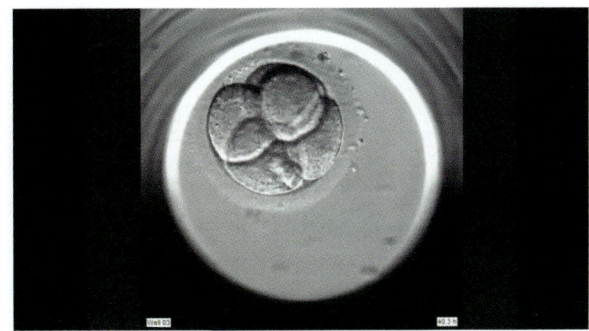

Video 7.11 Multinucleation on day 3.

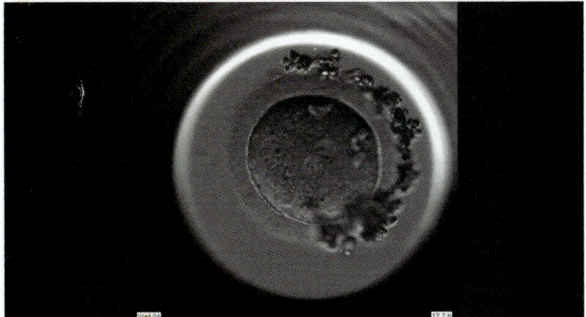

Video 7.12 Asymmetric blastomeres.

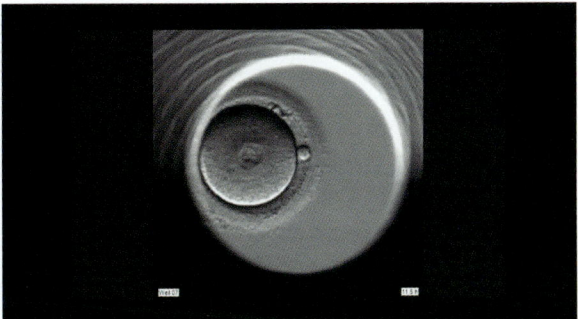

Video 7.13 Symmetric blastomeres.

the much smaller cell size, and therefore it would be less reliable (Video 7.11).

7.3.4. Blastomere size

When the first mitotic divisions take place, they are not always synchronic and the newly formed cells are not always of the same size (Video 7.12). An embryo undergoing synchronic and symmetric cleavage will present two, four, and eight cells of a similar size and so on (Video 7.13). Unevenly sized blastomeres are related to a reduction in the implantation potential of the embryo. It is evident that asynchronous cell divisions will lead to the co-existence of cells from two different cell cycles [13].

According to the consensus, embryos with two, four, eight, and 16 cells should present even-sized blastomeres, and embryos with an odd number of blastomeres should present two different sizes of blastomeres [14]. This means that inequalities in the sizes of the blastomeres should be compatible with a positive classification when they appear in the following way:

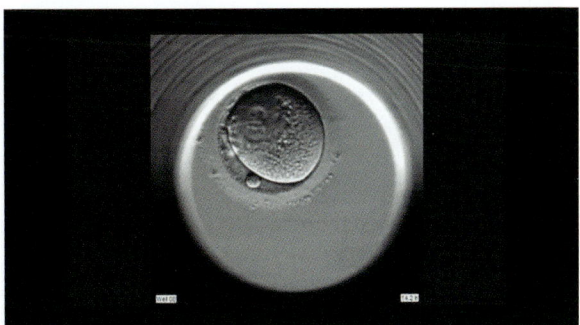

Video 7.14 Symmetry of three-cell embryo.

3 cells: 2 small + 1 large (Video 7.14)
5 cells: 2 small + 3 large (Video 7.15)
6 cells: 4 small + 2 large (Video 7.16)
7 cells: 6 small + 1 large (Video 7.17)

Different size combinations than the above mentioned imply an asymmetric division with an anomalous distribution of cytoplasm between the resulting cells (Video 18, Video 19, Video 20, Video 21).

107

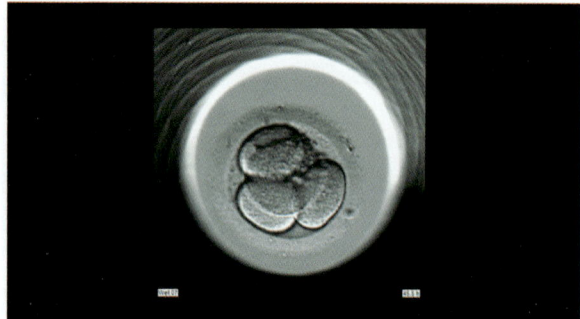

Video 7.15 Symmetry of five-cell embryo.

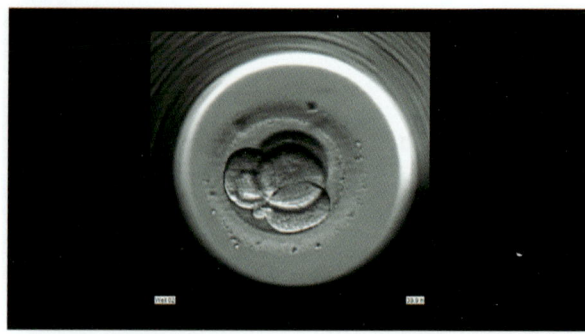

Video 7.16 Symmetry of six-cell embryo.

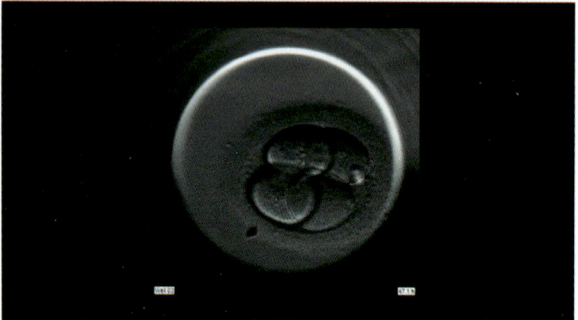

Video 7.17 Symmetry of seven-cell embryo.

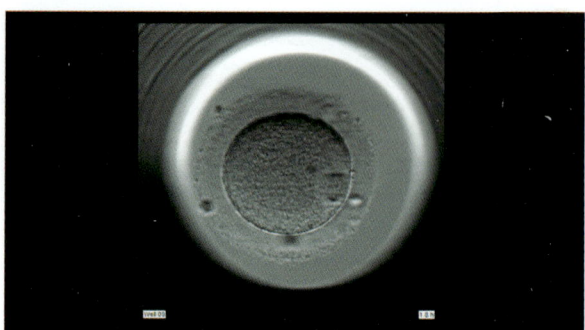

Video 7.18 Asymmetric division (example 1).

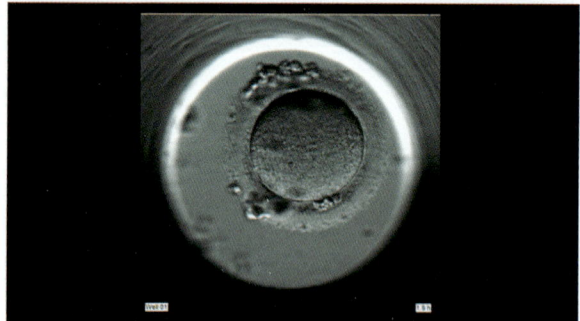

Video 7.19 Asymmetric division (example 2).

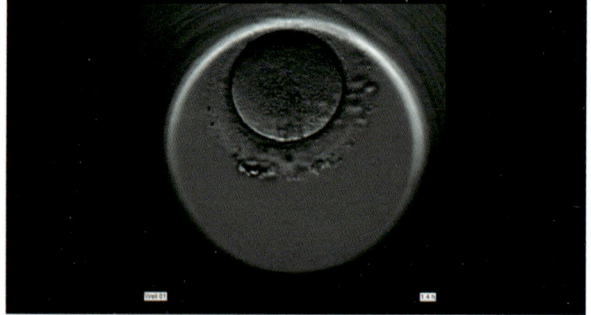

Video 7.20 Asymmetric division (example 3).

7.3.5. Other morphological features of day 2 and day 3 embryos

There are other morphological parameters that seem to be correlated with embryo quality, although these features can vary between the patient's embryos or even between patients. These morphological features can also be scored as part of the morphological assessment of day 2 and day 3 embryos.

According to the consensus there is no significant evidence to support a clear biological effect of these features on implantation potential. Therefore, we are going to review the most important features that could affect the implantation rates.

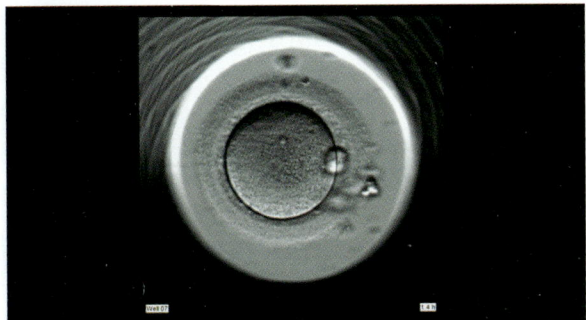

Video 7.21 Asymmetric division (example 4).

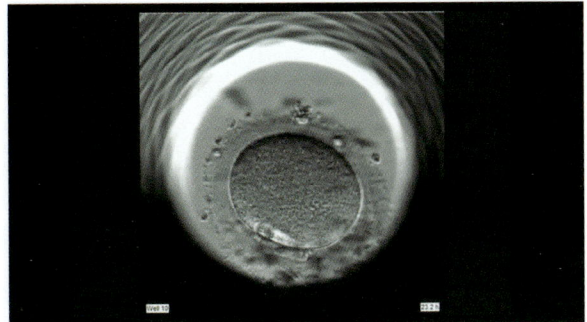

Video 7.22 Irregular-shape blastomeres.

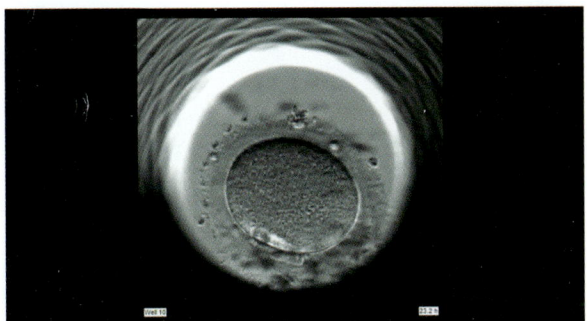

Video 7.23 Acytoplasmatic ring.

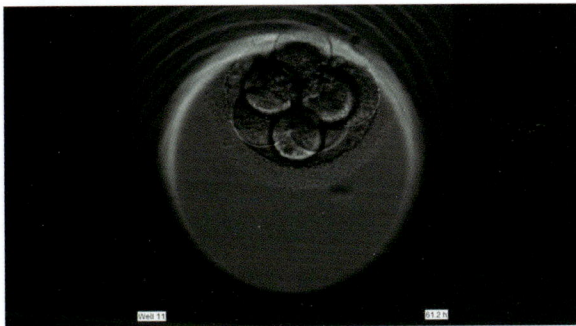

Video 7.24 Presence of vacuoles.

Irregular shape blastomeres may either have a physiological alteration or be undergoing division. It is difficult to evaluate the relationship of this characteristic with the potential for implantation and it has therefore not been included as a criterion for assigning embryo category (Video 7.22).

The presence of an acytoplasmic ring in the embryo has been related to a process of cellular lysis. This is characterized by the apparent contraction of the cytoplasm leaving a large translucent ring with no organelles close to the edge of the blastomere (Video 7.23).

The presence of vacuoles is also associated with a reduction in the developmental potential. However there is no evidence in the literature relating the exact number of vacuoles, their size or distribution in the blastomeres, with implantation rates. But it has been demonstrated that vacuoles with diameters of less than 5 μm may not compromise the development of the embryo [15] (Video 7.24).

Abnormalities of the zona pellucida are also associated with a low implantation rate probably due

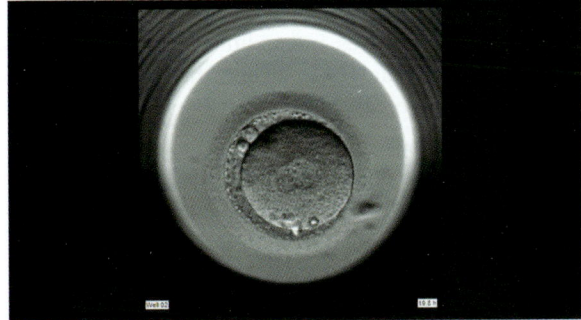

Video 7.25 Normal zona pellucida.

to the lack of hatching [15, 16]. A healthy zona pellucida may be described as having a round silhouette and a thickness of approximately 17 μm. In addition, it should not be homogeneous and present a double layer due to the partitioning of the inner zone. A normal zona pellucida should not present bulges and it should be translucent in appearance [17–19] (Video 7.25 and Video 7.26).

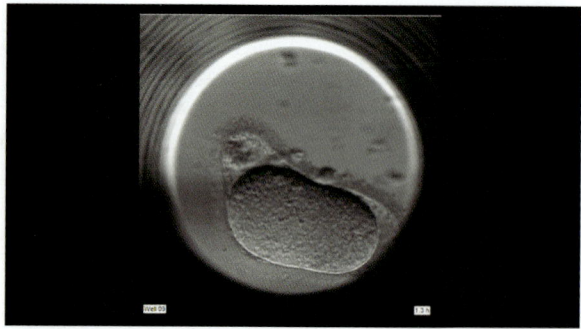

Video 7.26 Bulges in zona pellucida.

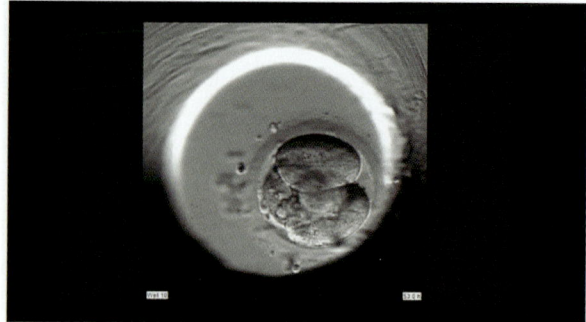

Video 7.27 Initiation of adhesion.

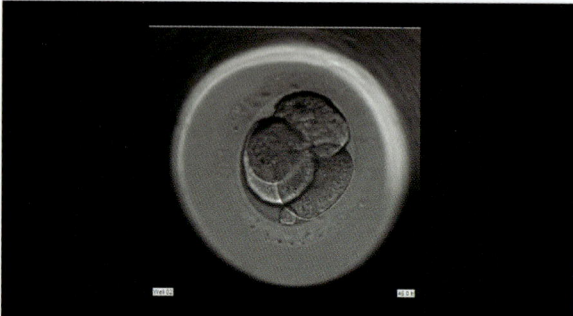

Video 7.28 Embryo compaction.

Early compaction is also a sign of embryo activation which involves intercellular contact, the formation of intercellular bonds and the start of embryo polarization. We can distinguish between two levels of compaction:

- The initiation of adhesion: the stage at which the cells can be individually identified, although the membranes are adjacent (Video 7.27).
- Compaction: the stage at which it is difficult to distinguish individual blastomeres (Video 7.28).

The references we reviewed did not provide sufficient evidence to allow us to include this parameter in an embryo classification system, possibly because it is highly influenced by the culture protocol and the culture medium utilized, although some recommendations can be suggested based on the level of adhesion and the observation day [20].

The last parameter has a strange appearance of orange peel which should not be confused with extensive vacuolization [20]. It is known as "pitting" and similarly to cell adhesion the appearance of this

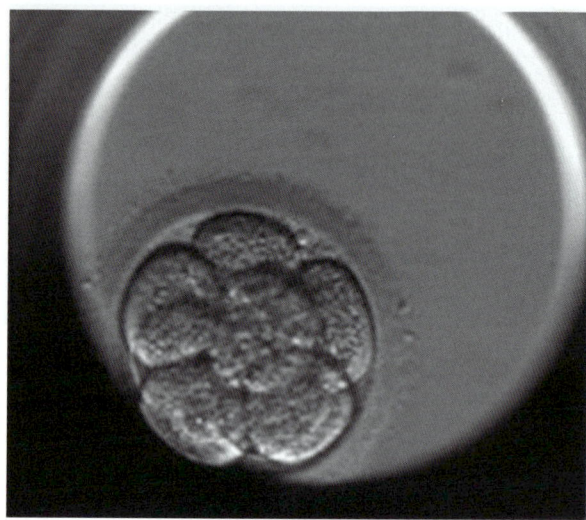

Figure 7.1 A phenomenon known as "pitting": a strange appearance of orange peel.

morphological parameter indicates that cytoplasmic activation has taken place and it should occur at the moment of physiological embryo activation at D+3 (Figure 7.1).

7.3.6. Cleavage-stage embryo scoring system

According to the consensus an optimal day 2 embryo (44 + 1 h post-insemination) should have four equally sized mononucleated blastomeres with < 10% fragmentation, and an optimal day 3 embryo (68 + 1 h post-insemination) should have eight equally sized mononucleated blastomeres, with < 10% fragmentation

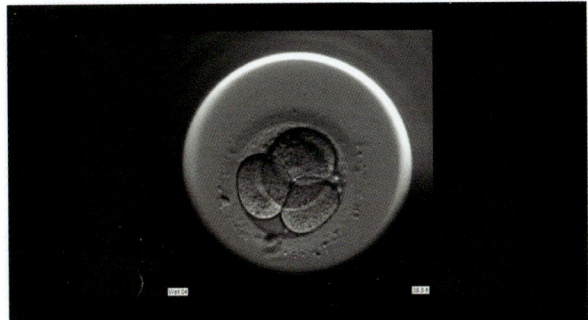

Video 7.29 Optimal day 3 embryo.

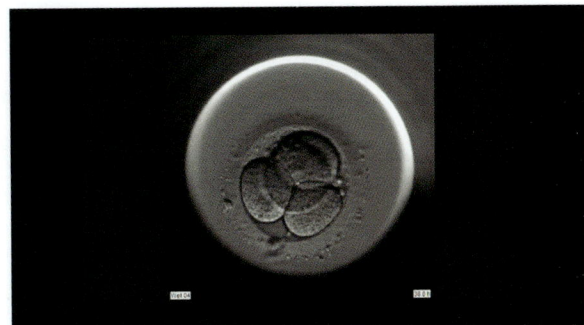

Video 7.30 Day 3 embryo Grade I.

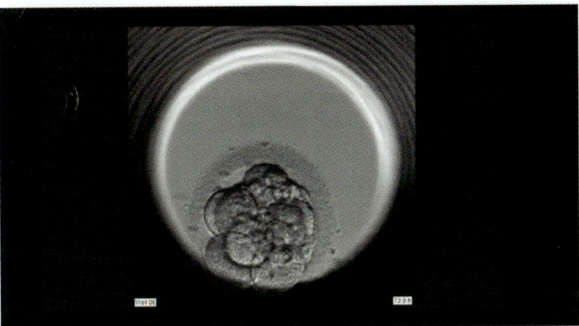

Video 7.31 Day 3 embryo Grade II.

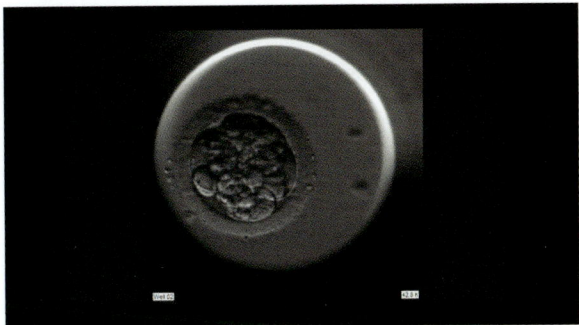

Video 7.32 Day 3 embryo Grade III.

(Video 7.29). The consensus scoring system for cleavage-stage embryos (based on cell number, grade and reason for the grade) is presented as follows:

Grade 1 (good quality): < 10% fragmentation, stage-specific cell size and no multinucleation (Video 7.30).

Grade 2 (fair quality): 10–25% fragmentation, stage-specific cell size for majority of cells and no evidence of multinucleation (Video 7.31).

Grade 3 (poor quality): Severe fragmentation (> 25%), cell size not stage specific and evidence of multinucleation (Video 7.32).

7.4. Morula and blastocyst classification

7.4.1. Morula stage

Cell compaction is usually observed on D + 4 of embryo culture although it initiates around the eight-cell stage. Visually speaking, it is characterized by the appearance of a compacted mass of cells (Video 7.33). According to

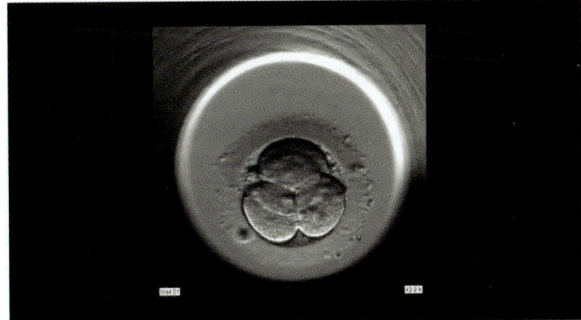

Video 7.33 Morula stage embryo.

the consensus, optimal embryos should be compacted or compacting at 92 + 2 h and they should have entered into a fourth round of cleavage. This compaction is the result of the tight bonds formed between cells thus preventing one from clearly distinguishing their shape. At this stage cells are no longer totipotent, corresponding to an embryo which has already become differentiated. Therefore, the presence of compaction on D + 4 is

111

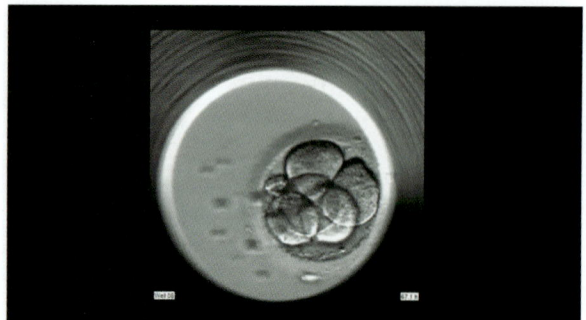

Video 7.34 Morula-stage embryo (partial compaction).

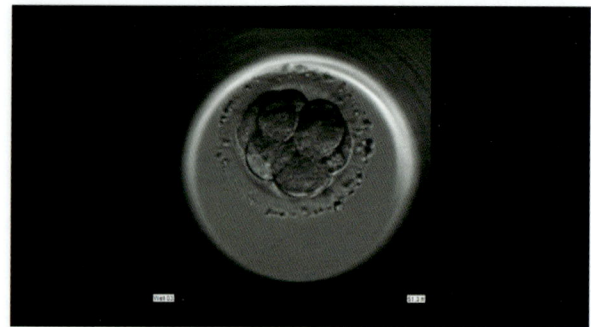

Video 7.35 Day 4 embryo Grade I.

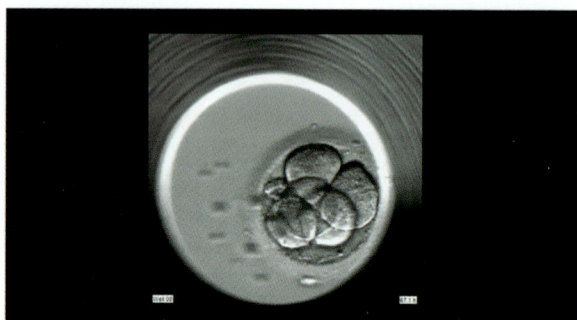

Video 7.36 Day 4 embryo Grade II.

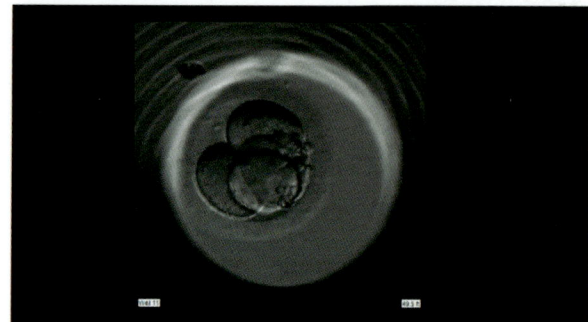

Video 7.37 Day 4 embryo Grade III.

a good prognosis for development. Compaction should affect all the embryo volume, but sometimes it only affects some of the cells, resulting in partial compaction (Video 7.34).

At present, reported evidence does not allow us to draw conclusions on the possible correlation between the type of compaction and the probability of embryo implantation, although the latest studies have shown that partial compaction may imply a lower chance of implantation.

When partial compaction occurs, the proportion of cells undergoing compaction will be correlated to the size of the resulting embryo. Embryos which are smaller than usual with more than half of the embryo excluded have a lower chance of implantation [21].

The consensus scoring system for day-4 embryos is based on a rating from good to poor quality, as follows:

Grade 1 (good quality): Entered into a fourth round of cleavage, evidence of compaction that involves virtually all the embryo volume (Video 7.35).

Grade 2 (fair quality): Entered into a fourth round of cleavage and compaction that involves the majority of the embryo (Video 7.36).

Grade 3 (poor quality): Disproportionate compaction involving less than half of the embryo, with two or three cells remaining as discrete blastomeres (Video 7.37).

7.4.2. Blastocyst classification

The embryo presents a complex structure at the blastocyst stage due to the large number of cells and their organization. Several authors have stated which are the key parameters for morphologically optimal blastocysts and how they correlate with higher implantation rates [22–27]. The following elements are the key parameters that should be taken into account for blastocysts:

-Blastocoel.
-Zona pellucida.
-Inner cell mass.
-Trophectoderm.

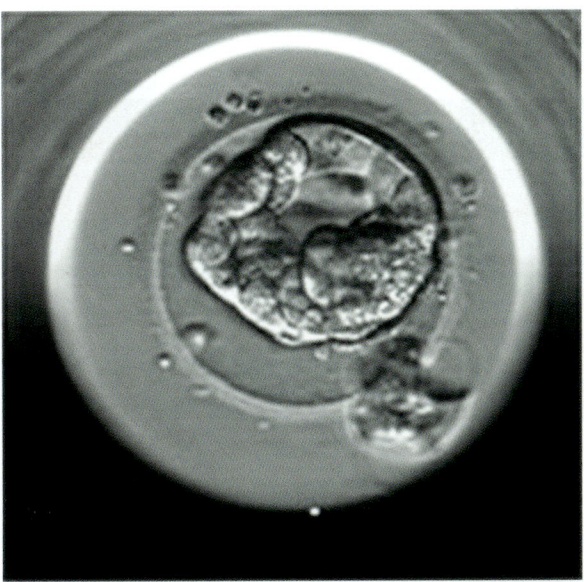

Figure 7.2 Contraction of blastocyst where at least the 50% of the blastocoel is collapsed.

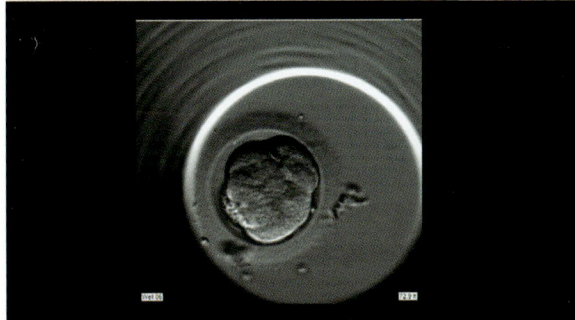

Video 7.38 Zona pellucida of expanded blastocyst.

7.4.3. Blastocoel expansion

The increase in the size of the blastocoel is very difficult to categorize because it is time-dependent and it may change during the blastocoel collapsing phases. In presence of this event (Figure 7.2), the evaluation of the embryo should be postponed for a few hours. Further re-expansion of the blastocoel will result in the thinning of the zona pellucida and once organized, the blastocoel normally occupies the majority of the volume of the embryo.

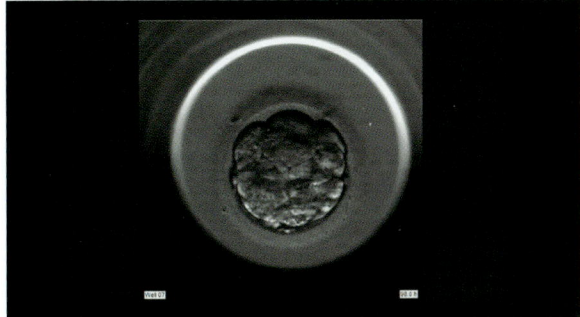

Video 7.39 Embryo hatching.

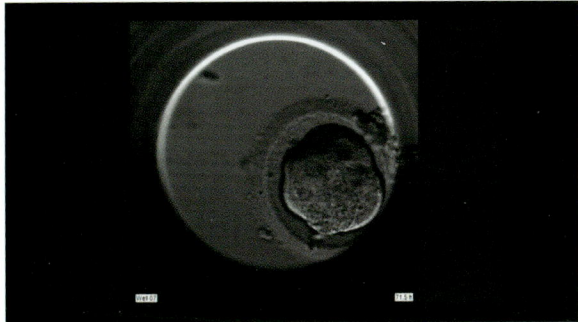

Video 7.40 Thinning of zona pellucida on day 5 embryo.

7.4.4. Zona pellucida

The thickness of the zona pellucida can greatly vary during blastocyst development. It will become thinner during blastocoel expansion (Video 7.38), reaching the minimum thickness when the blastocyst is completely expanded. The next step is embryo hatching and this begins with the fracture of the zona pellucida (Video 7.39). Some authors consider the thinning of the zona pellucida to be a favorable factor for implantation [22, 25]. Zona pellucida, from best to worst implantation rate:

-Thinning appears at D + 5 (Video 7.40)
-Thinning appears at D + 6 (Video 7.41)

7.4.5. Inner cell mass (ICM)

This is the most important parameter to define the various categories of blastocysts. The inner cell mass should be oval in shape and its cells should be compacted (Video 7.42).

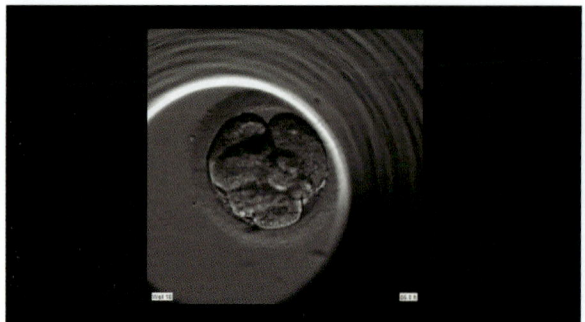

Video 7.41 Thinning of zona pellucida on day 6 embryo.

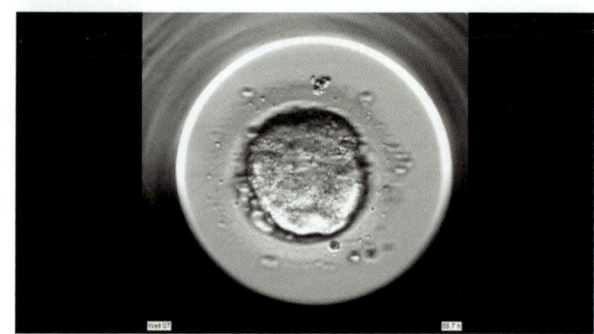

Video 7.42 Inner cell mass.

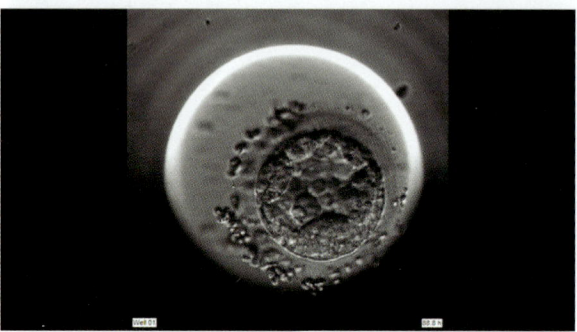

Video 7.43 Small inner cell mass.

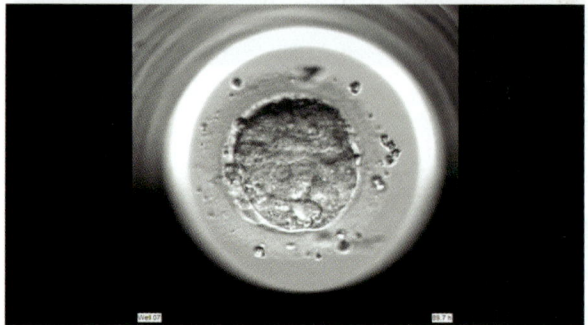

Video 7.44 Trophectoderm.

The most favorable size varies between 1,900 and 3,800 μm^2. Smaller sizes imply a lower implantation potential [26] (Video 7.43). An ICM of 3,800 μm^2 is comparable to the size of a blastomere in a four-cell stage embryo.

Inner cell mass, from best to worst implantation rate:

-ICM sizes from 1,900–3,800 μm^2, with an oval compacted appearance.

-ICM size < 1,900 μm^2, with a compacted non-oval appearance.

7.4.6. Trophectoderm (TE)

This structure presents a single layer of bonded cells which form the wall of the blastocoel or blastocyst cavity. The number, shape, and cohesion degree of these cells will allow us to classify the blastocyst into several categories (from best to worst implantation rate):

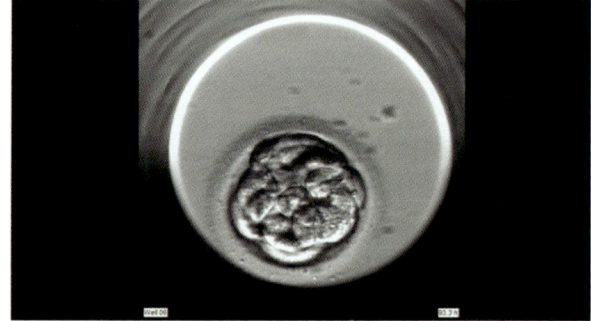

Video 7.45 Trophectoderm with irregular ephitelium.

-Homogeneous epithelium with elliptical cells (Video 7.44)

-Irregular epithelium (Video 7.45)

-Irregular epithelium with few cells (Video 7.46)

Blastocysts having a suboptimal trophoectoderm have good implantation rate as long as the inner cell mass has normal morphology [28].

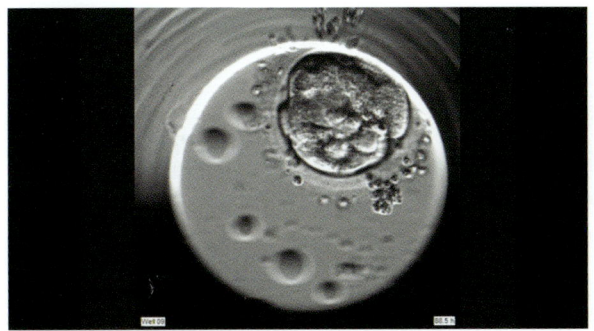

Video 7.46 Trophectoderm with few cells.

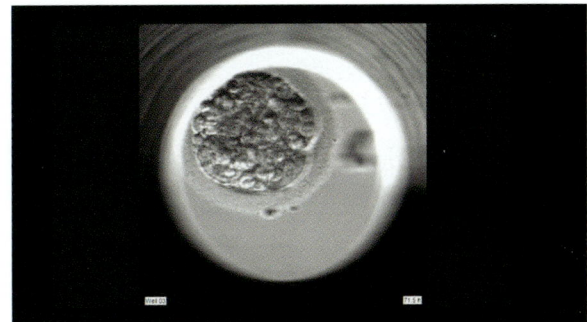

Video 7.47 Optimal blastocyst.

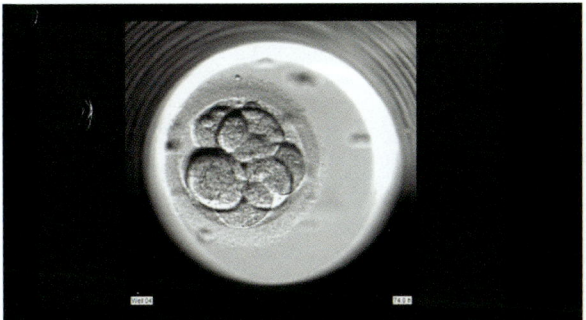

Video 7.48 Blastocyst development (early blastocyst).

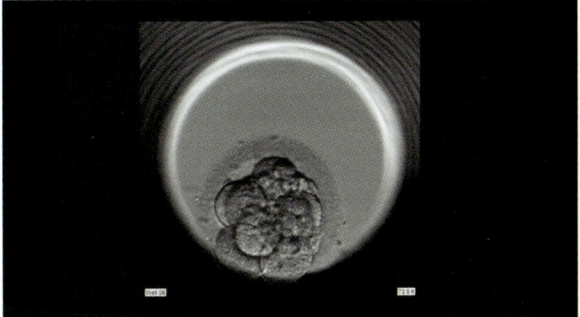

Video 7.49 Blastocyst development (cavitated blastocyst).

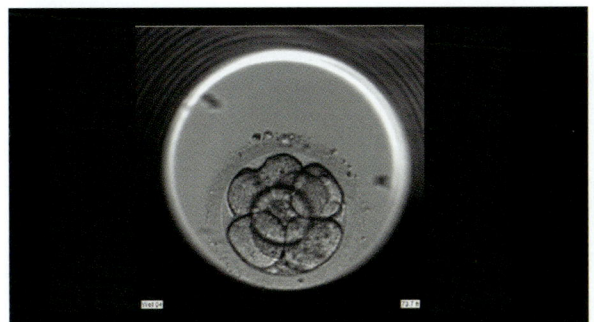

Video 7.50 Blastocyst development (expanded blastocyst).

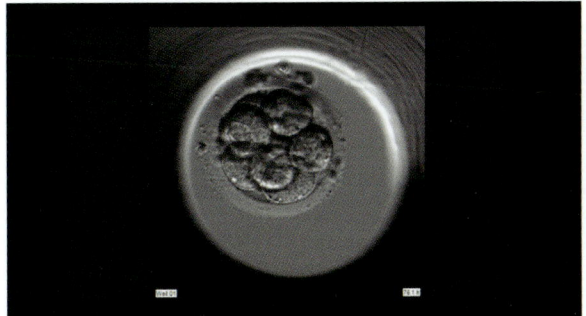

Video 7.51 Blastocyst development (hatching blastocyst).

Considering these aspects and according to this consensus an optimal blastocyst at this period of time (116 + 2 h) will be a fully expanded through hatched blastocyst with: an ICM that is prominent, easily discernible, and consisting of many compacted and tightly adherent cells, and with a TE that comprises many cells forming a cohesive epithelium (Video 7.47). It was agreed that while the ICM has a high prognostic value for implantation and fetal development, a functional TE is also essential.

The consensus for a blastocyst scoring system is a combination of the stage (expansion) and of the grade of ICM and of the TE, as follows.

Stage of development:

1 (early) (Video 7.48), 2 (blastocyst) (Video 7.49), 3 (expanded) (Video 7.50), 4 (hatched or hatching) (Video 7.51).

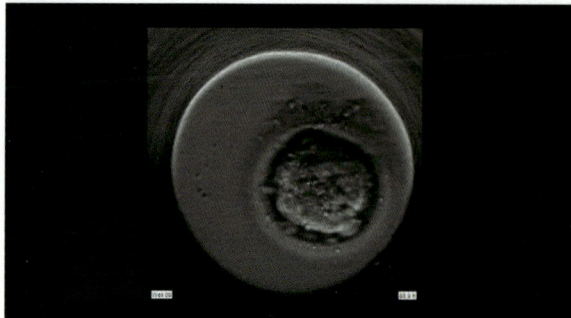

Video 7.52 Inner cell mass Grade I.

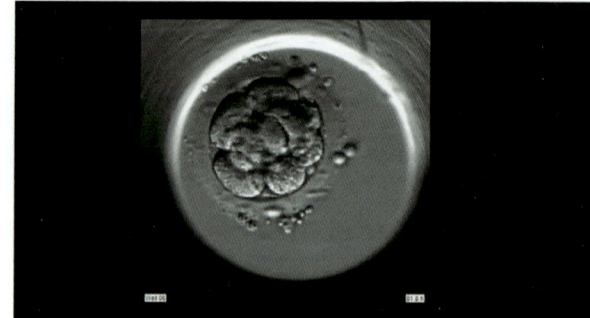

Video 7.53 Inner cell mass Grade II.

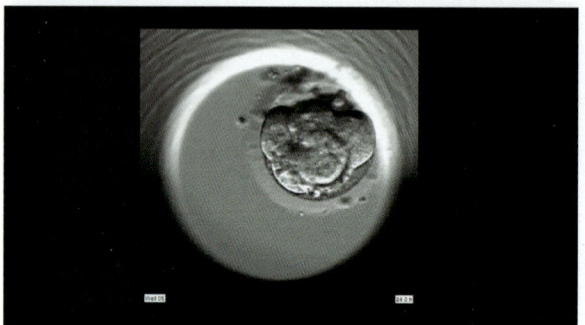

Video 7.54 Inner cell mass Grade III.

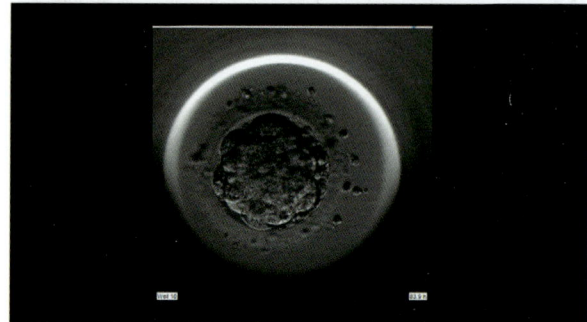

Video 7.55 Trophectoderm Grade I.

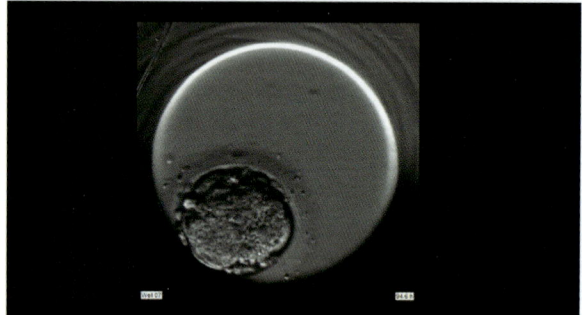

Video 7.56 Trophectoderm Grade II.

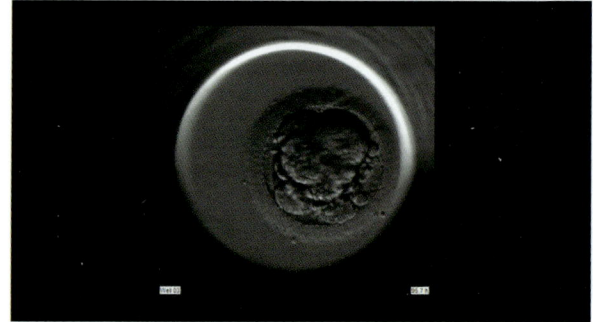

Video 7.57 Trophectoderm Grade III.

ICM:

Grade 1 (good quality): Prominent, easily discernible, with many cells that are compacted and tightly adhered together (Video 7.52).

Grade 2 (fair quality): Fairly easily discernible, with many cells that are loosely grouped together (Video 7.53).

Grade 3 (poor quality): Poor, difficult to discern, with few cells (Video 7.54)

TE:

Grade 1 (good quality): Many cells forming a cohesive epithelium (Video 7.55)

Grade 2 (fair quality): Few cells forming a loose epithelium (Video 7.56)

Grade 3 (poor quality): Poor. Very few cells (Video 7.57)

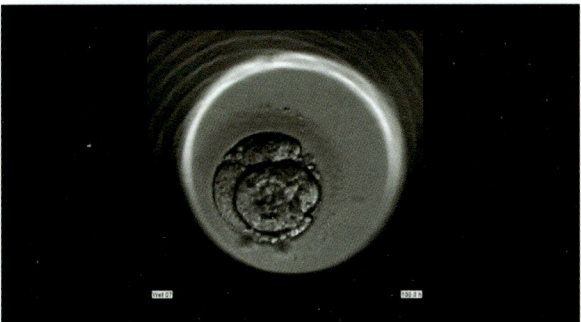

Video 7.58 Fragmentation and vacuolization.

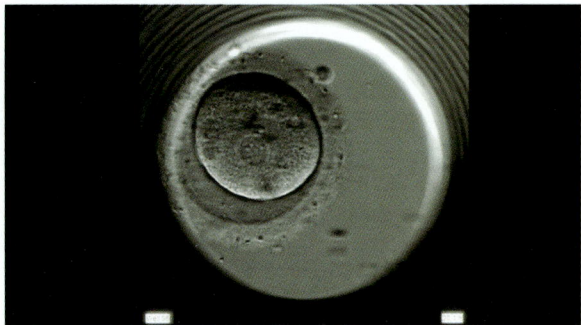

Video 7.59 Cell division annotations: t2–t8.

It was agreed that "hatching" is defined as the obvious emergence of the TE with enclosed blastocoel through a thin zona pellucida. It was noted that if a blastocyst is collapsed at the time of assessment, it cannot be graded reliably. These blastocysts should be re-evaluated 1–2 h later, as a regular cycle of collapse and re-expansion of blastocysts is normal.

7.4.7. Fragmentation and vacuolization

Both parameters indicate the start of apoptosis, however there are no reported references directly relating them to implantation failures.

Presence of vacuoles and fragmentation are incompatible with a correct morphology of the blastocyst and those presenting these abnormalities should be assigned to the lowest category of embryo quality (Video 7.58).

7.5. Embryo morphokinetics

The previous section of this chapter has focused on embryo quality based on standard morphological assessment. In this section we will cover embryo quality based on a different approach: embryo morphokinetics. This new concept is the result of the recent implementation of time-lapse systems (TMS) in the IVF laboratories and it combines the concept of embryo morphology with the kinetics of embryo development.

Embryo development is a dynamic process thus it makes complete sense to evaluate it in an uninterrupted manner. The current methods of embryo evaluation only allow intermittent observations of the embryos with the consequent loss of information and accuracy related to embryo divisions. In addition, the most relevant scoring criteria and strategies used in human embryology may need to be re-evaluated. For instance, the use of the "pronuclear score" to select embryos: it has been demonstrated that this score can change depending on the interval of hours at which it is analyzed. Likewise, the morphology of an embryo can also change within a span of only 4 hours leading once again to different scores [29].

During the past 5 years the use of TMS has increased dramatically, especially in Europe where legislation regarding the number of embryos transferred demands better selection methods. In fact, different groups have focused their efforts in identifying kinetic markers associated with embryo viability, implantation and live birth rates. Some of these markers are being used to "select" embryos while others have turned out to be useful to "deselect" embryos. It is our intention in this section of the chapter to summarize some of these studies and to describe embryo quality from a different perspective.

7.5.1. Kinetic parameters

According to the proposed guidelines on the nomenclature and annotation of dynamic human embryo monitoring [30] we can define the following morphokinetic "individual" variables (each timing defines the first time-lapse frame in which the expected phenomenon is observed or detected) (Video 7.59). Figure 7.3 describes some of the variables.

t0	Time of IVF or mid-time of micro/injection (ICSI/IMSI)
tPB2	The second polar body is completely detached from the oolemma
tPN	Fertilization status is confirmed
tPNa	Appearance of individual pronuclei; tPN1a, tPN2a; tPN3a. . .

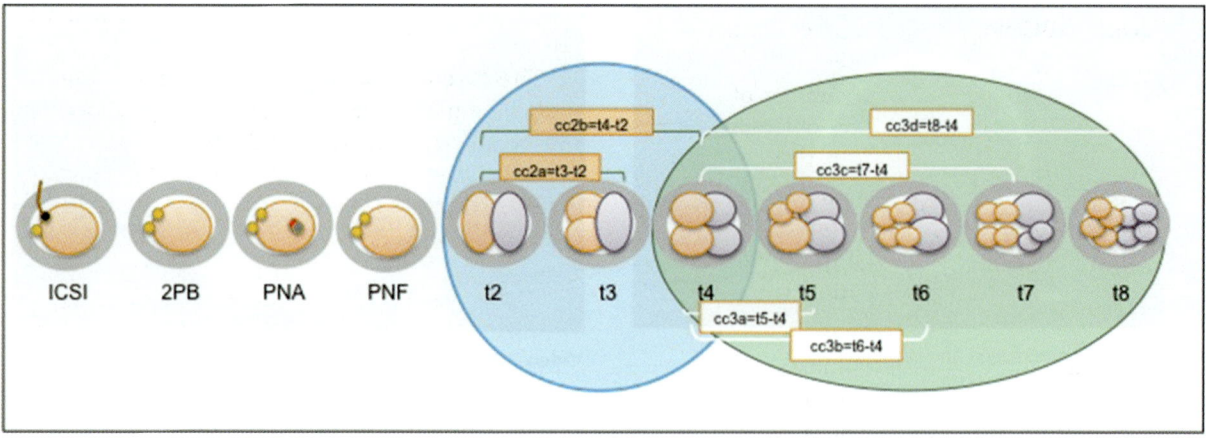

Figure 7.3 Graphical representation of kinetic variables up to the eight-cell stage.

tPNf	Time of pronuclei disappearance; tPN1f; tPN2f...	
tZ	Time of PN scoring	
t2 to t9	Time to two to nine discrete cells	
tSC	First evidence of compaction	
tMf/p	End of compaction (last frame before cavity formation) "f" corresponds to fully compacted; "p" corresponds to partial compaction	
tSB	Initiation of blastulation	
tByz	Full blastocyst (last frame before zona starts to thin) "y" corresponds to morphology of inner cell mass; "z" corresponds to morphology of trophectoderm cells	
tEyz	Initiation of expansion; first frame of zona thinning	
tHNyz	Herniation; end of expansion phase and initiation of hatching process	
tHDyz	Fully hatched blastocyst	

VP	tPNf–tPNa	PN Duration
ECC1	t2–tPB2	Duration of 1st cell cycle
ECC2	t4–t2	Duration of 2nd cell cycle
		Duration of single blastomere cycle: cc2a = t3–t2; cc2b = t4–t2
ECC3	t8–t4	Duration of 3rd cell cycle
		Duration of single blastomere cell cycle: cc3a = t5–t4; cc3b = t6–t4; cc3c = t7–t4; cc3d = t8–t4
s2	t4–t3	Synchronization of cell divisions
s3	t8–t5	Synchronization of cleavage pattern
dcom		Duration of compaction tMf–tSC (full comp.); tMp–tSC (partial comp.)
dB	tB–tSB	Duration of blastulation
dexp	tHN–tE	Duration of blastocyst expansion

In addition different "calculated" variables for dynamic monitoring of human embryo development have been defined (Figure 7.3). They usually represent the duration of a certain cell stage or the duration of a specific cell cycle as described:

dcol	tBCend (n)-tBCi(n)	Duration of blastocyst collapse; "n" is number of episodes of collapse and re-expansion
dre-exp	tre-exp end (n)-tre-expi(n)	Duration of re-expansion
dHN	tHN-tHD	Duration of herniation

The natural approach to describing embryo quality based on morphokinetics would be to correlate kinetic markers (associated with different stages of embryo development) with embryo viability and clinical outcomes and to identify the most relevant ones.

7.5.2. Kinetic markers associated with pronuclear dynamics and first cytokinesis

Several studies have correlated dynamics of pronuclear stage with embryo development, implantation and live birth rates.

Lemmen et al. (2008) [31] in a retrospective study of 102 fertilized oocytes observed that embryos that developed into $\geq$ four-cell embryos on day 2 of development had significantly earlier pronuclei disappearance and first cleavage than those that developed to three or two-cells. In addition, synchrony in appearance of nuclei after the first cleavage was significantly associated with pregnancy success ($p < 0.05$).

Wong et al. (2010) [32], in a retrospective study of 242 embryos derived from thawed oocytes, identified three time-lapse markers with distinct time windows that could predict blastocyst formation by the four-cell stage. One of them is associated with the first cytokinesis and is defined as P1 = duration of the first cytokinesis: 14.3 ± 6.0 minutes.

Azzarello et al. (2012) [33], in a prospective study, compared the time of PN breakdown between embryos resulting in live birth (n = 46) or no birth (n = 113). They found that the timing of PN breakdown was significantly longer in the live birth group than in the no birth group (24.9 ± 0.6 h vs. 23.3 ± 0.4 h; p = 0.022). In addition they found no live births in embryos whose PN fading happened before 20 h 45 min. and based on their findings the authors suggested PN breakdown as a novel exclusion criterion for embryo selection. In this study no PN morphological parameters were associated with live birth.

Chamayou et al. (2013) [34], in a retrospective study compared the time of PN appearance and PN breakdown between implanting (n = 72) and non-implanting (n = 106) embryos and observed no difference between the two groups for these parameters. Similar results were obtained by Kirkegaard et al. (2013) [35] in a prospective study that compared the same parameters between single blastocysts resulting in clinical pregnancy (n = 26) and no pregnancy (n = 58).

Aguilar et al. (2014) [36], in a retrospective study, compared the time for extrusion of the second polar body, time of PN breakdown, and length of the S-phase between implanting (n = 212) and non-implanting (n = 687) embryos. Significant differences were observed for the three parameters. However no differences were observed with respect to PN morphology, PN appearance and PN abuttal.

Kirkegaard et al. (2013) [35], in a prospective study, correlated the same parameter as Wong but taking implantation as an end point. The median duration of the first cytokinesis varied between implanting and non-implanting embryos (0.3 h, 95% CI 0–0.5 vs. 0 h, 95% CI 0–8.4 respectively, p= 0.04).

7.5.3. Kinetic markers associated with different cell stages

There is some inconsistency between the nomenclatures used by the different groups to define markers related to the different cell stages. However, they can be grouped in three categories: (i) kinetic markers associated with the time to **reach a specific cell stage**, (ii) kinetic markers associated with the **duration of a specific cell stage,** and (iii) kinetic markers associated with the **duration of a cell cycle** and its **synchronicity.**

7.5.4. Kinetic markers associated with the time to reach a specific cell stage

Several studies have correlated the time to reach a specific cell stage with implantation based on the analysis of embryos with known implantation data (KID embryos).

Meseguer et al. (2011) [37] in a retrospective analysis of 247 KID embryos determined that the times to reach the two-cell, three-cell, four-cell and

119

five-cell stages were shorter among implanted vs. non-implanted embryos and identified **t5** as the most effective parameter to discriminate between implanted and non-implanted embryos. An improved version of this study has been published recently by the same group, with a higher number of KID embryos (754) being analyzed [38]. In this case **t3** was identified as the most effective parameter to discriminate between implanted and non-implanted embryos.

Dal Canto *et al.* (2012) [39] in a retrospective analysis of 134 KID embryos found that implanted embryos (n = 19) developed to the eight-cell stage **(t8)** faster than non-implanted embryos (n = 115; 54.9 ± 5.2 vs. 58.0 ± 7.2 h, p = 0.035) and Freour *et al.* (2013) [40] observed that implanting embryos from active smokers reached the four-cell stage **(t4)** faster than non-implanting embryos (42 vs. 41 h respectively, p = 0.005).

In addition, one study has correlated individual kinetic parameters with embryo euploidy. Campbell *et al.* (2013) [41] observed that multiple aneuploid embryos were delayed at the initiation of compaction (tSC; median 85.1 hours post insemination (hpi); p = 0.02) compared with euploid embryos (tSC median 79.7 hpi). In addition embryos having single or multiple aneuploidy (median 103.4 hpi, p = 0.004, and 101.9 hpi, p = 0.006, respectively) had delayed initiation of blastulation compared with euploid embryos (median 95.1 hpi). Moreover, multiple aneuploidy embryos had delayed time to reach full blastocyst stage (tB; median 110.9 hpi, p = 0.01) compared with euploid embryos (tB median 105.9 hpi). All other timings tested in this study were not significantly different.

7.5.5. Kinetic markers associated with the duration of a specific cell stage or cell cycle

The correlation between the duration of different cell stages or cell cycles and embryo viability and implantation has been addressed in several studies. Interestingly a good number of studies coincide in the times mostly associated with implantation: the durations of the two-cell and three-cell stages.

Wong *et al.* (2010) [32] in a study involving thawed oocytes at the pronuclear stage observed that embryos that reached the blastocyst stage had a mean value for the duration of the two-cell stage (P2) of 11.1 ± 2.2 h and for the three-cell stage

(P3) of 1.0 ± 1.6 h, respectively. In addition to P1, the authors concluded that embryos that reached the blastocyst stage could be predicted, with a sensitivity and specificity of 94% and 93%, respectively, by having a P1 of 0–33 min, a P2 of 7.8–14.3 h, and a P3 of 0–5.8 h. Conversely, embryos that exhibited values outside of one or more of these windows were predicted to arrest.

Meseguer *et al.* (2011) [37] in a retrospective analysis of 247 KID embryos observed that the mean duration of the two-cell (11.8 + 1.2 vs. 11.8 + 3.3 h; p < 0.006) and the three-cell stage (0.78 + 0.73 vs. 1.77 + 2.83 h, respectively; p < 0.016) was significantly different between implanting and non-implanting embryos. In a second study by the same group, Cruz *et al.* (2012) [42] observed 834 embryos for 120 hours and confirmed that cc2 (duration od the two-cell stage) and s2 (duration of the three-cell stage) were statistically significant indicators of blastocyst development. Moreover, Rubio *et al.* (2012) [43] found that embryos with extremely short cc2 (< 5 h) had very low implantation rate (1.2%). This phenomenon was described as direct cleavage or "DC 2–3" (< 5 h), and based on these findings the authors suggested DC 2–3 < 5 h as a novel exclusion criterion for embryo selection.

Dal Canto *et al.* (2012) [39] also found that embryos that developed into expanded blastocysts had significantly shorter cc2 and s2 than those that developed into non-expanded blastocysts. Finally, another retrospective study by Hlinka *et al.* (2012) [44], analyzed images from 180 embryos and determined cc2, cc3, s2, and s3 among other variables as useful selection criteria for embryo outcome.

The correlation between euploidy and the duration of specific cell stages has been studied as well. Basile *et al.* (2014) [45] in a retrospective study of 504 embryos observed that embryos falling within the optimal ranges defined for cc3 (11.7–18.2 h) and t5-t2 (> 20.5 h) exhibited a significantly greater probablity of being chromosomally normal than those falling outside these ranges (33.4%, and 34.4% vs. 16.3%, and 10.4%, respectively).

In addition, Wong *et al.* (2010) [32], collected single embryos for gene expression analysis and revealed that embryos with time-lapse markers P1, cc2 (named by the authors as P2), and s2 (named by the authors as P3) outside of the optimal ranges exhibited abnormal RNA patterns for embryo cytokinesis, micro RNA (miRNA) biogenesis, and maternal mRNA reserve suggesting that embryo fate may be predetermined and inherited very early in development (by the four-cell stage).

Chavez *et al.* (2012) [46] subsequently observed that euploid embryos clustered tightly in the P1, cc2, or s2 window that was predictive of blastocyst formation according to Wong *et al.* (2010) [32]. Performing further molecular analysis, the authors discovered that fragmentation dynamics detected by time-lapse imaging, together with P1, cc2, and s2, could potentially distinguish euploid from aneuploid embryos at the four-cell stage, as the fragments contained nuclear DNA, kinetochore proteins, and whole chromosomes detected by fluorescence *in situ* hybridization (FISH).

7.5.6. Morphology dynamics

Continuous monitoring of embryo development has opened the window for the detection of morphological features that were undetectable with standard morphological assessment. Examples of these phenomena are: reabsorption of fragments (Video 7.60), irregular divisions (Video 7.61), direct cleavage (Video 7.61), three PNs turning into two PNs (Video 7.62), blastocyst collapsing (Video 7.63), etc.

Lemmen *et al.* (2008) [31] showed that embryos cultured in a time-lapse system can have dramatic differences in fragmentation grade and blastomere symmetry just prior to the time of embryo assessment on day 2. Meseguer *et al.* (2011) [37] observed that of the total 247 KID embryos, 48 (19.4%) exhibited one or more of the following morphological events: (i) direct cleavage (DC) from zygote to three blastomere embryo, defined as: cc2 = t3-t2 < 5 h (n = 8); (ii) uneven blastomere size (UBS) at the two-cell stage during the interphase where the nuclei are visible (n = 26). Blastomeres were considered uneven sized if the average diameter of the large blastomere was > 25% larger than the average diameter of the small

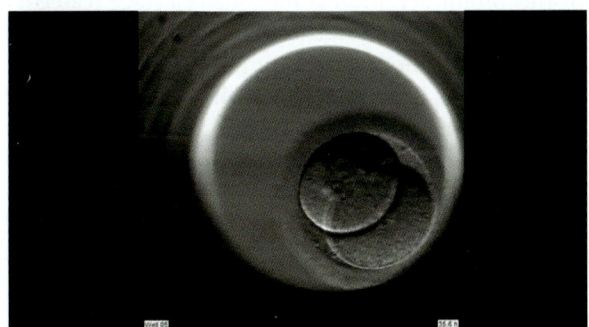

Video 7.60 Fragment reabsorption.

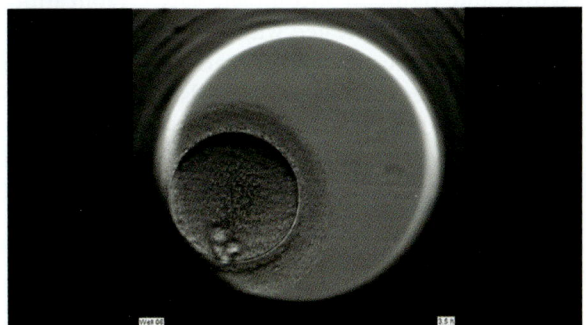

Video 7.61 Irregular division.

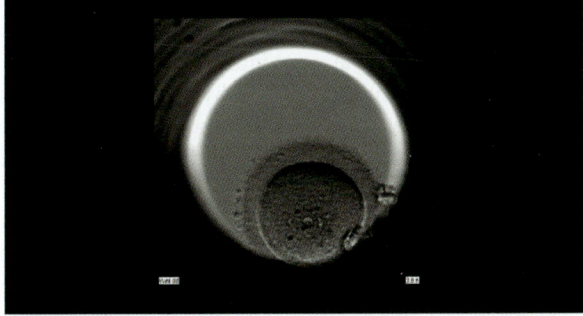

Video 7.62 PN disappearance: 3 PN to 2 PN.

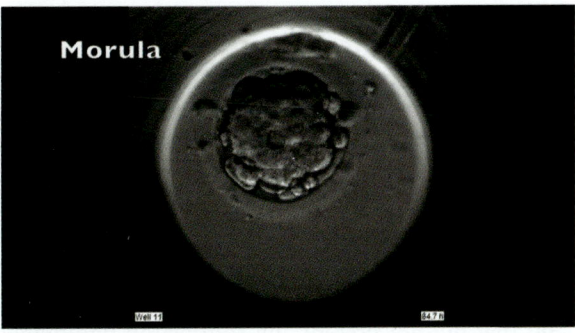

Video 7.63 Blastocyst collapse.

blastomere; (iii) multinucleation (MN) at the four-cell stage during the interphase where the nuclei are visible. Out of the 48 embryos, only four implanted (8%) (two with uneven blastomere size and two that were multinucleated) and therefore these three parameters have been proposed as exclusion criteria for embryo selection.

Following this study, Rubio *et al.* (2012) [43] performed a retrospective multicenter study and confirmed some of these findings, observing that embryos with DC2–3 (< 5 h) had a statistically significantly lower implantation rate than embryos with a normal cleavage pattern (1.2% vs. 20.2%). According to this study the incidence of this phenomenon among the cohort of transferred embryos (n = 1,659) was 6.6% (n = 109) therefore the authors suggest rejecting these embryos for transfer. It is known from published data that the duration of the cell cycle is around 10 to 12 hours. This interval is sufficient for the embryo to undergo two consecutive phases of cytokinesis and replicate the whole cell genome. By the use of TMS, abnormally short cell cycles of as little as 1.8 hours can be detected [43]. Extremely short cell cycles could be related to other factors and result in incomplete DNA replication, which might be associated with an unequal distribution of DNA to the blastomeres. According to previous studies with animal models, zygotes undergoing direct division from one cell to three blastomeres can have similar developmental ability and embryonic cell numbers to those with normal division, although with a high frequency of chromosomal abnormalities

Basile *et al.* (2014) [38], in a retrospective study of a larger data set (1616 KID embryos) confirmed that embryos presenting any of the three exclusion parameters proposed by Meseguer *et al.* (2011) [37] had significantly lower IR than those not presenting them: 2.9%, 17.8%, 18.1% vs. 28.7%, 28.1%, 27.9% for DC, UBS, and MN respectively.

Another important aspect of morphology dynamics is the blastocyst collapsing: defined as a phenomenon in which the trophectoderm is detached from the zona pellucida in more than 50% of its surface. This process can be detected by a time-lapse system (Video 7.63). According to the literature, blastocyst contraction negatively affects the natural process of blastocyst hatching. The transferred blastocysts (n = 438) that did not present any contraction had higher implantation rate (48.52%) than those (n = 64) that

had at least one contraction (35.11%) (data not published).

7.5.7. Morphokinetic models for embryo selection

In addition to the standard morphological assessment several groups have proposed different algorithms to select embryos based on kinetic data:

Meseguer *et al.* (2011) [37] developed a hierarchical model representing a classification tree, which subdivided embryos into six categories from A to F (Figure 7.4). Four of these categories (A–D) were further subdivided into two subcategories (+) or (-). The hierarchical classification procedure starts with a morphological screening of all embryos in a cohort to eliminate those embryos that are clearly *not* viable (i.e., highly abnormal, atresia or clearly arrested embryos). Those embryos that are clearly not viable are discarded and not considered for transfer (category F). Next step in the model is to exclude embryos that fulfill any of the three exclusion criteria: (i) uneven blastomere size at the two-cell stage, (ii) abrupt division from one to three or more cells, or (iii) multinucleation at the four-cell stage (category E). The subsequent levels in the model follow a strict hierarchy based on the binary timing variables $t5$, $s2$, and $cc2$. First, if the value of $t5$ falls inside the optimal range (48.8–56.6 h), the embryo is categorized as A or B. If the value of $t5$ falls outside the optimal range (or if $t5$ has not yet been observed at 64 h), the embryo is categorized as C or D. If the value of $s2$ falls inside the optimal range (≤ 0.76 h) the embryo is categorized as A or C depending on $t5$; similarly, if the value of $s2$ falls outside the optimal range, the embryo is categorized as B or D depending on $t5$. Finally, the embryo is categorized with the extra plus (+) if the value for $cc2$ is inside the optimal range (≤ 11.9 h) (A+/B+/C+/D+) and is categorized with a minus (-) as (A-/B-/C-/D-) if the value for $cc2$ is outside the optimal range (Figure 7.4).

The algorithm presented by this group has recently been validated in a randomized prospective controlled trial [47] comparing embryos cultured in a standard incubator (SI) with development evaluated only by morphology (control group) and embryos cultured in TMS with embryo selection based on the algorithm (study group). The ongoing pregnancy rate per treated cycle was statistically significantly increased to 51.4% (95% CI, 46.7–56.0) for the TMS

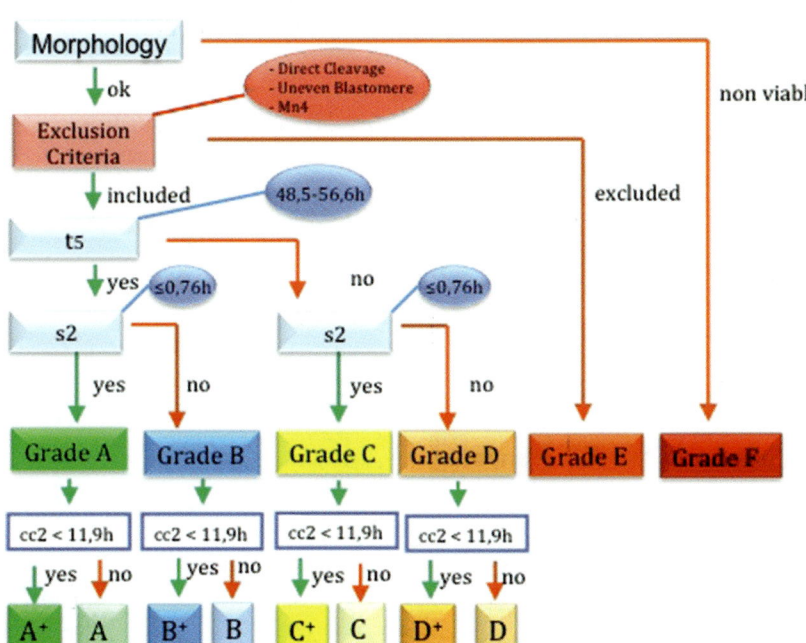

Figure 7.4 Embryo classification model based on exclusion and inclusion criteria from Meseguer *et al.* (2011).

group compared with 41.7% (95% CI, 36.9–46.5) for the SI group. For pregnancy rate, differences were not statistically significant at 61.6% (95% CI, 56.9–66.0) vs. 56.3% (95% CI, 51.4–61.0). The results per transfer were similar: statistically significant differences in ongoing pregnancy rate of 54.5% (95% CI, 49.6–59.2) vs. 45.3% (95% CI, 40.3–50.4) and not statistically significant for pregnancy rate at 65.2% (95% CI, 60.6–69.8) vs. 61.1% (95% CI, 56.2–66.1). Early pregnancy loss was statistically significantly decreased for the TMS group with 16.6% (95% CI, 12.6–21.4) vs. 25.8% (95% CI, 20.6–31.9). The implantation rate was statistically significantly increased in the SI group at 44.9% (95% CI, 41.4–48.4) vs. 37.1% (95% CI, 33.6–40.7). Based on these results we can conclude that the strategy of culturing and selecting embryos in a TMS improves reproductive outcomes.

Conaghan *et al.* (2013) [48] performed a two phase study to develop and validate an algorithm to predict blastocyst formation. In this case they used the Eeva prediction and cell-tracking software and observed a high probability of usable blastocyst formation when both P2 and P3 were within specific cell division timing ranges (P2, 9.33–11.45 hours; and P3, 0–1.73 hours) and a low probability when either P2 or P3 was outside the specific cell division timing ranges. The time between

cytokinesis 1 and 2 (P2) and the time between cytokinesis 2 and 3 (P3) dominated the prediction model, and the duration of first cytokinesis (P1) was of lesser statistical value; therefore, they proposed a prediction model based on P2 and P3 only. By using this model the authors observed that D3 embryos that became usable blastocysts could be predicted with a specificity of 84.2% (95% CI = 78.7%–88.5%), sensitivity of 58.8% (95% CI = 47.0%–69.7%), positive predictive value (PPV) of 54.1% (95% CI = 42.8%–64.9%), and negative predictive value (NPV) of 86.6% (95% CI = 81.3%–90.6%). By comparison, the same prediction based on morphology alone was achieved with a specificity of 52.1% (95% CI = 39.7%–64.6%), sensitivity of 81.8% (95% CI = 70.6%–92.9%), PPV of 34.5% (95% CI = 31.5%–37.5%), and NPV of 90.9% (95% CI = 87.3%–94.5%). The authors concluded that the use of this algorithm significantly improved the specificity (84.2% vs. 52.1%; p < 0.0001) and PPV (54.1% vs. 34.5%; p < 0.01) of usable blastocyst predictions and recommend the adjunctive use of this algorithm to improve embryo selection since it enables embryologists to better discriminate which embryos would be unlikely to develop to blastocyst.

Basile *et al.* (2014) [38] followed the study by Meseguer *et al.* (2011) [37] and published an

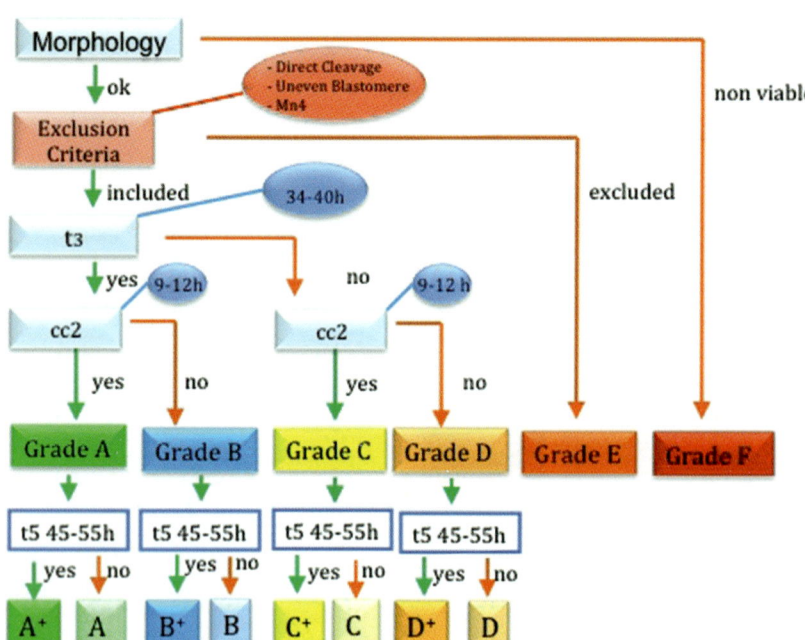

Figure 7.5 Embryo classification model based on exclusion and inclusion criteria from Basile *et al.* (2014).

improved version of the algorithm by studying a larger data set of embryos from four different IVF clinics (Figure 7.5). For that aim a sequential approach was adopted by the author. During Phase 1 of the study an algorithm was developed taking into consideration morphokinetic data of 754 KID embryos that were selected for transfer based only on conventional morphological criteria. The new algorithm included the variables t3, cc2, and t5 in combination with morphology and exclusion criteria (DC, UBS, and MN) and classified embryos from A to F according to implantation potential (Figure 7.5). Subsequently, during Phase 2 of the study, the predictive ability of this new algorithm was tested by applying it for embryo classification in a different group of IVF patients (885 cycles). Considering only cycles with known implantation (100% or 0% implantation, n = 1137), a significant decrease in IR was observed as embryos moved on from categories A to E. More specifically: "A" 32%, "B" 28%, "C" 26%, "D" 20%, and "E" the lowest 17%, $p < 0.001$.

Campbell *et al.* (2013) [41] elaborated an aneuploidy risk model based on the differences of tSB and tB between euploid and aneuploid embryos that had undergone trophectoderm biopsy. The model includes three categories: low risk, tB < 122.9 hpi, and tSB < 96.2 hpi; medium risk, tB < 122.9 hpi

and tSB 96.2 hpi; and high risk, tB < 122.9 hpi. The same group in a different study [49] applied this model to evaluate its effectiveness and potential clinical impact for unselected IVF patients without undergoing PGS after analyzing KID embryos. The study revealed significant differences in fetal heart rate (72.7; 25.5; 0) and live birth rate (61.1; 19.2; 0) between the three categories low, medium, and high respectively. This demonstrates that time-lapse imaging using defined morphokinetic data can be used to classify human pre-implantation embryos according to their risk of aneuploidy, without performing biopsy and PGS, and that this correlates well with clinical outcome. However it should never be taken as a replacement of genetic screening.

Basile *et al.* (2013) [45] also correlated morphokinetics with embryo aneuploidy on 77 patients undergoing genetic screening due to recurrent miscarriage or implantation failure. In this case embryo biopsy was performed on day 3 of development and the total number of embryos analyzed was 504. A logistic regression analysis was used to select and organize which observed timing events (expressed as binary variables inside or outside the optimal range) were most relevant to select embryos with higher probability of being chromosomally normal. The model identified t5–t2, OR = 2.853 (95% CI 1.763–4.616)

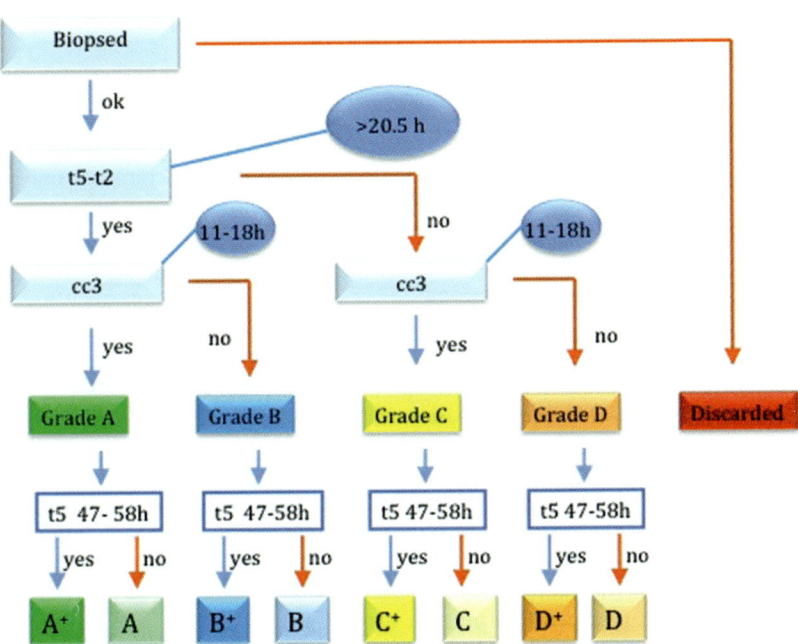

Figure 7.6 Embryo classification model based on two variables from Basile *et al.* (2013).

followed by cc3, OR = 2.095 (95% CI 1.356–3.238) as the most relevant variables related to normal chromosomal content. An algorithm for embryo selection based on these two variables classifies embryos from A to D with significant differences in the percentage of normal embryos as we move on from A to D. More specifically: A: 35.9%; B: 26.4%; C: 12.1%; D: 9.8% (p < 0.001) (Figure 7.6).

Freour *et al.* (2013) [40] elaborated a classification model based on data from 191 KID embryos (160 non-implanted embryos and 31 implanted embryos) identifying t4 and s3 as the most relevant kinetic parameters with significantly higher implantation rates in the first two quartiles (below median) than in the two last ones (32.3% vs. 18.1% for t4 and 32.1% vs. 15% for s3; p < 0.05 for both). According to this hierarchical classification model, embryos displaying t4 within reference range were graded as "A" and those outside the range were graded as "B." In addition, embryos displaying s3 within reference range were graded as "+" and those outside the range were graded as "-". An embryo associating t4 and s3 within range was then classified as "A+" corresponding to the best morphokinetic category. The authors validated this classification in a database including all transferred embryos observing implantation rates of 38.7%, 33.3%, 30.7%, and 15.3% for A+, A-, B+, and B-categories, respectively.

To conclude, the present chapter presents an overview of the current methods of embryo evaluation and selection based on two different approaches: 1. standard morphological assessment, and 2. embryo morphokinetics. It is our opinion that standard morphological assessment should remain the gold standard to initiate embryo evaluation; however, if possible, it should be complemented with the detection of kinetic markers known to improve clinical results.

References

1. Guerin P, El Mouatassim S, Menezo Y. Oxidative stress and protection against reactive oxygen species in the pre-implantation embryo and its surroundings. *Hum Reprod Update* 2001;7:175–89.

2. Wittemer C, Ohl J, Bailly M, Bettahar-Lebugle K, Nisand I. Does body mass index of infertile women have an impact on IVF procedure and outcome? *J Assist Reprod Genet* 2000;17:547–52.

3. Dechaud H, Anahory T, Reyftmann L, Loup V, Hamamah S, Hedon B. Obesity does not adversely affect results in patients who are undergoing in vitro fertilization and embryo transfer. *Eur J Obstet Gynecol Reprod Biol* 2006;127:88–93.

4. Edwards RG, Fishel SB, Cohen J, Fehilly CB, Purdy JM, Slater JM, et al. Factors influencing the success of in vitro fertilization for alleviating human infertility. *J In Vitro Fert Embryo Transf* 1984;1:3–23.

5. Jensen TK, Slama R, Ducot B, Suominen J, Cawood EH, Andersen AG, et al. Regional differences in waiting time to pregnancy among fertile couples from four European cities. *Hum Reprod* 2001;16:2697–704.

6. Gamiz P, Rubio C, de los Santos MJ, Mercader A, Simon C, Remohi J, et al. The effect of pronuclear morphology on early development and chromosomal abnormalities in cleavage-stage embryos. *Hum Reprod* 2003;18:2413–19.

7. Metwally M, Cutting R, Tipton A, Skull J, Ledger WL, Li TC. Effect of increased body mass index on oocyte and embryo quality in IVF patients. *Reprod Biomed Online* 2007;15:532–8.

8. Alikani M, Palermo G, Adler A, Bertoli M, Blake M, Cohen J. Intracytoplasmic sperm injection in dysmorphic human oocytes. *Zygote* 1995;3:283–8.

9. Krizanovska K, Ulcova-Gallova Z, Bouse V, Rokyta Z. Obesity and reproductive disorders. *Sb Lek* 2002;103:517–26.

10. Lintsen AM, Pasker-de Jong PC, de Boer EJ, Burger CW, Jansen CA, Braat DD, et al. Effects of subfertility cause, smoking and body weight on the success rate of IVF. *Hum Reprod* 2005;20:1867–75.

11. Metwally M, Ong KJ, Ledger WL, Li TC. Does high body mass index increase the risk of miscarriage after spontaneous and assisted conception? A meta-analysis of the evidence. *Fertil Steril* 2008;90:714–26.

12. Andreasen KR, Andersen ML, Schantz AL. Obesity and pregnancy. *Acta Obstet Gynecol Scand* 2004;83:1022–9.

13. Wang JX, Davies MJ, Norman RJ. Obesity increases the risk of spontaneous abortion during infertility treatment. *Obes Res* 2002;10:551–4.

14. Scott L. The biological basis of non-invasive strategies for selection of human oocytes and embryos. *Hum Reprod Update* 2003;9:237–49.

15. Nguyen RH, Wilcox AJ, Skjaerven R, Baird DD. Men's body mass index and infertility. *Hum Reprod* 2007;22:2488–93.

16. Metwally M, Tuckerman EM, Laird SM, Ledger WL, Li TC. Impact of high body mass index on endometrial morphology and function in the peri-implantation period in women with recurrent miscarriage. *Reprod Biomed Online* 2007;14:328–34.

17. Veeck LL. Oocyte assessment and biological performance. *Ann N Y Acad Sci* 1988;541:259–74.

18. Goyanes VJ, Ron-Corzo A, Costas E, Maneiro E. Morphometric categorization of the human oocyte and early conceptus. *Hum Reprod* 1990;5:613–18.

19. Gabrielsen A, Lindenberg S, Petersen K. The impact of the zona pellucida thickness variation of human embryos on pregnancy outcome in relation to suboptimal embryo development. A prospective randomized controlled study. *Hum Reprod* 2001;16:2166–70.

20. Desai NN, Goldstein J, Rowland DY, Goldfarb JM. Morphological evaluation of human embryos and derivation of an embryo quality scoring system specific for day 3 embryos: a preliminary study. *Hum Reprod* 2000;15:2190–6.

21. Tao J, Tamis R, Fink K, Williams B, Nelson-White T, Craig R. The neglected morula/compact stage embryo transfer. *Hum Reprod* 2002;17:1513–18.

22. Balaban B, Urman B, Sertac A, Alatas C, Aksoy S, Mercan R. Blastocyst quality affects the success of blastocyst-stage embryo transfer. *Fertil Steril* 2000;74:282–7.

23. Gardner DK, Lane M, Stevens J, Schlenker T, Schoolcraft WB. Blastocyst score affects implantation and pregnancy outcome: towards a single blastocyst transfer. *Fertil Steril* 2000;73:1155–8.

24. Schoolcraft WB, Gardner DK. Blastocyst culture and transfer increases the efficiency of oocyte donation. *Fertil Steril* 2000;74:482–6.

25. Racowsky C, Combelles CM, Nureddin A, Pan Y, Finn A, Miles L, et al. Day 3 and day 5 morphological predictors of embryo viability. *Reprod Biomed Online* 2003;6:323–31.

26. Richter KS, Harris DC, Daneshmand ST, Shapiro BS. Quantitative grading of a human blastocyst: optimal inner cell mass size and shape. *Fertil Steril* 2001;76:1157–67.

27. Shoukir Y, Campana A, Farley T, Sakkas D. Early cleavage of in-vitro fertilized human embryos to the 2-cell stage: a novel indicator of embryo quality and viability. *Hum Reprod* 1997;12:1531–6.

28. Kovacic B, Vlaisavljevic V, Reljic M, Cizek-Sajko M. Developmental capacity of different morphological types of day 5 human morulae and blastocysts. *Reprod Biomed Online* 2004;8:687–94.

29. Montag M, Liebenthron J, Koster M. Which morphological scoring system is relevant in human embryo development? *Placenta* 2011;32 Suppl 3:S252–6.

30. Ciray HN, Campbell A, Agerholm IE, Aguilar J, Chamayou S, Esbert

M, et al. Proposed guidelines on the nomenclature and annotation of dynamic human embryo monitoring by a time-lapse user group. *Hum Reprod* 2014;29:2650–60.

31. Lemmen JG, Agerholm I, Ziebe S. Kinetic markers of human embryo quality using time-lapse recordings of IVF/ICSI-fertilized oocytes. *Reprod Biomed Online* 2008;17:385–91.

32. Wong CC, Loewke KE, Bossert NL, Behr B, De Jonge CJ, Baer TM, et al. Non-invasive imaging of human embryos before embryonic genome activation predicts development to the blastocyst stage. *Nat Biotechnol* 2010;28:1115–21.

33. Azzarello A, Hoest T, Mikkelsen AL. The impact of pronuclei morphology and dynamicity on live birth outcome after time-lapse culture. *Hum Reprod* 2012;27:2649–57.

34. Chamayou S, Patrizio P, Storaci G, Tomaselli V, Alecci C, Ragolia C, et al. The use of morphokinetic parameters to select all embryos with full capacity to implant. *J Assist Reprod Genet* 2013;30:703–10.

35. Kirkegaard K, Kesmodel US, Hindkjaer JJ, Ingerslev HJ. Time-lapse parameters as predictors of blastocyst development and pregnancy outcome in embryos from good prognosis patients: a prospective cohort study. *Hum Reprod* 2013;28:2643–51.

36. Aguilar J, Motato Y, Escriba MJ, Ojeda M, Munoz E, Meseguer M. The human first cell cycle: impact on implantation. *Reprod Biomed Online* 2014;28:475–84.

37. Meseguer M, Herrero J, Tejera A, Hilligsoe KM, Ramsing N, Remohi J. The use of morphokinetics as a predictor of embryo implantation. *Hum Reprod* 2011:1–14.

38. Basile N, Vime P, Florensa M, Aparicio Ruiz B, Garcia Velasco JA, Remohi J, et al. The use of morphokinetics as a predictor of implantation: a multicentric study to define and validate an algorithm for embryo selection. *Hum Reprod* 2015;30(2): 276–83.

39. Dal Canto M, Coticchio G, Mignini Renzini M, De Ponti E, Novara PV, Brambillasca F, et al. Cleavage kinetics analysis of human embryos predicts development to blastocyst and implantation. *Reprod Biomed Online* 2012;25:474–80.

40. Freour T, Dessolle L, Lammers J, Lattes S, Barriere P. Comparison of embryo morphokinetics after in vitro fertilization-intracytoplasmic sperm injection in smoking and nonsmoking women. *Fertil Steril* 2013;99:1944–50.

41. Campbell A, Fishel S, Bowman N, Duffy S, Sedler M, Hickman CF. Modelling a risk classification of aneuploidy in human embryos using non-invasive morphokinetics. *Reprod Biomed Online* 2013;26:477–85.

42. Cruz M, Garrido N, Herrero J, Perez-Cano I, Munoz M, Meseguer M. Timing of cell division in human cleavage-stage embryos is linked with blastocyst formation and quality. *Reprod Biomed Online* 2012;25:371–81.

43. Rubio I, Kuhlmann R, Agerholm I, Kirk J, Herrero J, Escriba MJ, et al. Limited implantation success of direct-cleaved human zygotes: a time-lapse study. *Fertil Steril* 2012; 98(6):1458–63.

44. Hlinka D, Kalatova B, Uhrinova I, Dolinska S, Rutarova J, Rezacova J, et al. Time-lapse cleavage rating predicts human embryo viability. *Physiol Res* 2012;61:513–25.

45. Basile N, Nogales Mdel C, Bronet F, Florensa M, Riqueiros M, Rodrigo L, et al. Increasing the probability of selecting chromosomally normal embryos by time-lapse morphokinetics analysis. *Fertil Steril* 2014;101:699–704.

46. Chavez SL, Loewke KE, Han J, Moussavi F, Colls P, Munne S, et al. Dynamic blastomere behaviour reflects human embryo ploidy by the four-cell stage. *Nat Commun* 2012;3:1251.

47. Rubio I, Galan A, Larreategui Z, Ayerdi F, Bellver J, Herrero J, et al. Clinical validation of embryo culture and selection by morphokinetic analysis: a randomized, controlled trial of the EmbryoScope. *Fertil Steril* 2014; 102(5):1287–94.

48. Conaghan J, Chen AA, Willman SP, Ivani K, Chenette PE, Boostanfar R, et al. Improving embryo selection using a computer-automated time-lapse image analysis test plus day 3 morphology: results from a prospective multicenter trial. *Fertil Steril* 2013;100:412,9.e5.

49. Campbell A, Fishel S, Bowman N, Duffy S, Sedler M, Thornton S. Retrospective analysis of outcomes after IVF using an aneuploidy risk model derived from time-lapse imaging without PGS. *Reprod Biomed Online* 2013; 27(2):140–6.

Time-lapse implementation in a clinical setting: management of laboratory quality

Dean E. Morbeck

8.1. Introduction

Every IVF program should ask itself "how do we know we are good?" and from there "how can we get better?" These two questions form the foundation of the science of quality in healthcare in general and provide the framework for this chapter on the role of time-lapse in quality management in the IVF laboratory.

"Quality" for a process such as assisted reproduction can be measured at many stages and is influenced by many factors, making it often difficult to compare outcomes among clinics [1]. Like many indicators of quality in healthcare, success for an IVF program – a high rate of healthy singleton deliveries per oocyte retrieval or frozen embryo transfer (FET) – varies depending on risk factors such as age, ovarian reserve, duration of infertility, and various socio-economic factors [2, 3]. Implantation rate for young patients (< 35 years old) provides the most robust measure of program quality since it typically has the largest sample size of good prognosis patients and, unlike delivery rate, corrects for number of embryos transferred, thus providing an outcome measure that requires little risk adjustment.

Using implantation rate (IR) as a quality measure, how is the field of assisted reproduction doing? The Society for Assisted Reproduction and Technology (SART) has published clinic specific outcomes for > 20 years, most recently 2012, providing a robust data set for analysis of this question. Optimal implantation rates for good prognosis patients vary depending on day of embryo transfer [4], ranging from 30 to 40% for cleavage stage transfers to > 50% for blastocyst transfers. Thus, a quality threshold of 30% IR for patients < 35 is a conservative minimum benchmark, one that one-third of IVF clinics reporting to SART

fail to achieve (Figure 8.1). This graph illustrates that quality varies considerably among IVF clinics, variation that could be due to patient selection, clinical management, laboratory quality or embryo transfer technique. For the IVF laboratory, we return to the question "how do we know we are good?", which shifts the focus from clinical outcomes to laboratory-specific quality indicators.

Time-lapse imaging brings the embryology laboratory into the era of Big Data, providing the tools necessary for comprehensive quality management. This chapter will cover two uses of time-lapse imaging in the clinical IVF laboratory: quality indicators as markers of quality of the culture environment and the application of TL with mouse embryos for quality control testing of laboratory products.

8.2. Time-lapse for quality management of IVF laboratories

Quality indicators are an essential component of a laboratory's quality management plan and are referred to as key performance indicators (KPIs;[5, 6]). An example of a common KPI is fertilization rate: an objective, readily obtained indicator that reflects proper patient selection for IVF, optimal sperm concentration or ICSI technique, and quality of the gamete preparation and insemination process. Another common KPI used by IVF laboratories is embryo utilization rate, defined as percentage of fertilized oocytes that are transferred or cryopreserved. Embryo utilization rate is a reflection of the quality of the culture system and thus is the most commonly used and useful KPI for gauging laboratory quality because all laboratories track number of embryos used (transferred or frozen). While this KPI is readily available, it is influenced by number of embryos

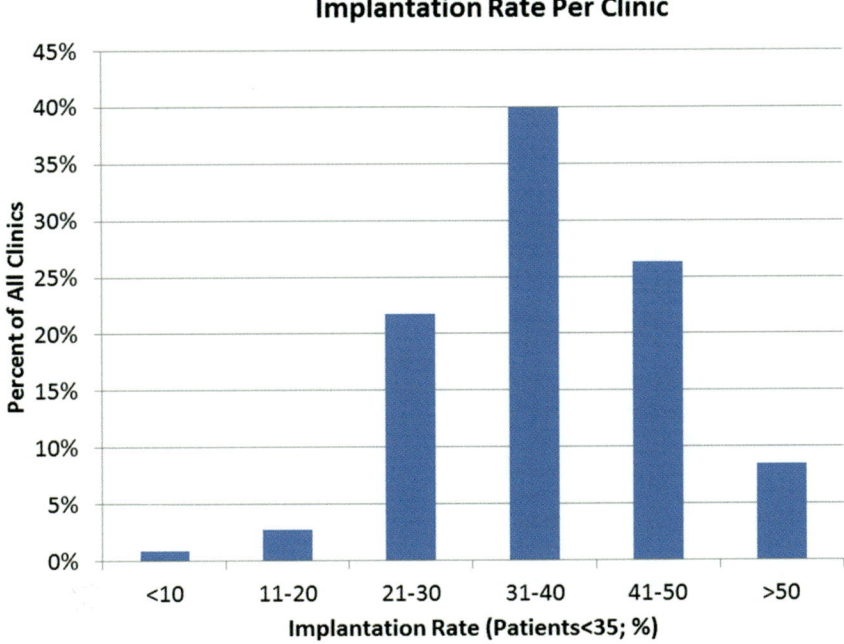

Implantation Rate Per Clinic

Figure 8.1 Distribution of clinics in the United States based on implantation rates for patients less than 35 years old. (Source: www.sart.org.)

transferred per patient and stage of cryopreservation (i.e., more embryos are frozen on day 2/3 than day 5), two laboratory controlled factors that limit the utility of this KPI as an industry standard. Other than KPIs for cryopreservation developed by the Alpha Scientists in Reproduction (Alpha), the field of embryology has not embraced the use of KPIs for quality management of the laboratory.

Useful KPIs must meet at least three criteria: they must be informative (i.e., meaningful), objective (i.e., not influenced by an hourly schedule or technician variation), and easily obtained. Informative KPIs can facilitate process control as well as identify quality improvement opportunities by providing an early detection system for process changes. The quality of the culture environment dictates embryo development, indicating that patterns of embryo development have potential as KPIs. Examples of shifts noted after process changes demonstrate that embryo development is sensitive to air quality [7–9], variable product quality [10–13] or variation in laboratory practice [14, 15]. Embryo development thus presents opportunities for informative KPIs, though currently internationally recognized KPIs with established benchmarks are limited to cryopreservation [6].

Time-lapse technology provides the tools necessary to meet two other criteria for KPIs: objectivity and ease of collection. Early cleavage [16] and

four-cell rate on day 2 [5] are examples of KPIs that are objective, yet suffer from two significant limitations: variation in timing when assessment occurs and disruption of culture in order to make the assessment. Time-lapse not only overcomes both of these limitations, it adds an important third element – exact timing of the division event as well as *how* the embryo got to that stage. Thus time-lapse provides comprehensive, objective data in an easily analyzed form without disturbing the embryo's environment.

8.3. Clinical examples of time-lapse quality management

Much of the focus of time-lapse technology to date has been on its potential for improving embryo selection, yet TL has many complementary benefits including its role in quality management. An early example of the power of TL for quality management is shown in Figure 8.2, where cell cycle timings were significantly delayed with a change in culture media. Since this same laboratory did not detect any difference in cell cycle timings with two different culture media [17], the findings in Figure 8.2 suggest that culture quality may have varied due to differences in lots of media, protein or oil. Culture media composition varies considerably [18] and the impact of these differences may be related to the amount of oxygen in use as well as the

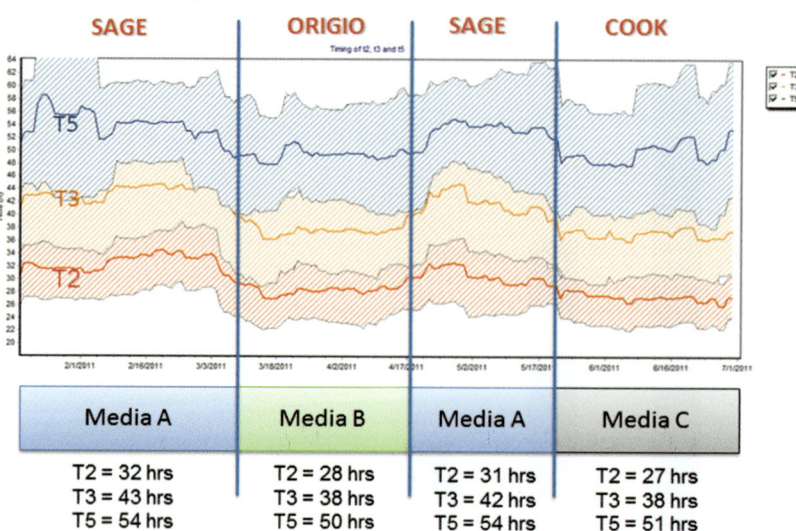

Average timing of Cell Divisions

Figure 8.2 Running average over 2 weeks of cell timings (t2, t3, and t5) during four periods with different culture media. (Figure courtesy of Marcos Meseguer, Ph.D.)

Media A	Media B	Media A	Media C
T2 = 32 hrs	T2 = 28 hrs	T2 = 31 hrs	T2 = 27 hrs
T3 = 43 hrs	T3 = 38 hrs	T3 = 42 hrs	T3 = 38 hrs
T5 = 54 hrs	T5 = 50 hrs	T5 = 54 hrs	T5 = 51 hrs

(Period: Feb-2011 to June 2011, Running average over 2 weeks)

protein supplement [19]. This example of a shift in timings due to culture conditions suggests that cell cycle timings could be useful for monitoring not only the quality of culture medium, but lot number changes of disposables used for embryo culture.

The IVF laboratory at Mayo Clinic has extensive experience using embryo development on day 2 as a quality indicator. When first tracked in 2006 to 2007, the average number of embryos at the four-cell stage on day 2 (42–44 h) was 34% (Figure 8.3A, quarterly averages). This quality indicator was the basis for quality improvement initiatives that included additional volatile organic compound (VOC) filtration in the air handler (HVAC) in 2008, change from big box cell culture incubators (Forma) to benchtop incubators (Planer) in 2011, and a change to culture in the EmbryoScope in 2013. An improvement in four-cell rate was achieved after adding the ultraviolet photo-catalytic oxidation system to the air-handler (48%) and again after introduction of TL in 2013 (58%). What is not apparent from this figure is a shift in embryo development that occurred during quarter 4 of 2011 and quarter 1 of 2012. Routine tracking of four-cell rate showed a decrease in 2 months in early 2012 but closer scrutiny of the pattern of embryo development demonstrated that there was a significant shift in the percent of embryos with > four cells on day 2 (Figure 8.3B) that persisted for exactly 6 months. The last 3 months of this period corresponded with introduction into clinical practice of an EmbryoScope with time-lapse imaging, though on a limited basis, providing a window into specific cell timings. Figure 8.4 illustrates that the increase in embryos with > four cells on day 2 during the first quarter of 2012 was due to both short two-cell and four-cell cycles. Of note, pregnancy and implantation rates were slightly reduced during these two quarters, shifts that may have been detected sooner with more sensitive quality indicators.

The power of TL for quality management is in the ability to track changes in cell division kinetics that occur in response to changes in lot numbers of products that come into direct contact with embryos, signals that can provide an early indication of a shift in quality that might otherwise not be detected for several weeks until pregnancy confirmation occurs. Detailed analysis of the different products used during the period in 2011 to 2012 is presented in Figure 8.5. Culture medium is not presented on the graph because a new lot was used monthly. A specific lot of protein, in this case SSS, had a direct overlap with the affected period, suggesting that this lot of protein was responsible for the shift observed. Had KPIs derived from TL been available at the time of the change in protein, the effects on development and outcomes could have been avoided.

With the power of TL, statistical process control using quality indicators can be a routine component

(A)

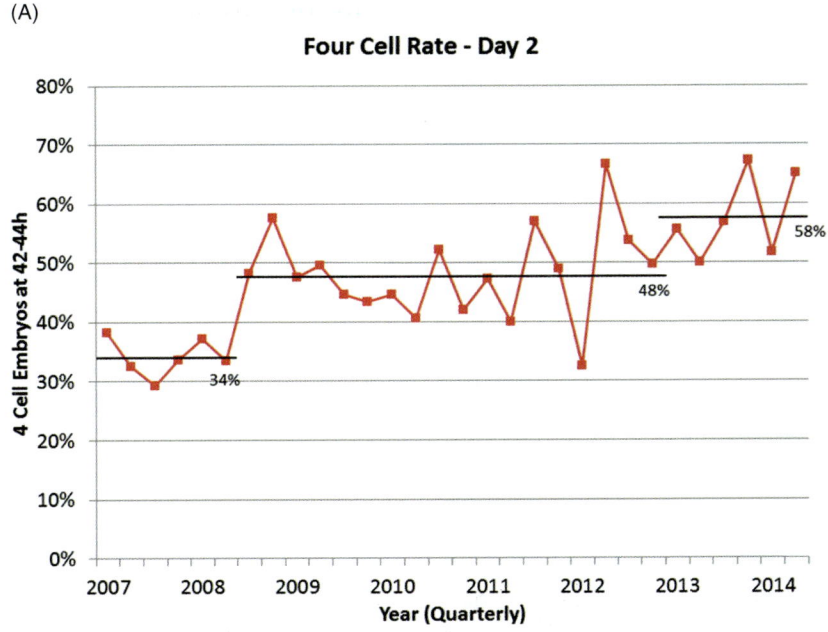

(B)

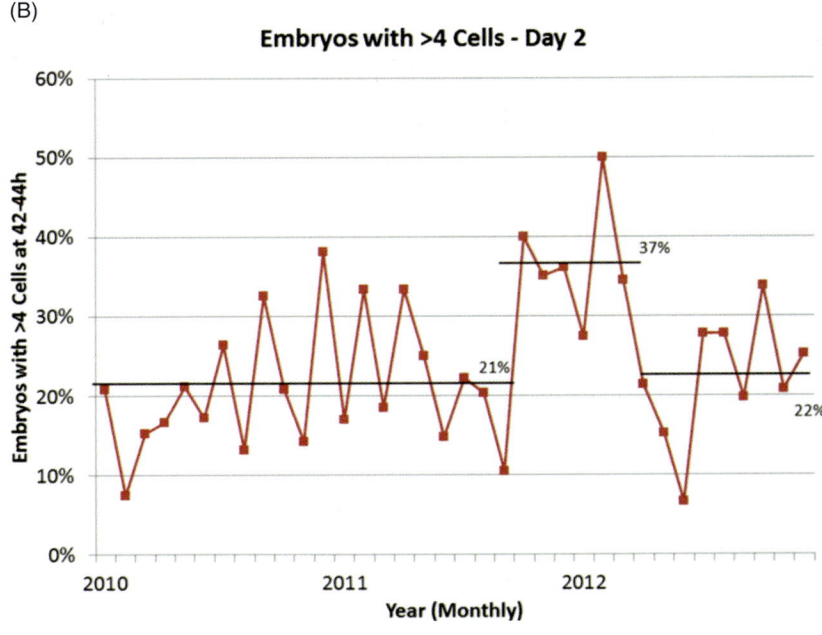

Figure 8.3 A: Quarterly averages; B: Increase in embryos with four cells on day 2.

of the quality management system for embryology, with many yet-to-be determined KPIs. Given their historical role in traditional embryo assessment, appropriate development on days 2, 3, and 5 can continue to be tracked with TL, yet with more accuracy, less disturbance, and less effort. In addition, features of embryo development that cannot be captured with static measures, such as direct cleavage of blastomeres [20], reverse cleavage [21], atypical development [22], and multinucleation [23], may be indicators of shifts in the quality of the embryo culture environment.

Air quality is one component of the culture environment that is poorly defined, can vary considerably,

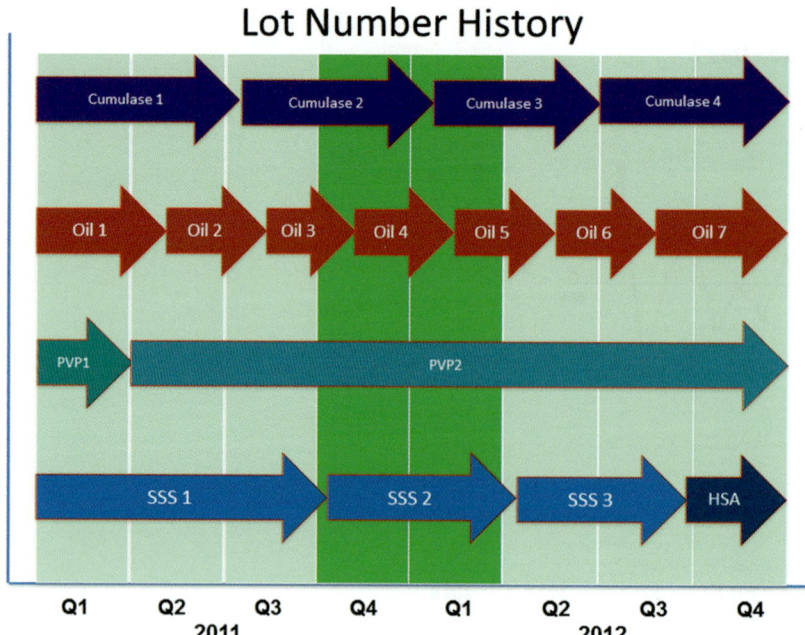

Figure 8.4 The increase in embryos with > four cells on day 2 during the first quarter of 2012 was due to both short two-cell and short four-cell cycles.

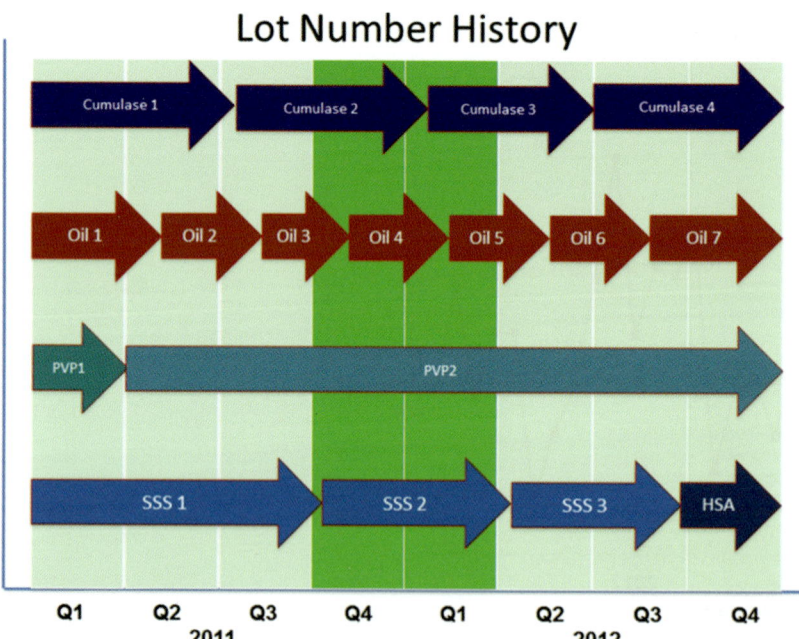

Figure 8.5 Graphical representation of changes in lot numbers of key reagents used for human IVF during 2011 and 2012. The green shaded area represents a period when a high percentage of embryos had short cell cycles resulting in > 30% of embryos having > four cells on day 2.

and can adversely affect embryo development and clinical outcomes [24]. Routine process control with several KPIs may detect subtle changes in air quality that currently go unnoticed. To test this hypothesis, we performed an experiment with a short exposure to acrolein, a common volatile organic compound. Using a standard mouse embryo assay, exposure of < 1.0 ppm acrolein for 24 h at the zygote stage was toxic (Video 8.1); however, addition of protein provided a protective effect that allowed acceptable blastocyst formation with standard protein concentrations (5 mg/ml HSA; Video 8.3) but not with a reduced amount of protein (1 mg/ml HSA; Video 8.2). These results provide evidence that embryo morphokinetics can be used to monitor air quality.

Protein-Free + Acrolein

Video 8.1 Mouse embryos (FVB female X CF1 male) cultured in HTF without protein in an EmbryoScope with 0.5 ppm acrolein for 24 hours. Four representative embryos that lyse < 30 hours of culture.

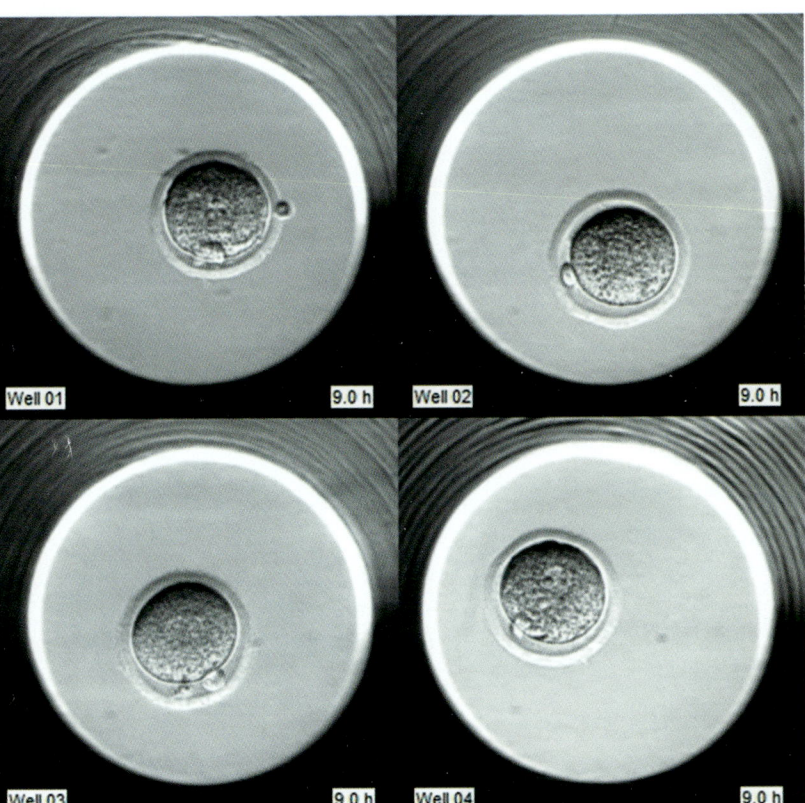

8.4. Time-lapse as a sensitive assay for quality control

Lot-to-lot variation in product quality remains a significant issue in the assisted reproduction laboratory. Quality control testing by manufacturers involves the use of a bioassay as a surrogate for human oocytes, sperm, and embryos. While most products are tested with a one-cell mouse embryo assay (MEA; [25]), its sensitivity has been brought into question as it failed to detect toxicity of several lots of mineral oil that ultimately were recalled by the manufacturers [12, 13]. This relatively poor sensitivity has led to numerous attempts to develop an assay that can provide assurance that toxicity is not present. Comparisons of sperm QC assays [26, 27], ICM outgrowth assays [28], and different mouse strains [29] demonstrate

little or no improvement and represent methods that are labor intensive.

Time-lapse imaging of mouse embryo development is a sensitive bioassay that may provide a substantial improvement over methods in routine use by manufacturers. In 2011 there were at least two lots of mineral oil recalled from clinical use due to toxicity. Both lots passed the standard one-cell MEA with > 80% blastocyst formation [13], illustrating clearly that the human embryo is more sensitive to oil toxicity than standard culture of mouse embryos in groups. In contrast, development of a multifactorial model incorporating variables that differed significantly from control embryos was many times more sensitive than blastocyst rate [13]. An example of how this model works is presented in Figure 8.6. While all five embryos shown developed to the

133

HSA (1 mg/ml) + Acrolein

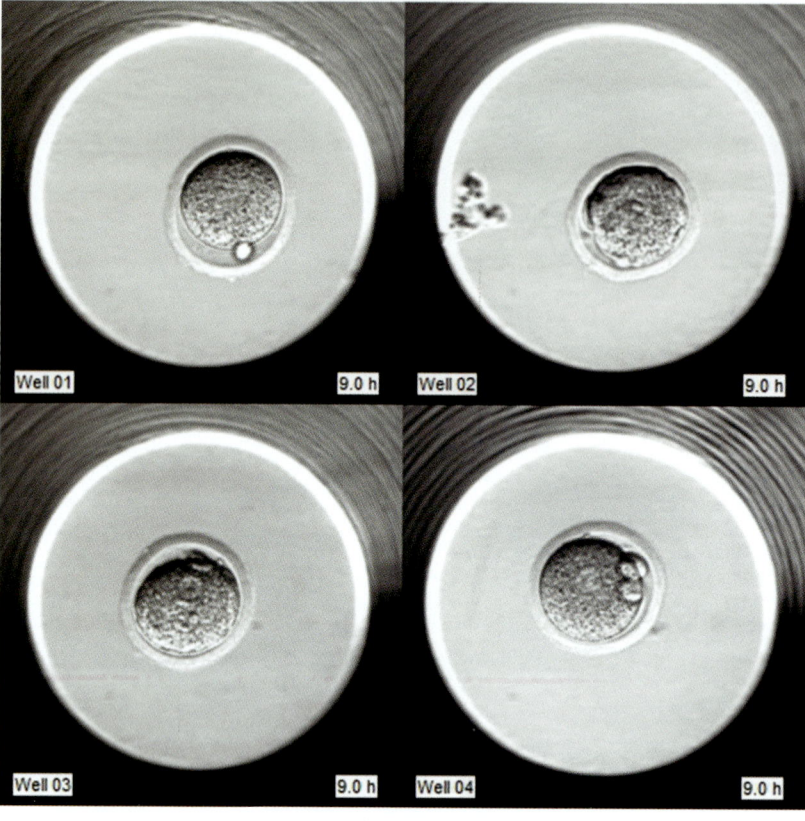

Video 8.2 Mouse embryos (FVB female X CF1 male) cultured in HTF with HSA (1 mg/ml) in an EmbryoScope with 0.5 ppm acrolein for 24 hours. Four representative embryos that develop to the morula stage, arrest, and lyse.

blastocyst stage, only embryo 3 met the timing limits established by the control group. Using this approach, we showed that two contaminants found in oil, TX-100 (Video 8.4–8) and peroxide (Video 8.9–13) did not affect blastocyst development at low concentrations but altered morphokinetics [13]. Given the precise nature of mouse embryo cell divisions and the small number of samples needed to detect a difference, this process could easily be automated and a sensitive QC score would provide an objective measure of differences in development between lots of a given product. In addition to markers that are considerably more objective than blastocyst formation, the ability to use morphokinetics from control embryos for each assay performed further increases sensitivity. For example, the tSB (63.9 h) for the control group in the TX-100 study (Video 8.4) was more than

3 hours earlier than for the peroxide treatment group (67.1 h; Video 8.9). Thus, rather than using a set benchmark for passing a product, such as 80% blastocyst rate, TL is more powerful by comparing the results obtained to exact timings for the controls that can change for each assay.

While routine testing with TL may not be practical, certain products with known lot-to-lot variability may benefit from additional layers of testing. For instance, mineral oil is a petroleum product that is largely undefined and has resulted in the most recalls of all products used in the IVF laboratory. The method of testing oil deserves attention, since group culture may provide a protective affect that limits the sensitivity of the MEA. This is illustrated in Video 8.14, where mouse embryos are cultured both individually and as a group

HSA (5 mg/ml) + Acrolein

Video 8.3 Mouse embryos (FVB female X CF1 male) cultured in HTF with HSA (5 mg/mL) in an EmbryoScope with 0.5 ppm acrolein for 24 hours. Four representative embryos; three that develop to the blastocyst stage within 96 hours of culture.

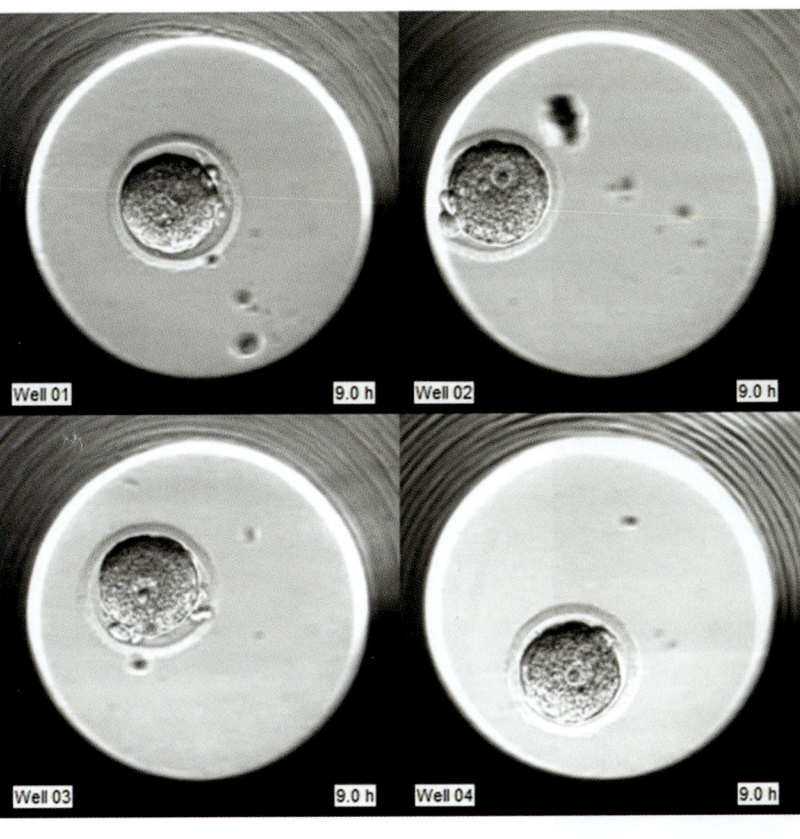

Morphokinetic Model

Embryo	cc2	t5	tSB	Classification
Upper 95%	<23.1	<39.4	<69.3	Control
Embryo 1	22.2	40.2	65.0	Delayed
Embryo 2	23.4	41.2	71.6	Delayed
Embryo 3	22.9	36.8	66.4	Normal
Embryo 4	21.1	36.5	76.5	Delayed
Embryo 5	24.2	39.0	67.2	Delayed

Figure 8.6 Illustration of the utility of a morphokinetic model for detecting embryotoxicity in the mouse embryo assay. The green shaded row represents the average cc2, t5, and tSB for the control embryos and the 95% CI. Embryos 1–5 are examples of embryos exposed to a test item for toxicity. Even though all five embryos develop to the blastocyst stage, note that only Embryo 3 meets all of the criteria established by the control group.

TX-100 Control

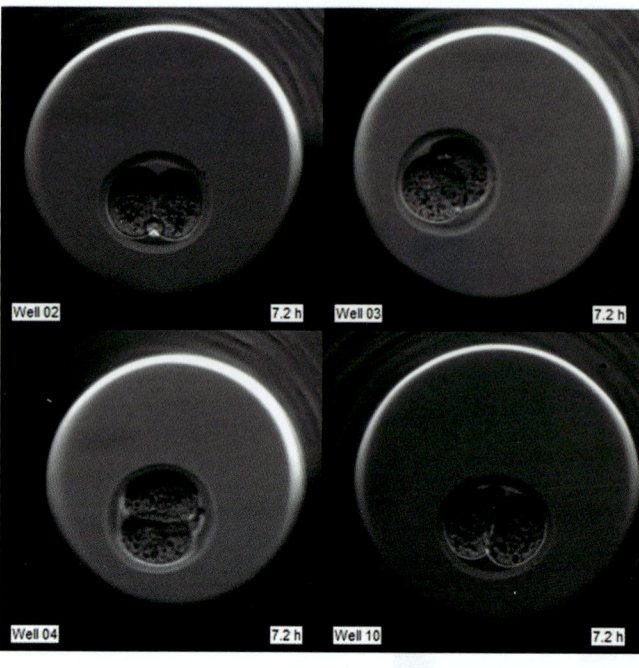

Video 8.4–8 Time-lapse mouse embryo assay (MEA) with thawed one-cell embryos from hybrid mice (see [13]) in protein-free HTF containing TX-100 at 0% (v4), 0.0004% (v5), 0.0008% (v6), 0.0012% (v7), and 0.0016% (v8).

TX-100 0.0008%

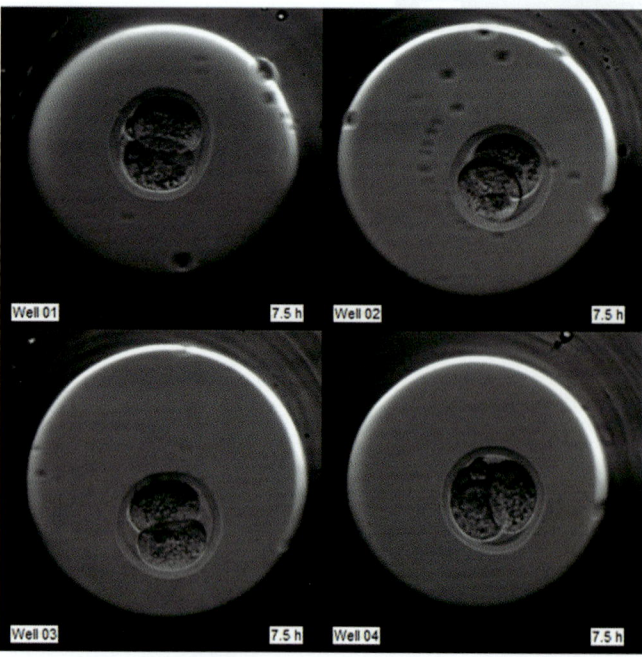

TX-100 0.0004%

Video 8.4–8 *(cont.)*

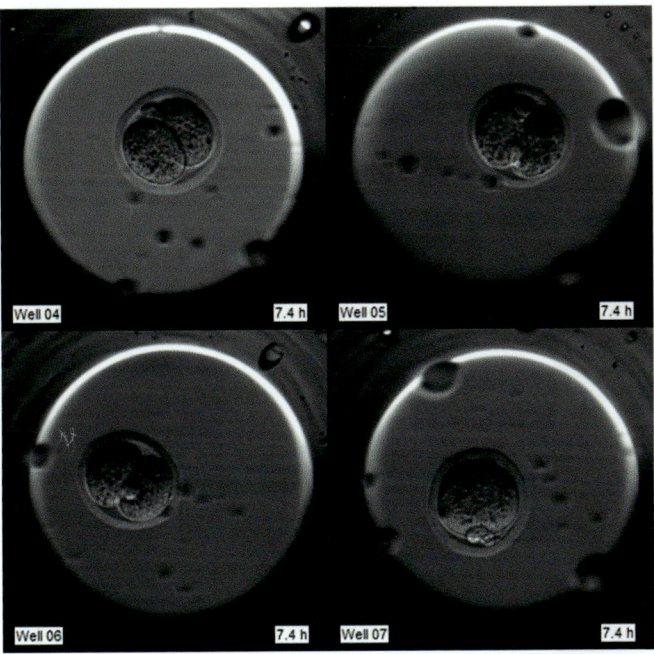

TX-100 0.0012%

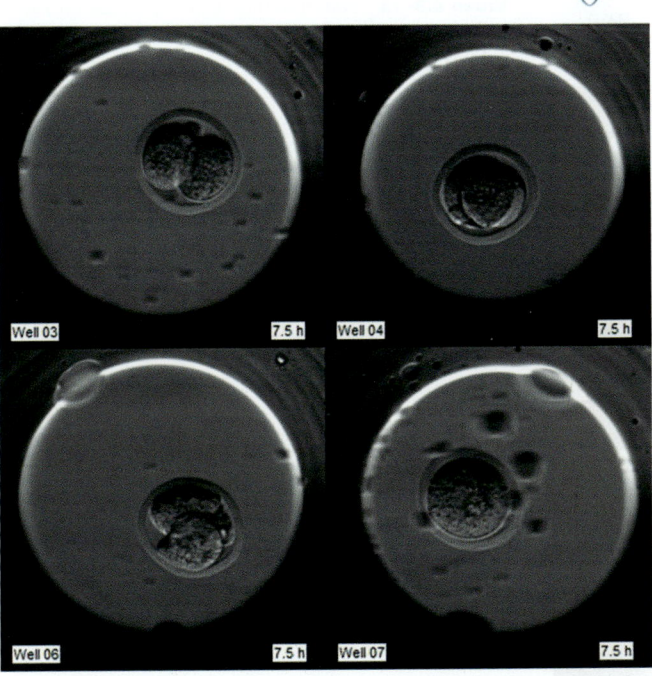

TX-100 0.0016%

MAYO
CLINIC

Video 8.4–8 (*cont.*)

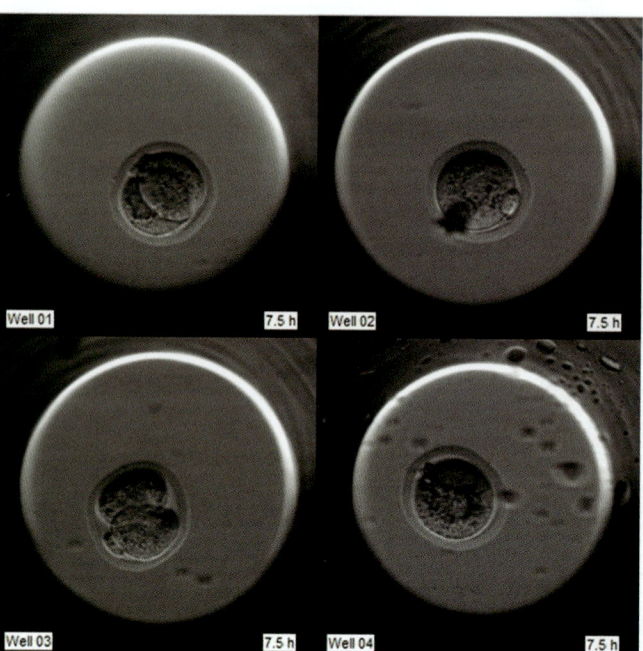

Cumene Peroxide Control

MAYO
CLINIC

Video 8.9–13 Time-lapse mouse embryo assay (MEA) with thawed one-cell embryos from hybrid mice (see [13]) in protein-free HTF containing cumene peroxide at 0 (v9), 2 μM (v10), 4 μM (v11), 6 μM (v12), and 8 μM (v13).

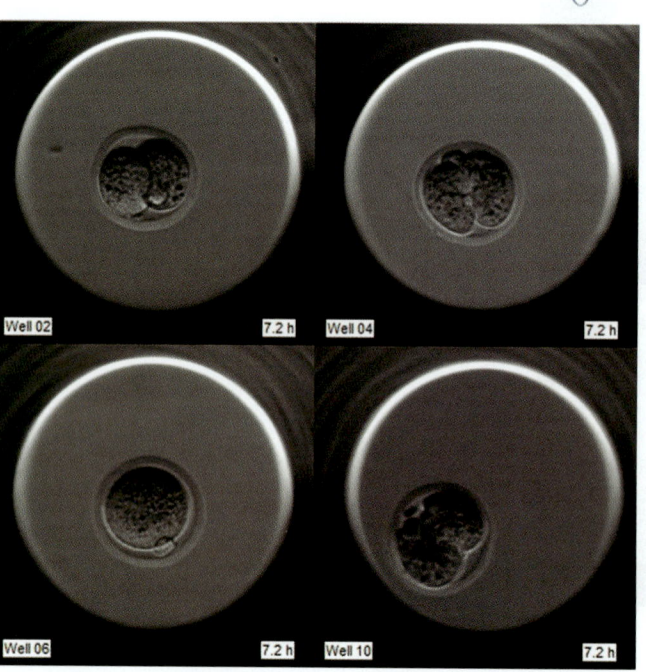

Cumene Peroxide 2 μM

Video 8.9–13 *(cont.)*

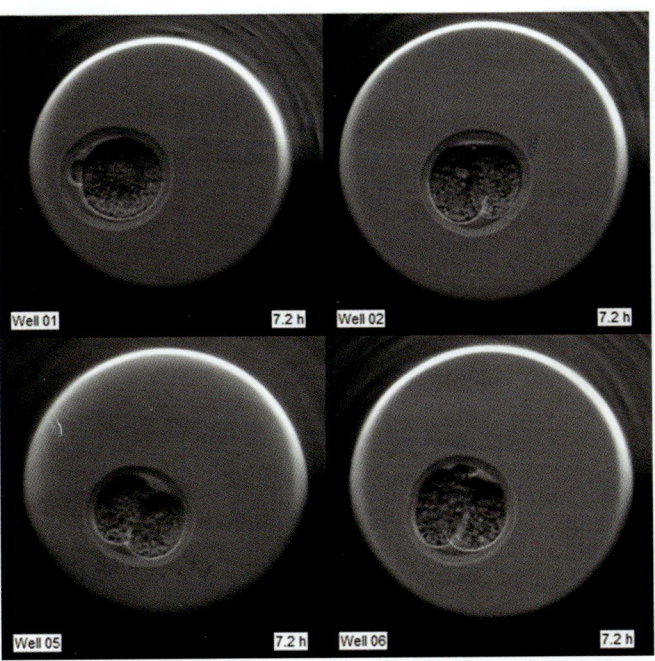

Cumene Peroxide 4 μM

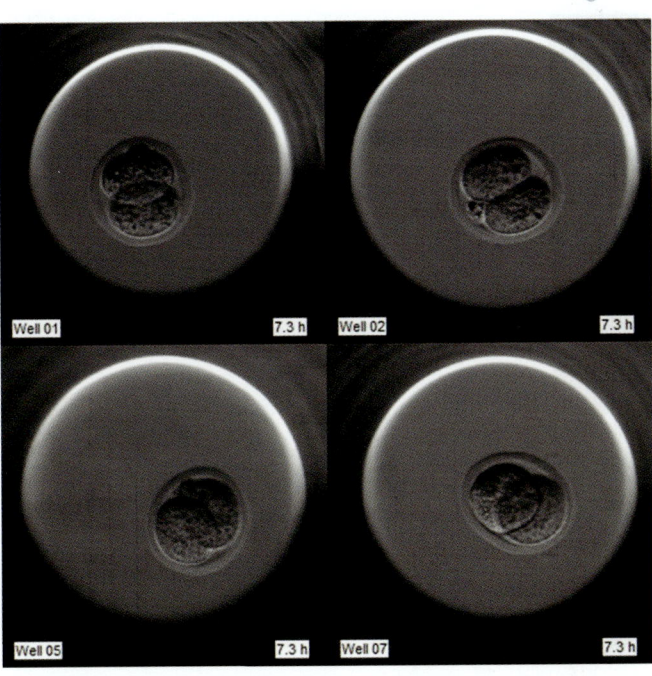

Cumene Peroxide 6 μM

Video 8.9–13 *(cont.)*

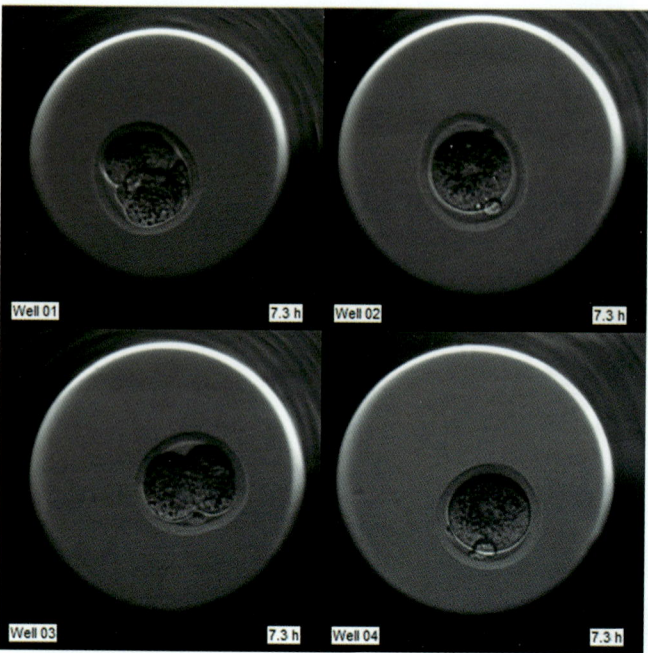

Cumene Peroxide 8 μM

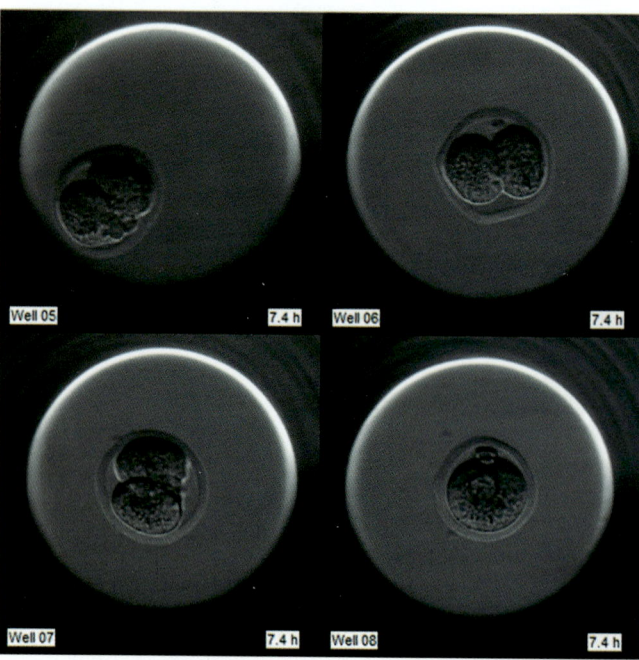

Oil Lot 2 (Wolff et al., 2013)

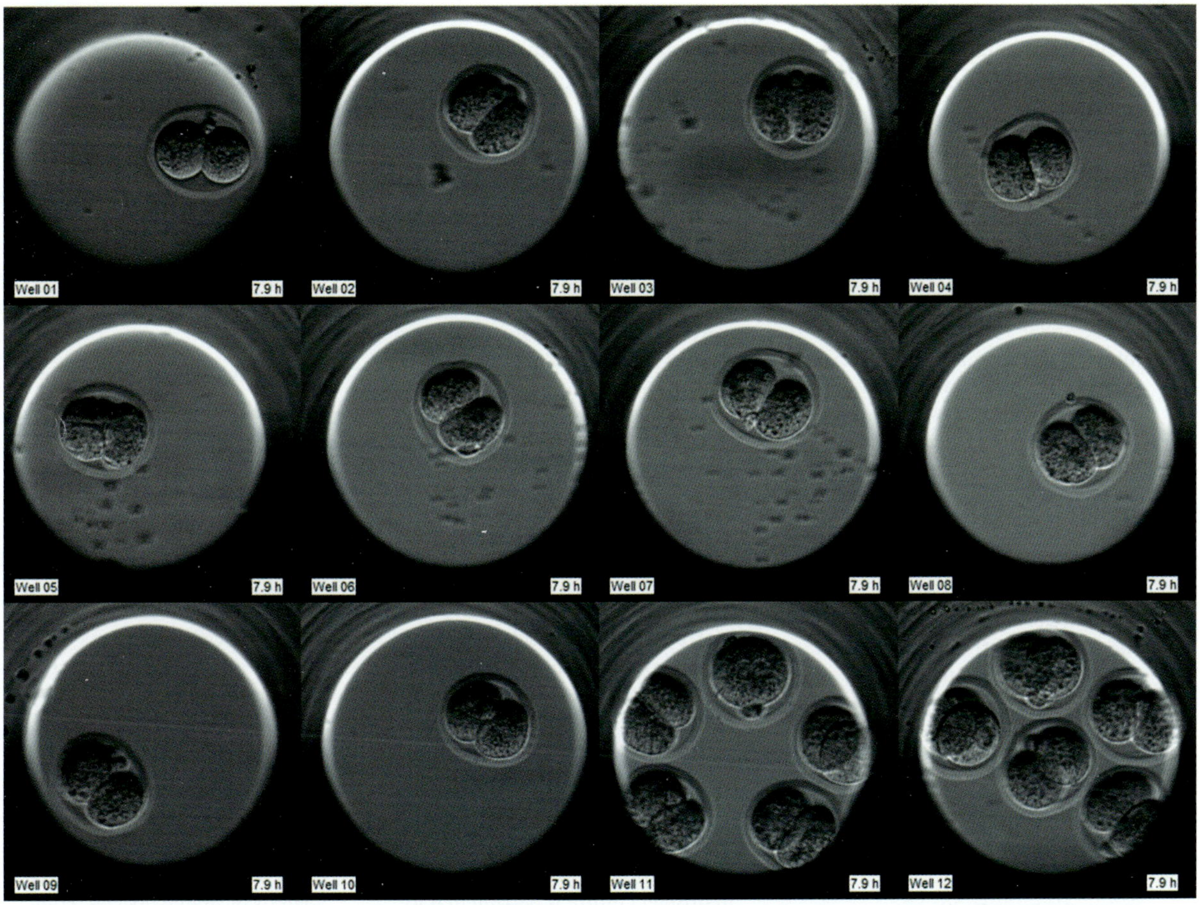

Video 8.14 Time-lapse mouse embryo assay (MEA) with thawed one-cell embryos from hybrid mice (see [13]) in protein-free HTF with peroxide containing mineral oil (Lot 2). Wells 1–10 contain one embryo each, wells 11 and 12 contain five embryos each. The latter embryos illustrate the protective effect of group culture during quality control testing since they remain expanded and in some cases completely hatch, whereas embryos cultured individually regress after 96 h.

(wells 11 and 12) in oil containing peroxide (Lot 2 from [13]). Timing of development and development to the blastocyst stage is improved in the group culture well, illustrating that group culture confers a protective effect that decreases the sensitivity of the MEA. In addition, embryos cultured in groups remained expanded and even hatched, whereas nine of the ten embryos cultured individually regressed after 96 hours of culture.

Protein, mostly as albumin purified from human serum (HSA), also demonstrates significant lot-to-lot variation, due in part to its highly undefined and variable composition [19] as well as the additive octanoic acid [30]. If manufacturers do not elect to test these products further, end users could screen lots as necessary. Time-lapse quality control testing provides end users with the most sensitive QC assay available for routine use.

8.5. Summary

"Measurement is the first step that leads to control and eventually to improvement. If you can't measure something, you can't understand it. If you can't understand it, you can't control it. If you can't control it, you can't improve it." This quote, by IBM's quality guru H. James Harrington, captures the essence of what time-lapse offers for quality management in the IVF laboratory. Time-lapse provides an opportunity to standardize clinical practice and develop robust key performance indicators. KPIs can then be used as aspirational benchmarks for the field of embryology, benchmarks that allow us to quantify how good we are and how much better we can become.

Quality management, at its core, requires robust data collection. Time-lapse technology is ideally suited to fill a gap in quality management in the IVF laboratory: the ability to obtain large amounts of objective data. The most impressive feature of this application of TL is that rather than requiring *more* handling to obtain additional data, it actually requires *less*!

References

1. Alper MM, Brinsden PR, Fischer R, Wikland M. Is your IVF programme good? *Hum Reprod.* 2002 Jan;17(1):8–10.

2. Johnson A, El-Toukhy T, Sunkara SK, Khairy M, Coomarasamy A, Ross C, et al. Validity of the in vitro fertilisation league tables: influence of patients' characteristics. *BJOG* 2007 Dec;114(12):1569–74.

3. Luke B, Brown MB, Wantman E, Stern JE, Baker VL, Widra E, et al. A prediction model for live birth and multiple births within the first three cycles of assisted reproductive technology. *Fertil Steril.* 2014 Sep;102(3):744–52.

4. Glujovsky D, Blake D, Farquhar C, Bardach A. Cleavage stage versus blastocyst stage embryo transfer in assisted reproductive technology. *The Cochrane database of systematic reviews.* 2012;7: CD002118.

5. Mortimer D, Mortimer S. *Quality and Risk Management in the IVF Laboratory.* Cambridge, UK: Cambridge University Press; 2005.

6. Alpha. The Alpha consensus meeting on cryopreservation key performance indicators and benchmarks: proceedings of an expert meeting. *Reprod Biomed Online.* 2012 Aug;25(2):146–67.

7. Boone WR, Johnson JE, Locke AJ, Crane MMt, Price TM. Control of air quality in an assisted reproductive technology laboratory. *Fertil Steril.* 1999 Jan;71(1):150–4.

8. Esteves SC, Bento FC. Implementation of air quality control in reproductive laboratories in full compliance with the Brazilian Cells and Germinative Tissue Directive. *Reprod Biomed Online.* 2013 Jan;26(1):9–21.

9. Khoudja RY, Xu Y, Li T, Zhou C. Better IVF outcomes following improvements in laboratory air quality. *J Assist Reprod Genet.* 2013 Jan;30 (1):69–76.

10. Sifer C, Pont JC, Porcher R, Martin-Pont B, Benzacken B, Wolf JP. A prospective randomized study to compare four different mineral oils used to culture human embryos in IVF/ICSI treatments. *Eur J Obstet Gynecol Reprod Biol.* 2009;147:52–6.

11. Otsuki J, Nagai Y, Chiba K. Peroxidation of mineral oil used in droplet culture is detrimental to fertilization and embryo development. *Fertil Steril.* 2007 Sep;88(3):741–3.

12. Turner T. The identification of a toxic substance in the in vitro fertilization laboratory: the value of inter-laboratory communication. *Fertil Magazine.* 2010;12:64–5.

13. Wolff HS, Fredrickson JR, Walker DL, Morbeck DE. Advances in quality control: mouse embryo morphokinetics are sensitive markers of in vitro stress. *Hum Reprod.* 2013 Jul;28(7):1776–82.

14. Zhang JQ, Li XL, Peng Y, Guo X, Heng BC, Tong GQ. Reduction in exposure of human embryos outside the incubator enhances embryo quality and blastulation rate. *Reprod Biomed Online.* 2010 Apr;20 (4):510–5.

15. Miller KA. Optimizing culture conditions. In: Quinn P, editor. *Culture Media, Solutions, and Systems in Human ART.* Cambridge: Cambridge University Press; 2014.

16. Sakkas D, Shoukir Y, Chardonnens D, Bianchi PG, Campana A. Early cleavage of human embryos to the two-cell stage after intracytoplasmic sperm injection as an indicator of embryo viability. *Hum Reprod.* 1998 Jan;13 (1):182–7.

17. Basile N, Morbeck D, Garcia-Velasco J, Bronet F, Meseguer M. Type of culture media does not affect embryo kinetics: a time-lapse analysis of sibling oocytes. *Hum Reprod.* 2013 Mar;28(3):634–41.

18. Morbeck DE, Krisher RL, Herrick JR, Baumann NA, Matern D, Moyer T. Composition of commercial media used for human embryo culture. *Fertil Steril.* 2014 Sep;102(3):759–66 e9.

19. Morbeck DE, Paczkowski M, Fredrickson JR, Krisher RL, Hoff HS, Baumann NA, et al. Composition of protein supplements used for human embryo culture. *J Assist Reprod Genet.* 2014 Sep 27.

20. Rubio I, Kuhlmann R, Agerholm I, Kirk J, Herrero J, Escriba MJ, et al. Limited implantation success

of direct-cleaved human zygotes: a time-lapse study. *Fertil Steril.* 2012 Aug 24.

21. Liu Y, Chapple V, Roberts P, Matson P. Prevalence, consequence, and significance of reverse cleavage by human embryos viewed with the use of the Embryoscope time-lapse video system. *Fertil Steril.* 2014 Sep 12.

22. Athayde Wirka K, Chen AA, Conaghan J, Ivani K, Gvakharia M, Behr B, et al. Atypical embryo phenotypes identified by time-lapse microscopy: high prevalence and association with embryo development. *Fertil Steril.* 2014 Jun;101(6):1637–48 e1–5.

23. Ergin EG, Caliskan E, Yalcinkaya E, Oztel Z, Cokelez K, Ozay A, et al. Frequency of embryo multinucleation detected by time-lapse system and its impact on pregnancy outcome. *Fertil Steril.* 2014 Oct;102(4):1029–33 e1.

24. Cohen J, Gilligan A, Esposito W, Schimmel T, Dale B. Ambient air and its potential effects on conception in vitro. *Hum Reprod.* 1997 Aug;12(8):1742–9.

25. Morbeck DE. Importance of supply integrity for in vitro fertilization and embryo culture. *Semin Reprod Med.* 2012 Jun;30(3):182–90.

26. Claassens OE, Wehr JB, Harrison KL. Optimizing sensitivity of the human sperm motility assay for embryo toxicity testing. *Hum Reprod.* 2000 Jul;15(7):1586–91.

27. Hughes PM, Morbeck DE, Hudson S, Fredrickson J, Walker DL, Coddington CC. Peroxides in mineral oil used for in vitro fertilization: defining limits of standard quality control assays *J Assist Reprod Genet.* 2010;27 (2–3):87–92.

28. Gada RP, Daftary GS J, Walker DL, Lacey JM, Matern D, Morbeck DE. Potential of inner cell mass outgrowth and amino acid turnover as markers of quality in the in vitro fertilization laboratory. *Fertil Steril.* 2012;98 (4):863–869.

29. Khan Z, Wolff HS, Fredrickson JR, Walker DL, Daftary GS, Morbeck DE. Mouse strain and quality control testing: improved sensitivity of the mouse embryo assay with embryos from outbred mice. *Fertil Steril.* 2013 Mar 1;99 (3):847–54 e2.

30. Leonard PH, Charlesworth MC, Benson L, Walker DL, Fredrickson JR, Morbeck DE. Variability in protein quality used for embryo culture: embryotoxicity of the stabilizer octanoic acid. *Fertil Steril.* 2013;100(2):544–549.

Time-lapse implementation in a clinical setting: outcome results

Nikica Zaninovic, Qiansheng Zhan, and Zev Rosenwaks

9.1. Introduction

Time-lapse microscopy (TLM) of embryos has been a revolutionary innovation in human IVF. TLM represents one of the most important improvements in clinical embryology since the implementation of ICSI. The introduction of the time-lapse systems in clinical embryology has enhanced our knowledge of embryo development and embryokinetics. With TLM, human embryology has attained a fourth dimension: time (4-D embryology). It has enhanced our knowledge of cleavage dynamics as delineated by cytokinesis, karyokinesis, temporal cellular events, polarity, and much more. Evaluation of the dynamic processes of embryo division has enabled a better understanding of embryonic well-being and has provided critical information about normal versus abnormal embryonic developmental patterns with the goal of identifying healthy embryos with the greatest implantation potential.

At Weill Cornell Medical College, we first studied mouse and human embryo development using the time-lapse system in 2005. We used a custom made time-lapse chamber (Tokai hit, Japan) that was mounted on the inverted microscope under stable gas and temperature conditions. Images were acquired by a digital camera and processed by computer digital imaging software. The resulting videos were used only as an academic teaching tool and were not designed for the clinical setting, as the system was only able to evaluate one to two embryos at a time.

In November 2011, we obtained our first clinical TLM-EmbryoScope® (Fertilitech, Vitrolife, Denmark). Elaborate testing and validation of the EmbryoScope was performed, starting first as an incubator and then as an embryo selection or deselection tool. First, we tested mouse embryo development and compared it to the standard incubator. In human embryo studies, we tested

embryo development of sibling oocytes between the EmbryoScope and the standard incubator (77 patients). Next, we tested different oxygen concentrations (20% vs. 5%) on sibling oocytes between two EmbryoScopes (28 patients). The EmbryoScope showed superior embryo development and more stable incubator conditions compared to the standard incubators, especially in reduced oxygen conditions (3E) (Figure 9.1). Few previous studies compared culture in standard incubators versus time-lapse incubators such as the EmbryoScope. No differences were found in embryo development between incubators, possibly due to the small number of subjects, incubation in ambient oxygen, and the duration of the culture. Comparison of clinical outcome between incubators was not evaluated in these studies [1, 2, 3].

After thorough evaluation of our embryo development results, we decided to apply the EmbryoScope with reduced oxygen (5%) for all IVF patients. Our decision was based on superior embryo development (higher number of four-cell embryos on day 2 and eight-cell embryos on day 3) (Figure 9.1 A), higher ET selection rate, higher blastulation rate and increased blastocyst freezing rate (Figure 9.1 B).

The effect of oxygen concentration as a key factor in embryo development has been well documented [4]. In mammalian embryos, oxygen regulates amino acid turnover and metabolic equilibrium [5]. In a recent Cochrane review, the success rate of IVF or ICSI cycles increased in embryos cultured in low oxygen conditions. This benefit is especially evident when assessing live birth rates [6]. In TLM studies, differences in oxygen concentrations alter developmental rates of embryos, as evidenced by a lower percentage of early blastocysts developed in ambient oxygen conditions [7].

In 2012, we used the EmbryoScope on 15% of IVF patients, followed by 70% use in 2013 and 80%

Time-Lapse Microscopy in In-Vitro Fertilization, ed. Marcos Meseguer. Published by Cambridge University Press.
© Cambridge University Press 2016.

(A)

Number of 4c and 8c by incubator

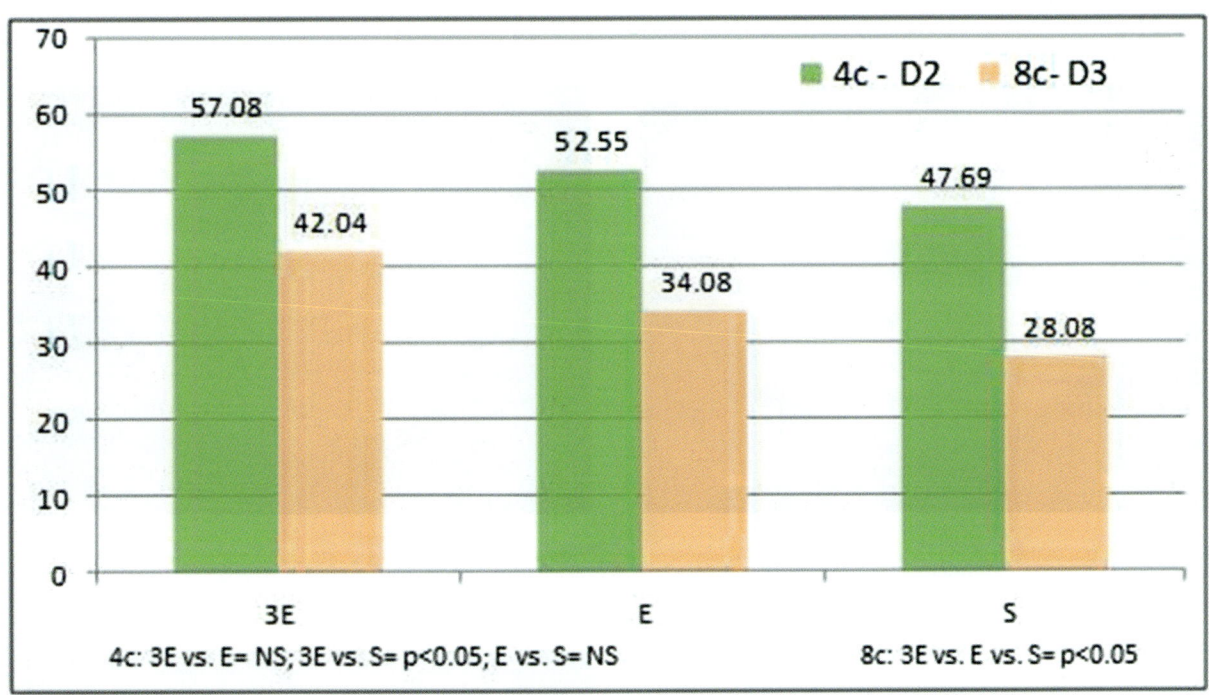

(B)

ET and BL cryo by incubator

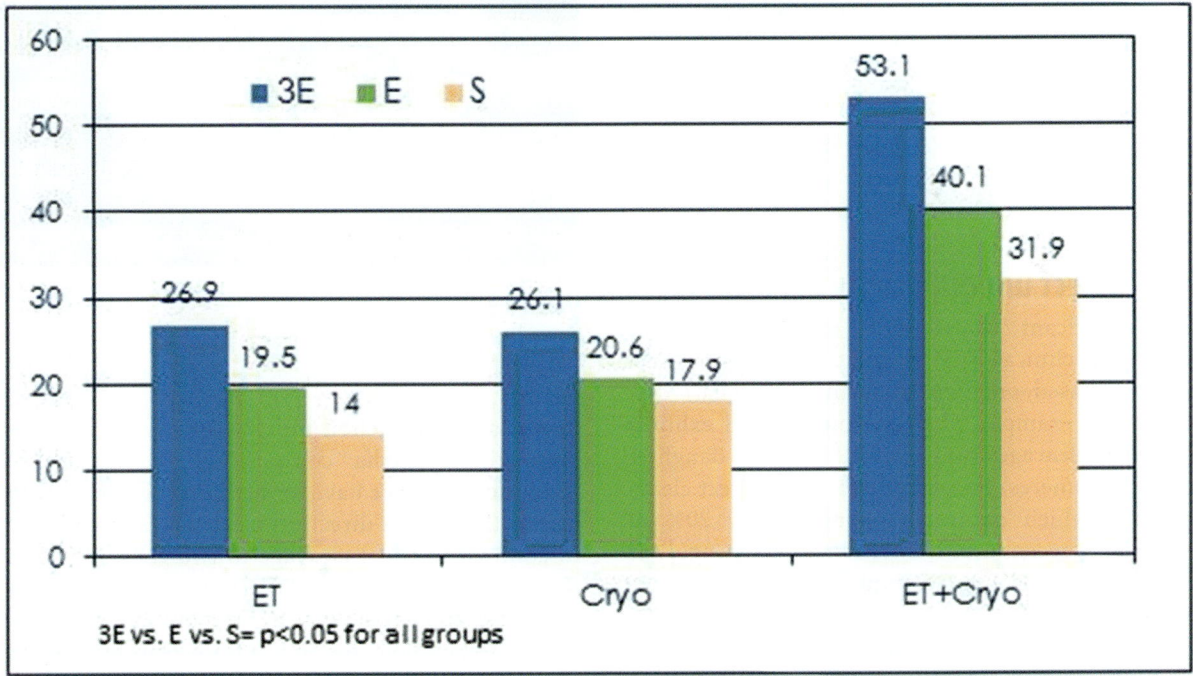

Figure 9.1 Testing and validation of EmbryoScopes TLM on sibling oocytes. (A) Number of four-cell (day 2) and eight-cell (day 3) embryos in triple gas EmbryoScope 5% O_2 (3E), EmbryoScope 20% O_2 (E) and standard incubator (S). 3E have a significantly higher number of four- and eight-cell embryos. (B) Percent of ET and BL cryoembryos by incubator. 3E have higher transfer and freezing rates compared to E or S. Chi-square test statistic.

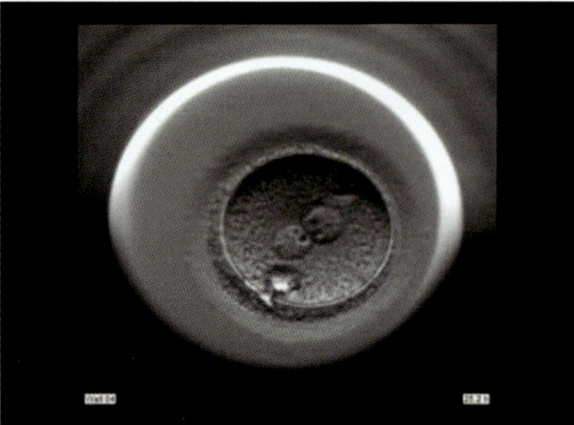

Video 9.1 Development of the DUC-1 embryo showing direct cleavage from one cell to three cells at the first division. No blastocyst formation was observed.

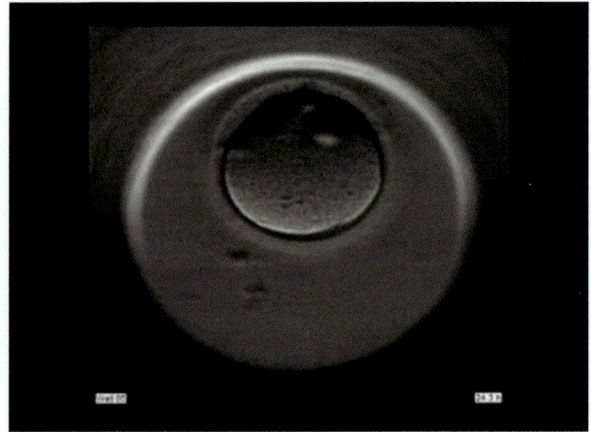

Video 9.2 Development of the DUC-2 embryo showing direct cleavage from one cell to three cells of one of the two blastomeres at the two-cell stage (second division).

utility in 2014. In 2015, we plan to expand to ten EmbryoScopes, which will allow us to use TLM on all patients, which comprise more than 3,000 cycles per year. We chose the EmbryoScope (E) time-lapse system as our center's primary incubator system due to its stable and unsurpassed incubator features; uninterrupted culture and constant monitoring; customizable O_2 and CO_2 gas concentrations; direct involvement of the embryologist in evaluation, annotation and embryo selection; the robust results reported throughout the world; and the ability to design and apply embryo selection models.

9.2. Direct unequal cleavages (DUCS)

One significant contribution of the time-lapse system to clinical IVF is the ability to deselect embryos based on their abnormal (unequal) divisions. For example, when the mother cell exhibits abnormal cleavage from one cell to three daughter cells (blastomeres) instead of two cells (direct cleavage) [8], which occurs in approximately 20% of embryos. A recent multicenter retrospective analysis [9] demonstrated limited implantation of embryos exhibiting one-to-three cell cleavage at the time of first mitosis (designated DUC-1) (Video 9.1). This observation could only be identified with time-lapse microscopy and would be missed by static observations. The occurrence of DUCS has been reported by many IVF laboratories, suggesting a

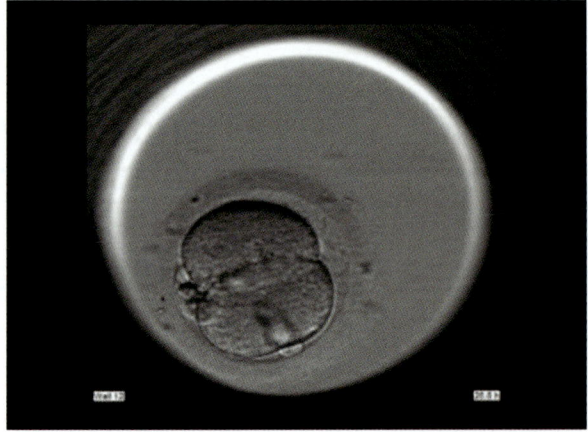

Video 9.3 Development of the DUC-3 embryo showing direct cleavage from one cell to three cells of one of the four blastomeres after the four-cell stage (third division).

universal phenomenon unrelated to specific patient populations, media composition or fertilization technique [10]. We have termed direct cleavage from one to three cells, direct unequal cleavages (DUCS), in contrast to equal cleavage to two cells. It is clear that embryos exhibiting DUCS have limited developmental and implantation potential and while the underlying etiology for DUCS is unclear, the transfer of these embryos should be avoided.

We have observed that DUCS can also occur during the second and third cleavage divisions (DUC-2, DUC-3), and not in all blastomeres (Videos 9.2 and 9.3) [11]. Our

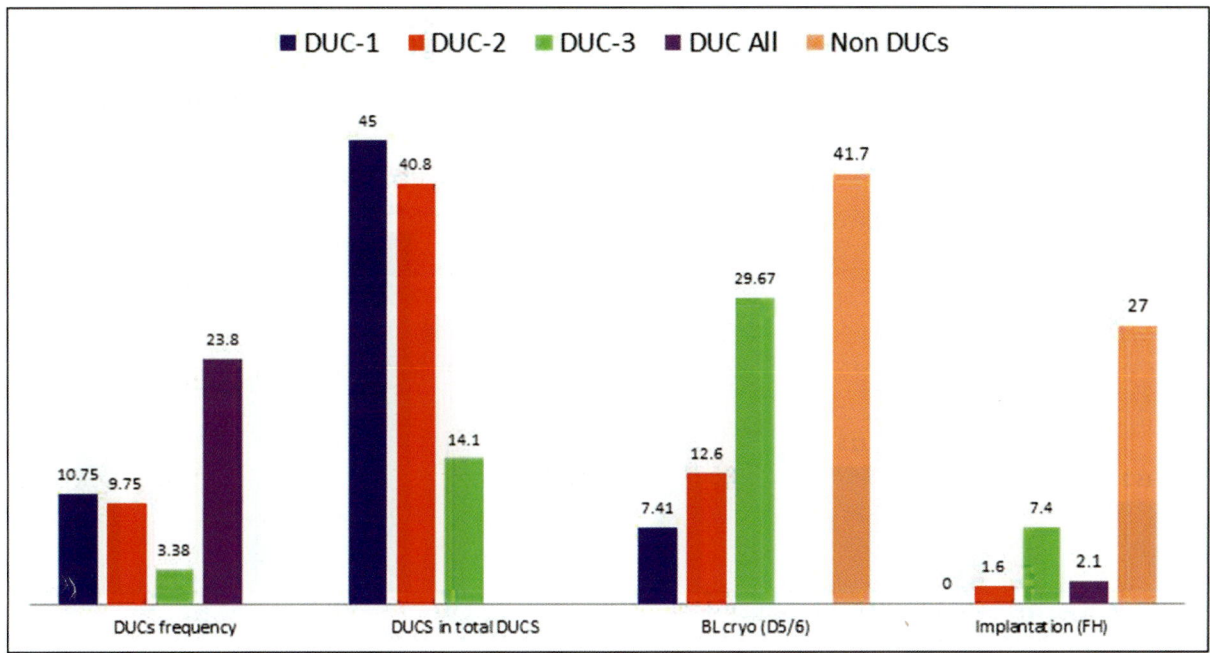

Figure 9.2 Frequency and outcome of direct unequal cleavages (DUCS) in 8,113 IVF embryos (%). DUC-1: first cleavage; DUC-2: second cleavage, and DUC-3: third cleavage.

analysis indicates that DUCS at early cell stages have a greater negative impact on subsequent embryo development than later stage DUCS, although overall development is impaired by the presence of DUCS at any stage (Figure 9.2). Underlying causes of DUCS embryos and whether chromosomal abnormalities are associated with their incidence remain in question. Chromosomal analysis of the DUCS embryos from PGS patients clearly indicated an absence of chromosomal triploidy as a main cause of DUCS. PGS analysis (biopsy day 3 or 5), however, detected chromosomal abnormalities in 84% of DUCS embryos, albeit different patterns of chromosomal aberrations of both maternal and paternal origin. DUCS-1 embryos exhibited extremely high aneuploidy rate, mostly complex aneuploidies (more than one aneuploidy) (Figure 9.3). Aneuploidy rates were lower when abnormal divisions occurred at second (DUC-2) and third (DUC-3) division (Figure 9.3). Possible mechanisms involved in DUCS might include: silent triploidy caused by non-extrusion of the second polar body, use of diploid sperm or diploid (giant) oocytes (extremely rare), abnormalities of the centrosome, and subsequent mitotic spindle, and/or postzygotic diploidization [12, 13, 14].

9.3. Clinical outcome data using TLM

Outcome data presented here include all IVF or DER (donor egg recipient) patients from 2011 (standard incubator only – S), 2012 (15% of E cycles), and 2013 (70% of E cycles). Data are grouped based on the type of incubator: S = standard incubator (Forma Scientific, 20% O_2, 5.8% CO_2), S 2011 or S 2012–2013; E = EmbryoScope (Fertilitech, Vitrolife, 5% O_2, 5.8% CO_2) E 2012–2013. Excluded cycles were: EmbryoScope testing cycles on sibling oocytes; cycles with mix ET (S and E); PGD/PGS cycles; endometrial co-culture cycles; cryo upfront cycles; and immature or 1PN ET cycles. No exclusions were performed based on infertility diagnosis, stimulation protocols, sperm origin, and maternal or paternal age. All outcome data are based on fetal sacs (clinical pregnancy – CP) or live birth (LB) rate per ET (ongoing pregnancy – OP, Implantation, and live birth cycles). Data are stratified by maternal age, fertilization techniques (ICSI or insemination), ET day (D3 or D5), and IVF or DER. Embryos in all groups were cultured in in-house made sequential medium (very similar to G1/2) with media changed on day 3 for BL cultures.

For the standard and EmbryoScope groups, embryo selections for day 3 ET were based on cell

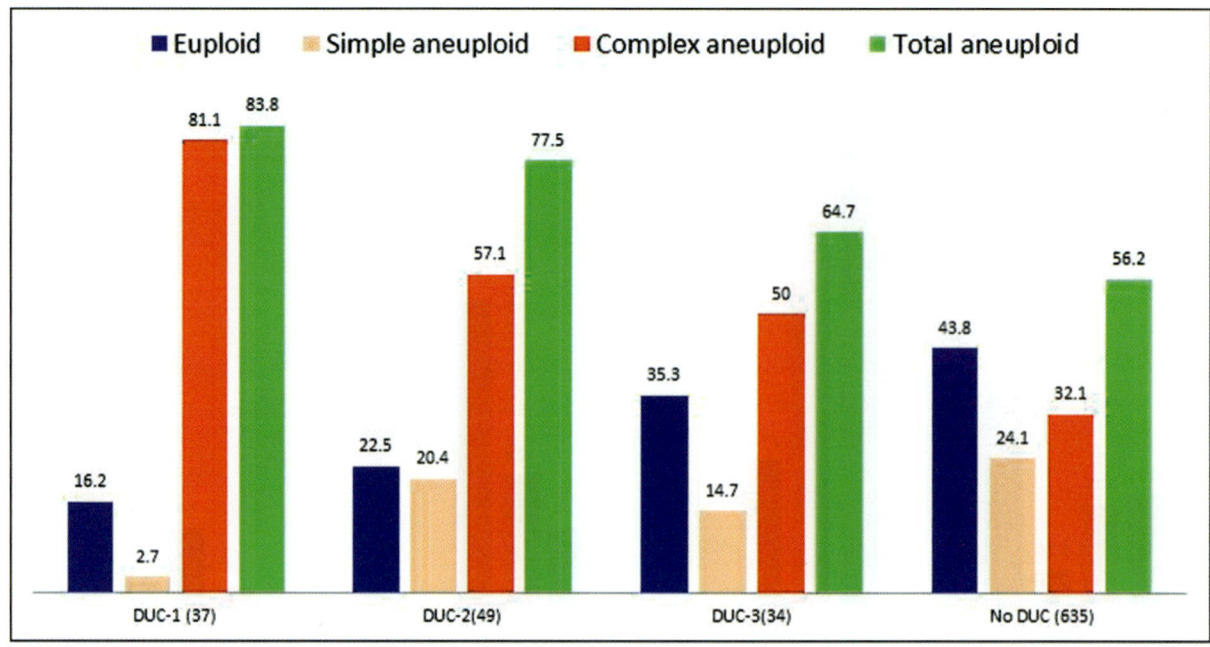

Figure 9.3 DUCS and chromosomal analysis in 132 PGS patients (%). Overall DUCS incidence of 16.9% (218/1293). High aneuploidy rates were observed in DUCS embryos compared to non-DUCS embryos. DUC-1 embryos express extremely high percentage of complex aneuploidies.

number and morphology, without using specific embryo selection models but eliminating DUCS embryos in E cultured patients. On day 5, blastocysts for ET were selected based on blastocoel expansion, ICM and TE morphology [15] without using morphokinetic data. Only high-quality blastocysts (2BB or higher on day 5 or 6) were selected for freezing.

Overall characteristics of the IVF patients are presented in Table 9.1. On average, similar numbers of embryos were transferred in all three groups. Higher numbers of BL transfers were observed in the EmbryoScope group (19.3%) indicating better growth of embryos in TLM and improved selection of patients for BL culture. The increase in BL-ET was primarily in the under-40 patient population, with an apparent tendency towards an increased D5-ET in older populations as well (Figure 9.4). In 2014, we increased the percentage of BL transfers to over 30%. There was also a conscious attempt to increase the number of single embryo transfer (SET) blastocysts in 2012 to 2013. Improvement in embryo culture (in our opinion mainly due to reduced O_2) resulted in doubling the number of cycles with BL freezing (day 5 or 6) with the EmbryoScope compared to the standard incubator, and on average one more BL was frozen per cycle (Table 9.1).

Overall clinical outcome data showed a significant increase in clinical pregnancy (CP), ongoing pregnancy (OP), implantation (number of LB per number of embryos ET), and live birth ET cycles when the EmbryoScope was used compared to the standard incubator in 2011 and 2012 to 2013 (Figure 9.5). When clinical outcome data were analyzed by maternal age, the most independent determining factor for clinical outcome, there was a clear increase in CP and OP rate, especially for patients younger than 38 years old (Figure 9.6 and Figure 9.7). This increase is even more evident when implantation rate is analyzed, indicating that embryos cultured in the EmbryoScope resulted in higher number of live births and higher number of cycles resulting in LB (Figure 9.8 and Figure 9.9). It is worth noting that the trend of higher clinical outcome in the EmbryoScope group is also evident in patients with advanced maternal age. Recent multicenter retrospective analyses, correlating EmbryoScope culture combined with morphokinetic embryo selection model and embryo culture in standard incubator with standard morphological evaluation, report an overall 20% increase in CP rate in favor of TLM culture [16]. Interestingly, embryos in both culture systems utilized 20% O_2;

Table 9.1 Overall patient population characteristics between 2011 standard (S) and 2012–2013 standard (S) and EmbryoScope (E) groups

IVF cycles	2011 (S)	2012–2013 (S)	2012–2013 (E)
Number of patients	1,231	1,199	1,318
Number of ET cycles	1,459	1,421	1,505
Number of oocytes (avg./cycle)	15,888 (avg. 12.9)	12,961 (avg. 10.8)	17,263 (avg. 13.1)
Number of MII (Avg.)	12,965 (Avg. 10.5)	10,465 (Avg. 8.7)	14,308 (Avg. 10.9)
Number of 2PN (Avg.)	9,646 (Avg. 7.8)	7,401 (Avg. 6.2)	10,918 (Avg. 8.3)
Number of embryos ET (Avg.)	3,907 (Avg. 2.7)	3,664 (Avg. 2.6)	3,846 (Avg. 2.6)
Number of D3-ET	1,321	1,324	1,214
Number of D5-ET (%) (few D2, D4 or D6 ET)	136 (9.3%)	91 (6.4%)	290 (19.3%)
Number of SET (D5)(%)	20 (14.7%)	22 (24.2%)	70 (24.1%)
Number of cycles with BL cryo (%/cycles)	401 (27.5%)	310 (21.2%)	633 (42.1%)

it will be of great interest to assess the impact of reduced oxygen in the same groups.

The improvement in clinical outcome in our E group was also followed by the decrease in miscarriage rates (Figure 9.10). This was also observed in larger retrospective and prospective studies by other groups [16, 17, 18].

To further evaluate the impact of TLM on the day of embryo transfer, we analyzed our data separately by ET day: day 3 (D3) or day 5 (D5). For D3 ET, we observed a significant increase in CP and OP in the EmbryoScope group compared to the standard group. When divided by maternal age, the improvement in clinical outcome is more evident in patients less than 40 years old (Figure 9.11 and Figure 9.12). This increase is further evidenced by higher implantation

rate and live birth rate per ET (Figure 9.13 and Figure 9.14). The improvement can be attributed to better culture conditions and to better embryo selection and especially deselection for ET (see DUCS section above). For day 5 ET cycles, there is a trend towards higher pregnancy and implantation rates using the EmbryoScope (Figure 9.15 to 9.18). We did not expect much of an increase in this group, as embryos that have reached the blastocyst stage have self-declared themselves as better embryos. The significant benefit of the TLM is attributed to better selection of patients and embryos for BL culture resulting in the increase of D5-ET cycles (Table 9.1). Retrospective and prospective studies comparing TLM with standard incubator demonstrate an improved clinical outcome in the TLM group. Interestingly, in retrospective analyses,

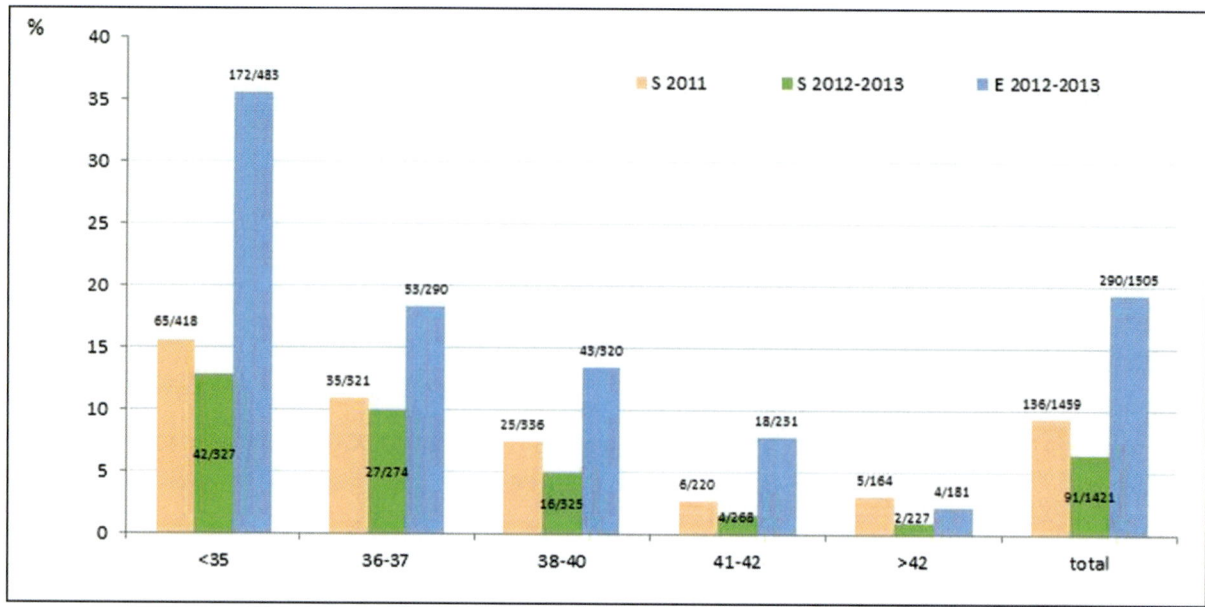

Figure 9.4 Percent of day 5 ET in correlation with maternal age.

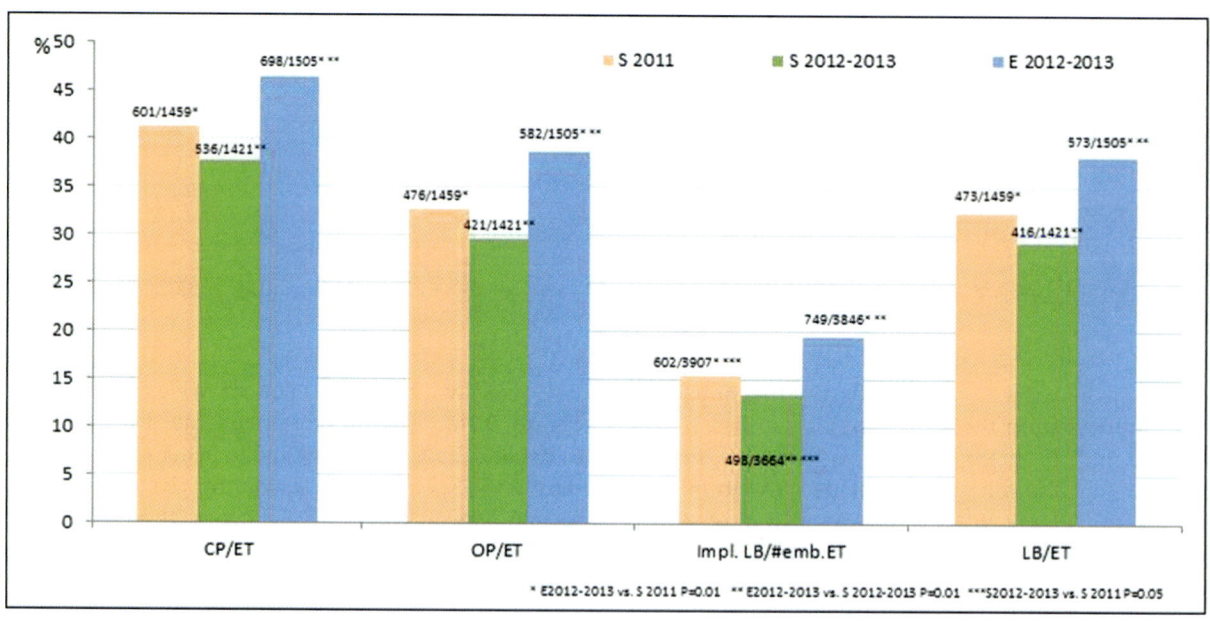

Figure 9.5 Overall clinical outcome data for clinical pregnancy (CP, sac seen), ongoing pregnancy (OP, per LB), implantation (number of LB per ET) and live birth cycles (LB/ET) between 2011 S, 2012–13 S and 2012–13 E cycles. Chi-square test statistic.

the improved results were more evident for day 3 ET; while in a prospective study, a significant increase was observed in ongoing pregnancy rates in day 5 ET, and implantation rates were significantly increased for day 3 ET [16, 18].

Although several groups have suggested that there is no need for TLM when embryos at the blastocyst stage are selected for 24-chromosome analysis (PGS), we believe that TLM is a complementary technique that aids efforts to reduce multiple pregnancies by

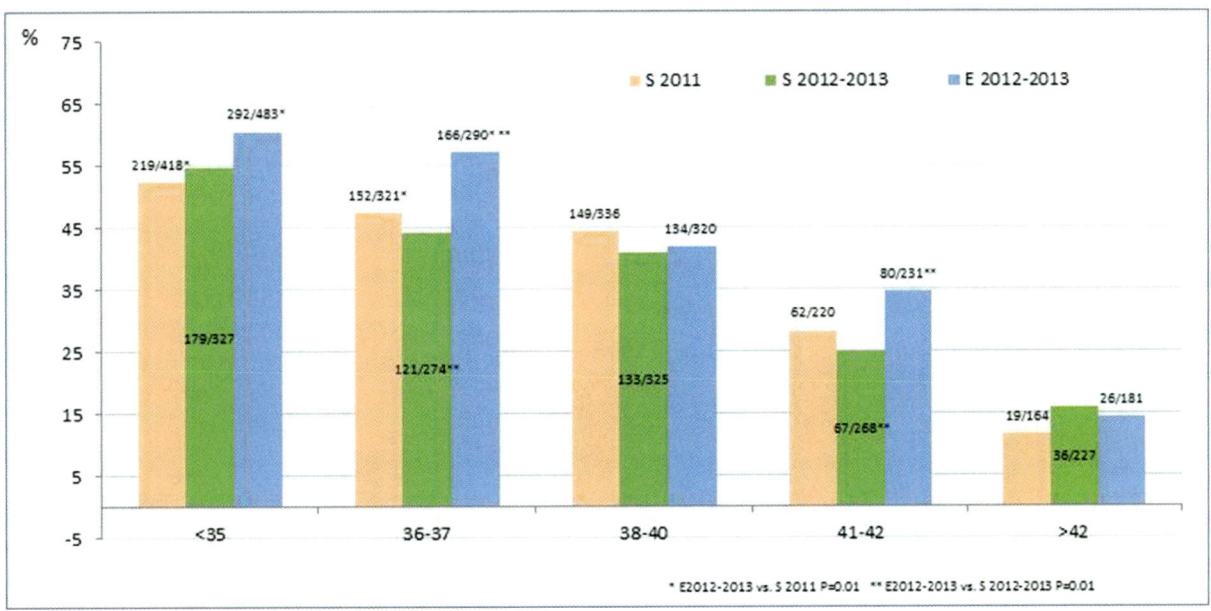

Figure 9.6 Clinical pregnancies per ET (sac seen) by maternal age in S 2011, S 2012–13 and E 2012–13 groups. Chi-square test statistic.

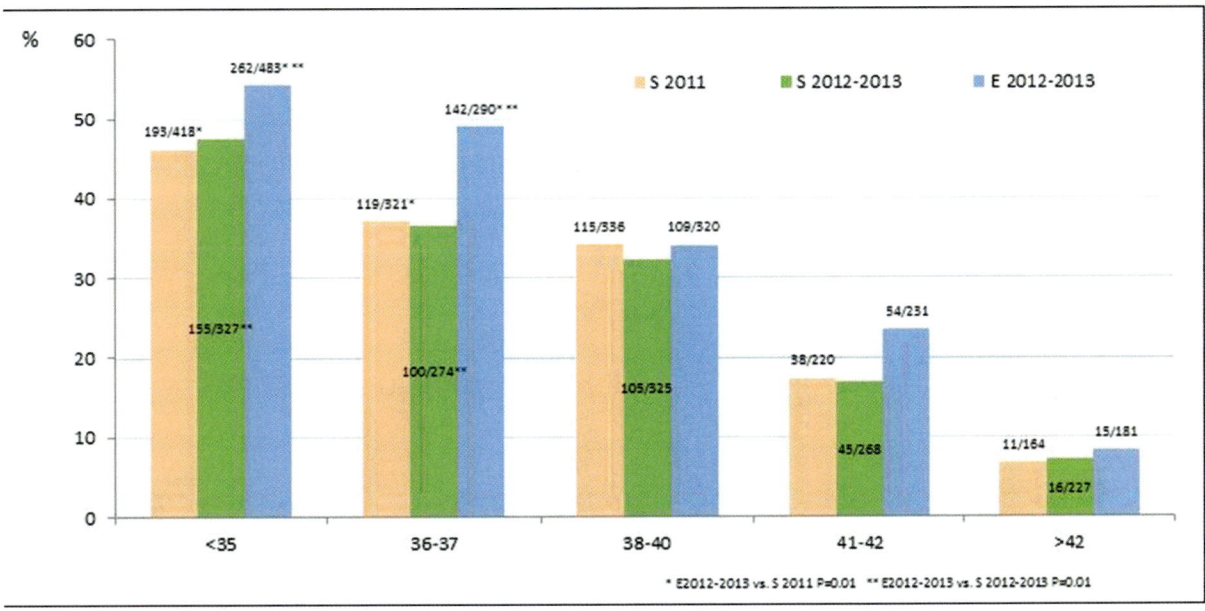

Figure 9.7 Ongoing pregnancies per ET (LB/ET) by maternal age in S 2011, S 2012–13 and E 2012–13 groups. Chi-square test statistic.

single-blastocyst transfer. In the elegant study by Yang *et al.* [19] it was demonstrated that we need both technologies, and they can complement each other. This study also demonstrates higher clinical outcome using TLM (EmbryoScope) cultured embryos compared to the standard incubator, when euploid embryos were transferred. Nonetheless, more studies are needed to ascertain whether PGS truly improves pregnancy rates.

To further delineate the influence of the fertilization method on clinical outcome using standard or EmbryoScope®, we analyzed ICSI and insemination

151

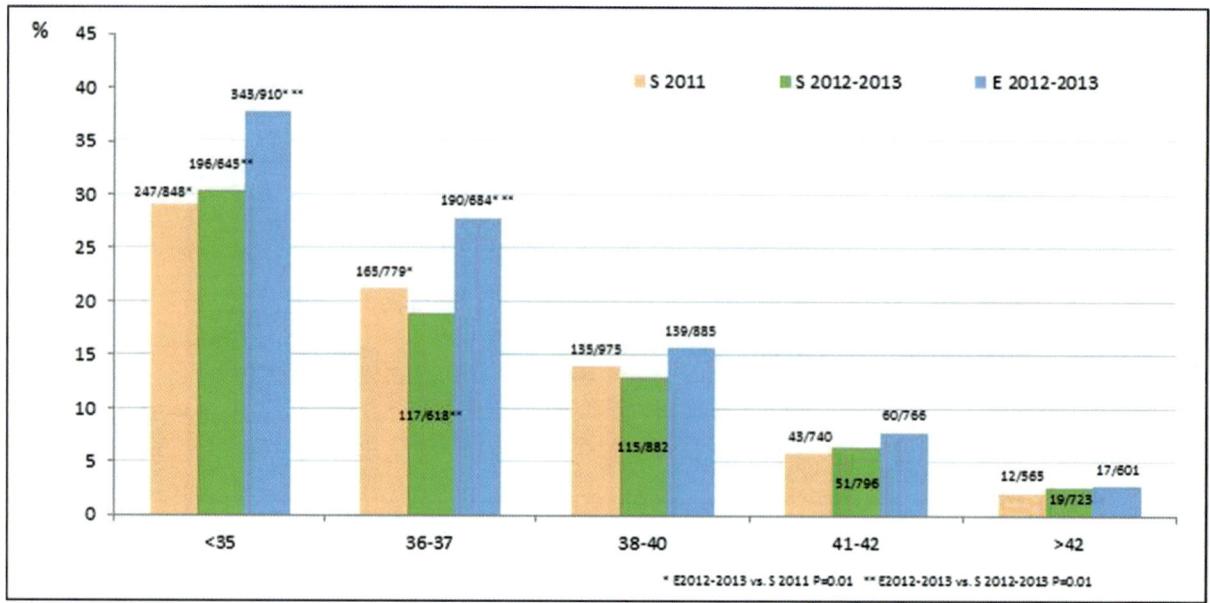

Figure 9.8 Implantation rate (number of LB/number of embryos ET) by maternal age in S 2011, S 2012–13 and E 2012–13 groups. Chi-square test statistic.

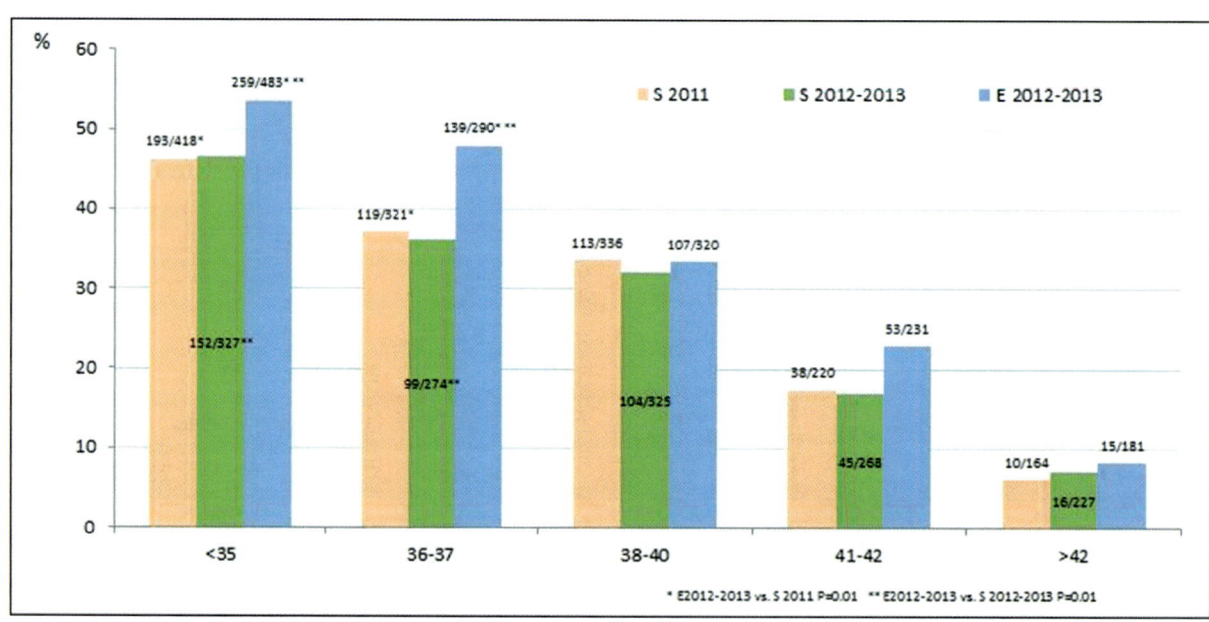

Figure 9.9 Live birth cycles (LB/ET) by maternal age in S 2011, S 2012–13 and E 2012–13 groups. Chi-square test statistic.

cycle data separately. Over 70% of our IVF cycles use ICSI. Clinical results are clearly improved when the EmbryoScope is utilized for ICSI cycles as evidenced by the increase of CP and OP (Figure 9.19 and Figure 9.20). Even more evident are the significantly higher implantation and live birth rates with the EmbryoScope compared to standard incubators (Figure 9.21 and Figure 9.22). More day 5 ET cycles were performed in the EmbryoScope group (18.7%) as compared to the standard group

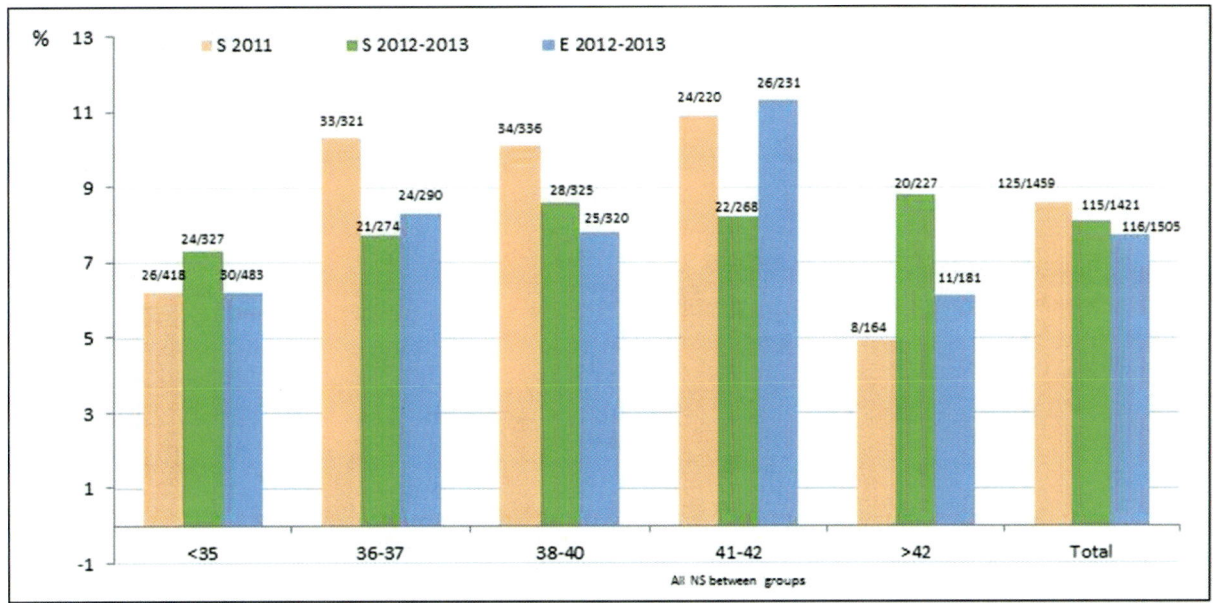

Figure 9.10 Miscarriage cycle rate by maternal age in S 2011, S 2012–13 and E 2012–13 groups. Chi-square test statistic.

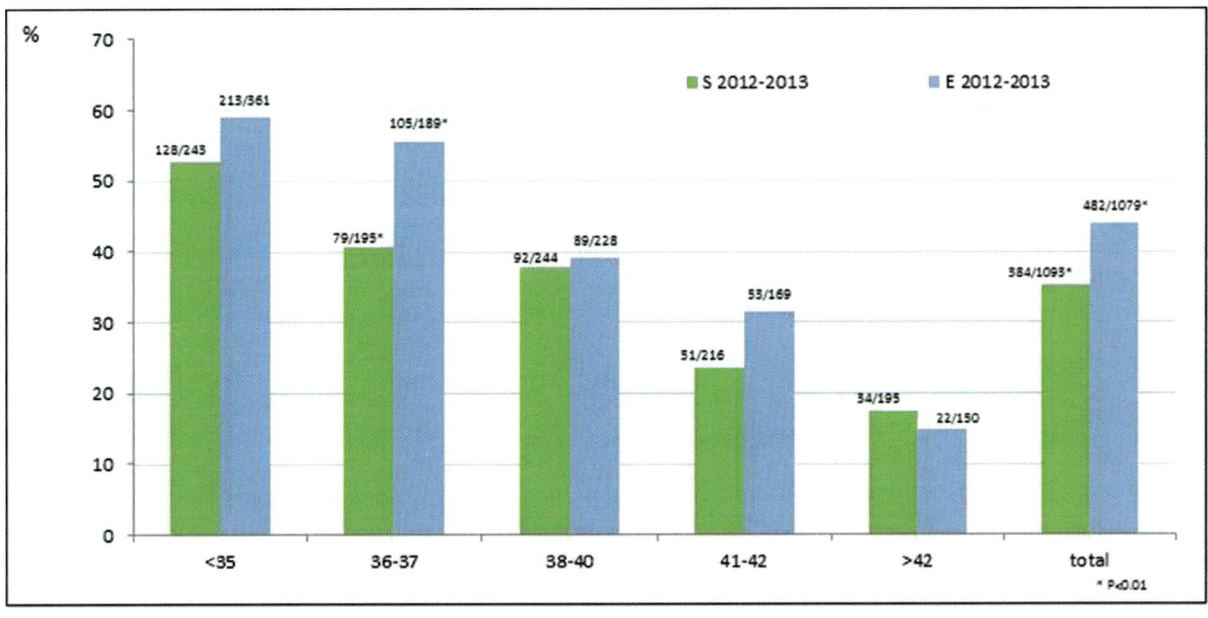

Figure 9.11 Clinical pregnancies per ET (sac seen) in ICSI cycles by maternal age: S 2012–13 vs. E 2012–13 groups. Chi-square test statistic.

(5.1%). Conventional insemination cycles represent approximately 30% of all IVF cycles at CRM but overall pregnancy and implantation results exceed those of ICSI cycles. This can be attributed to the potential deleterious effect of spermatozoal defects associated with the need to perform ICSI. We see improvements in pregnancy and implantation rates in insemination cycles with the Embryo-Scope use. This is particularly evidenced by the increased implantation rates (Figure 9.23 through Figure 9.26). Twenty percent of insemination EmbryoScope cycles were day 5 ET compared to

153

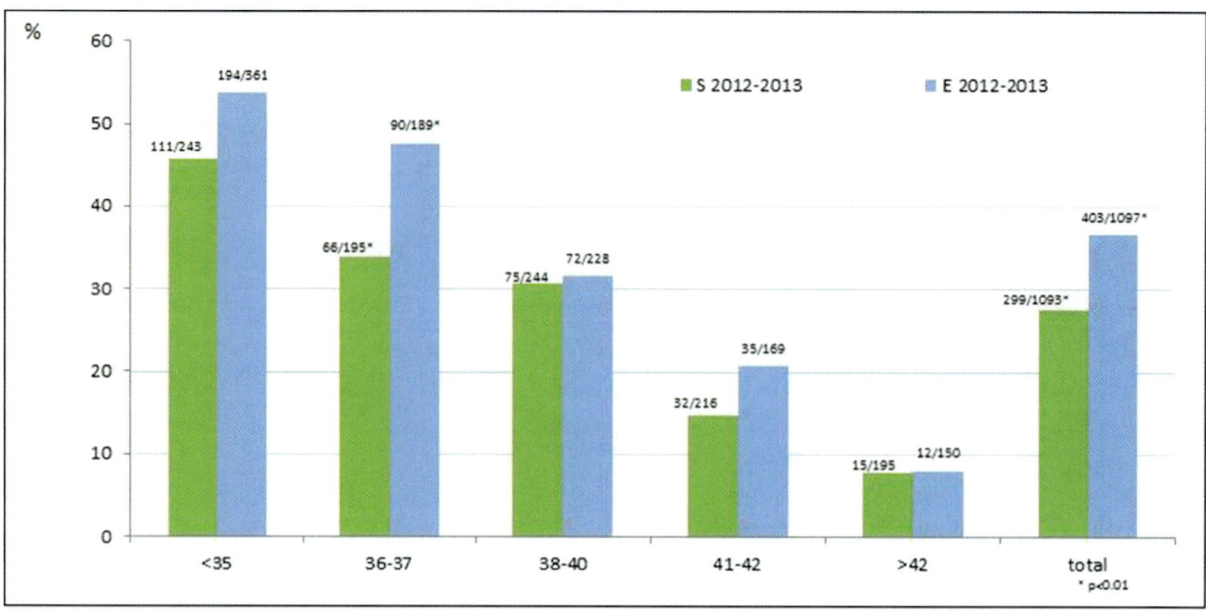

Figure 9.12 Ongoing pregnancies per ET (LB) in ICSI cycles by maternal age: S 2012–13 vs. E 2012–13 groups. Chi-square test statistic.

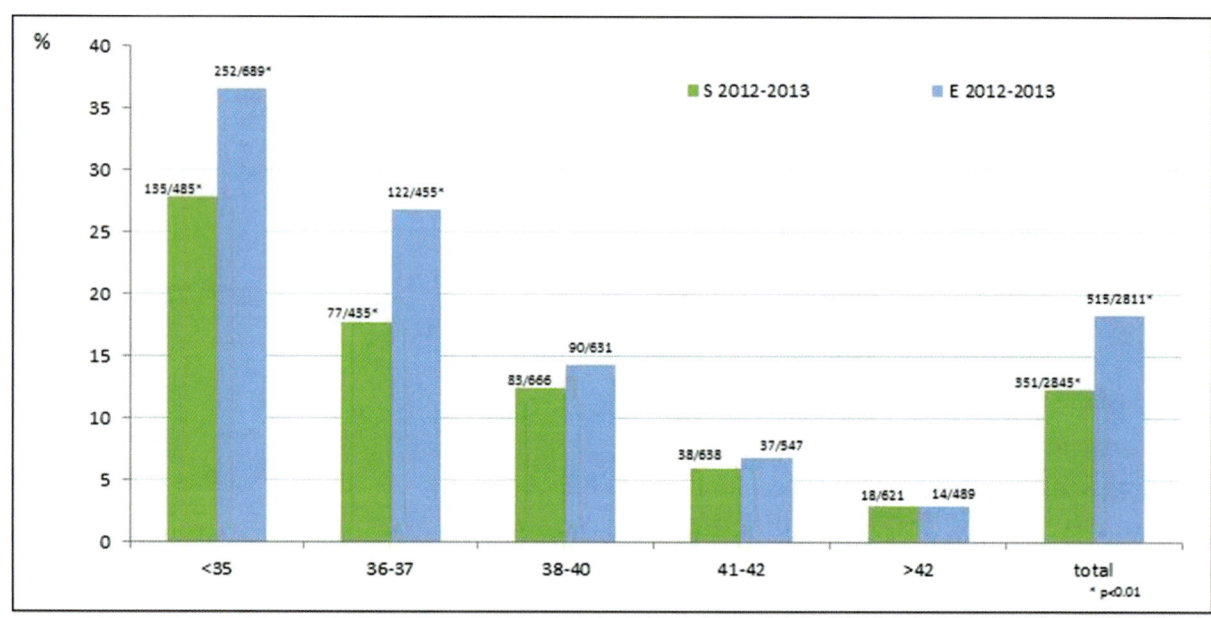

Figure 9.13 Implantation rate (number of LB/number of embryos ET) in ICSI cycles by maternal age: S 2012–13 vs. E 2012–13 groups. Chi-square test statistic.

10.7% in the standard group. An additional explanation for the clear benefit to ICSI cycles lies with the way that ICSI embryos are cultured. ICSI oocytes were put in the EmbryoScope immediately after injection, while inseminated zygotes were put in on the following day after fertilization check. The difference of one day might not seem significant, but it must be kept in mind that the pronuclear formation and the transition from meiosis to mitosis are probably one of the most important periods in zygote/embryo development. While conventional insemination is not possible using the EmbryoScope slides, we nonetheless

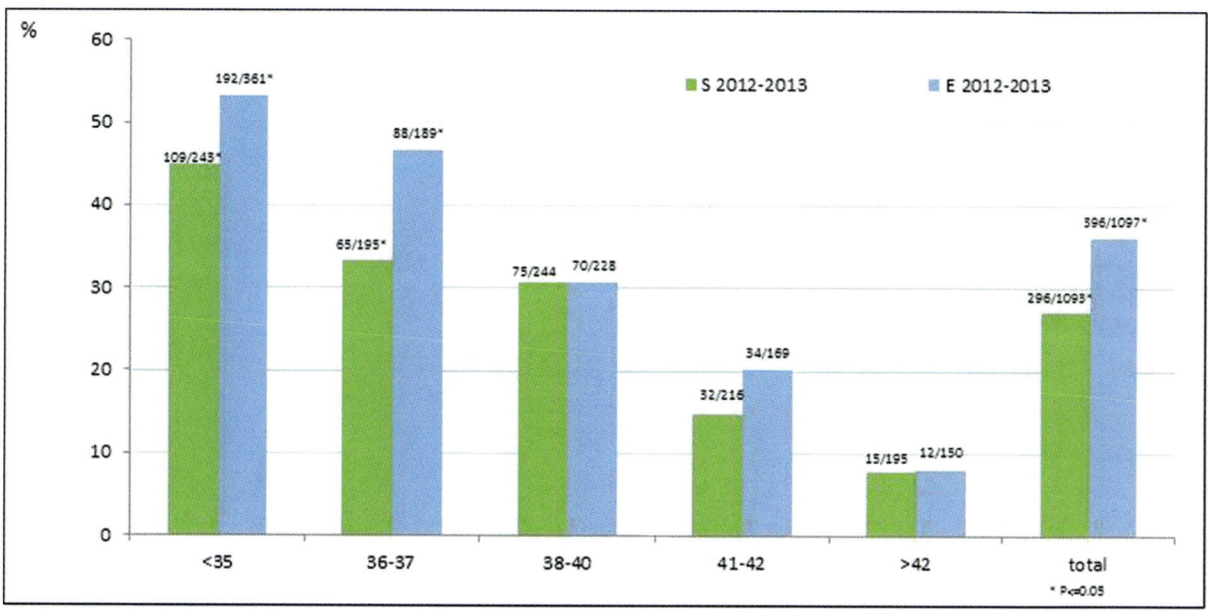

Figure 9.14 Live birth cycles (LB/ET) in ICSI cycles by maternal age: S 2012–13 vs. E 2012–13 groups. Chi-square test statistic.

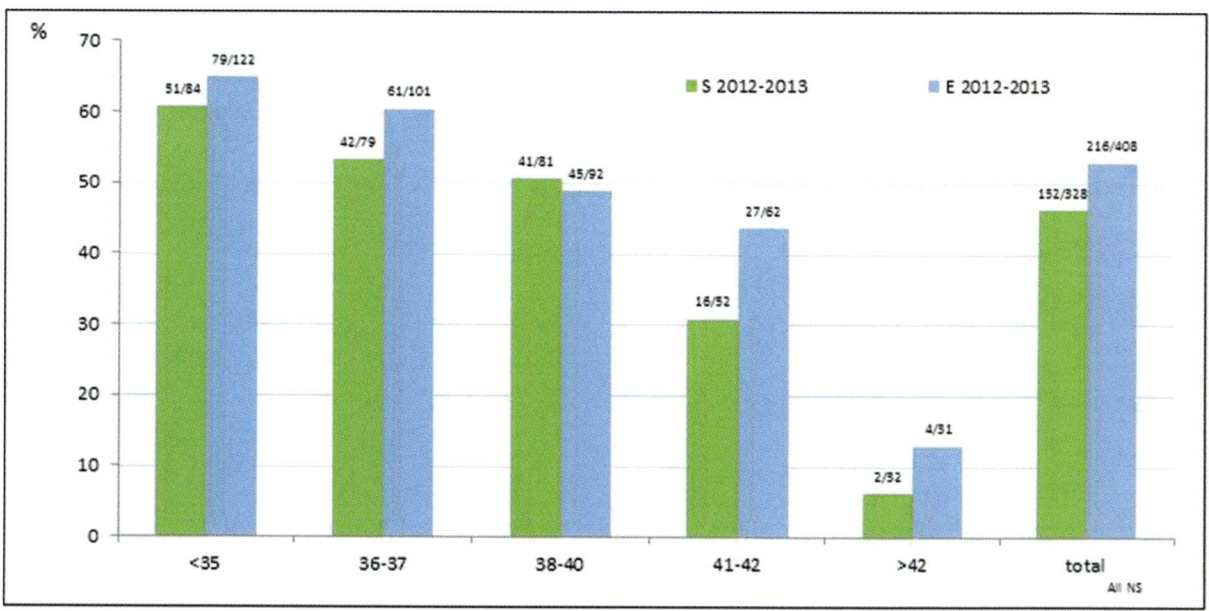

Figure 9.15 Clinical pregnancies per ET (sac seen) in inseminated cycles by maternal age: S 2012–13 vs. E 2012–13 groups. Chi-square test statistic.

believe that embryos can benefit if the insemination process occurs in triple gas standard incubators before transferring zygotes to TLM. Fertilization techniques have been reported to be related to the morphokinetic parameters, i.e., insemination of oocytes showed delayed starts compared to ICSI [10]. However, if start time is stratified, subsequent embryo development between ICSI and insemination embryos was similar in an oocyte donation study [10]. In that study, clinical outcome data were not different

155

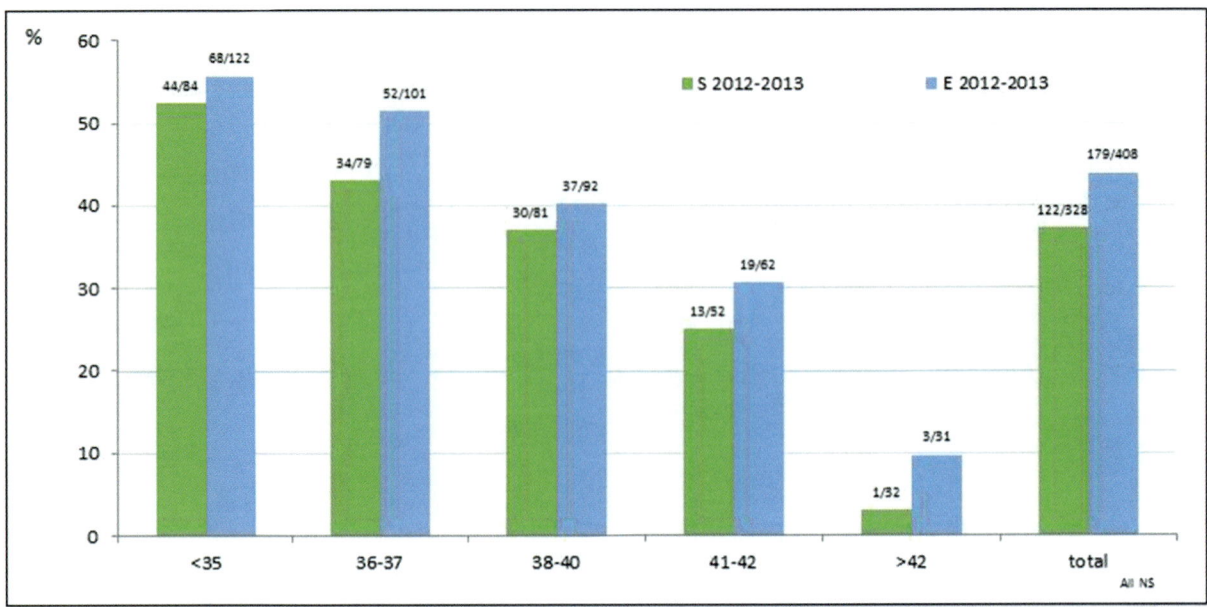

Figure 9.16 Ongoing pregnancies per ET (LB) in inseminated cycles by maternal age: S 2012–13 vs. E 2012–13 groups. Chi-square test statistic.

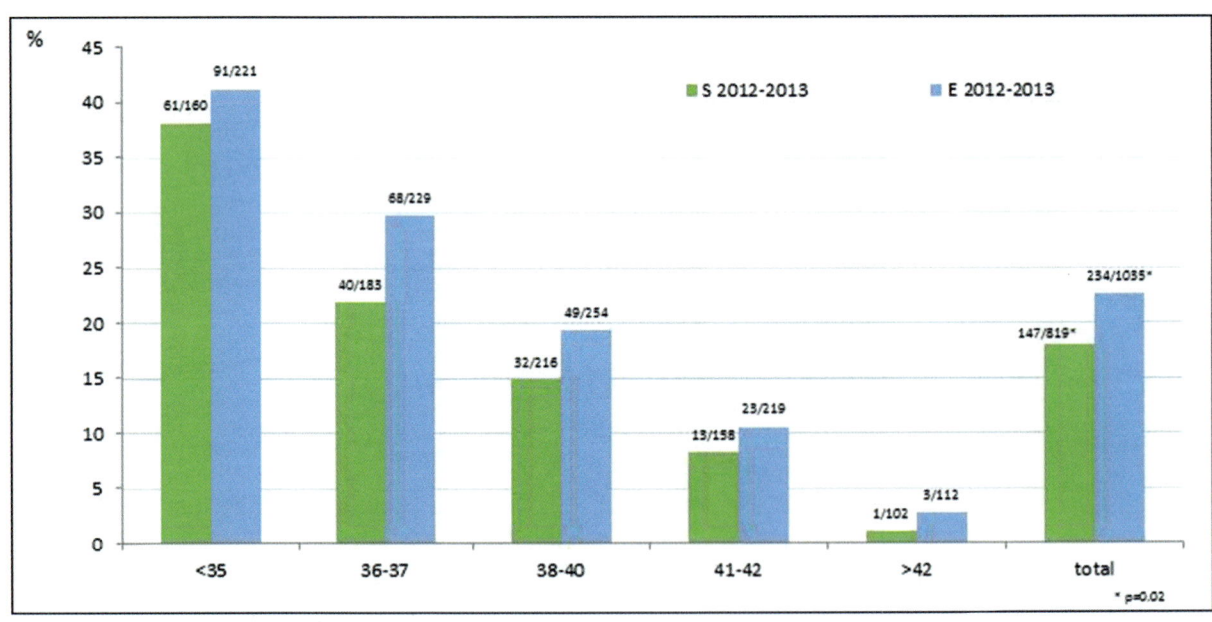

Figure 9.17 Implantation rate (number of LB/number of embryos ET) in inseminated cycles by maternal age: S 2012–13 vs. E 2012–13 groups. Chi-square test statistic.

between ICSI and insemination DER patients, as over 80% of patients had day 5 ET [10].

Our data using the EmbryoScope for fresh donor egg recipient (DER) transfers indicate an increased trend in clinical outcome and implantation rates (Figure 9.27). This is also exemplified in other studies using fresh or frozen donor oocytes [16, 18]. Even a slight percent increase in this patient population clearly

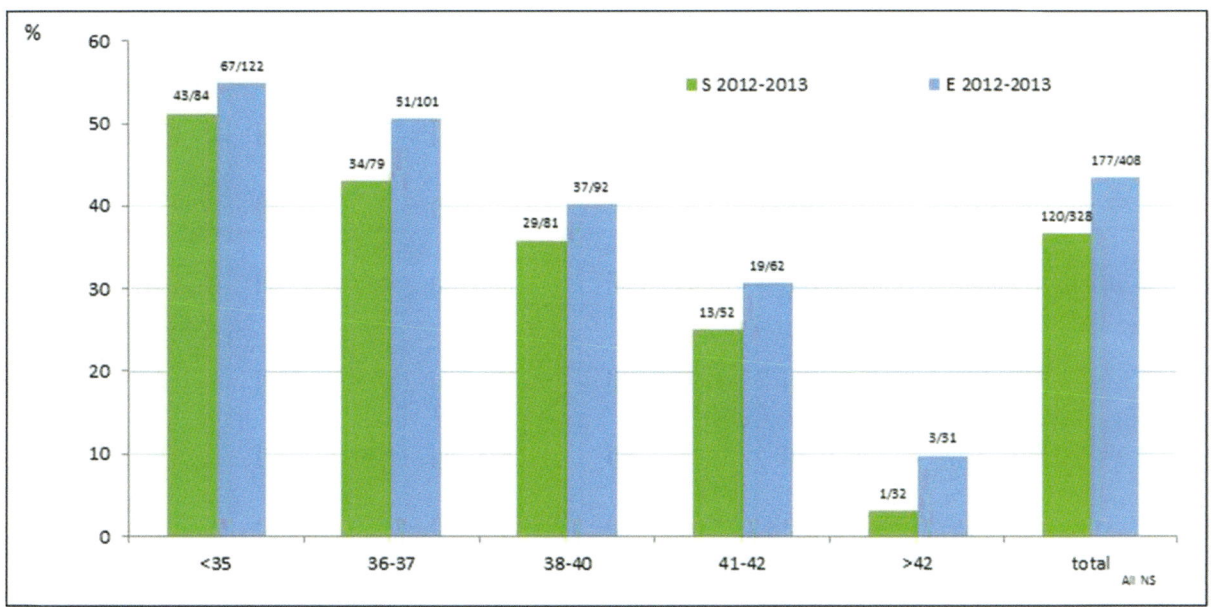

Figure 9.18 Live birth cycles (LB/ET) in inseminated cycles by maternal age: S 2012–13 vs. E 2012–13 groups. Chi-square test statistic.

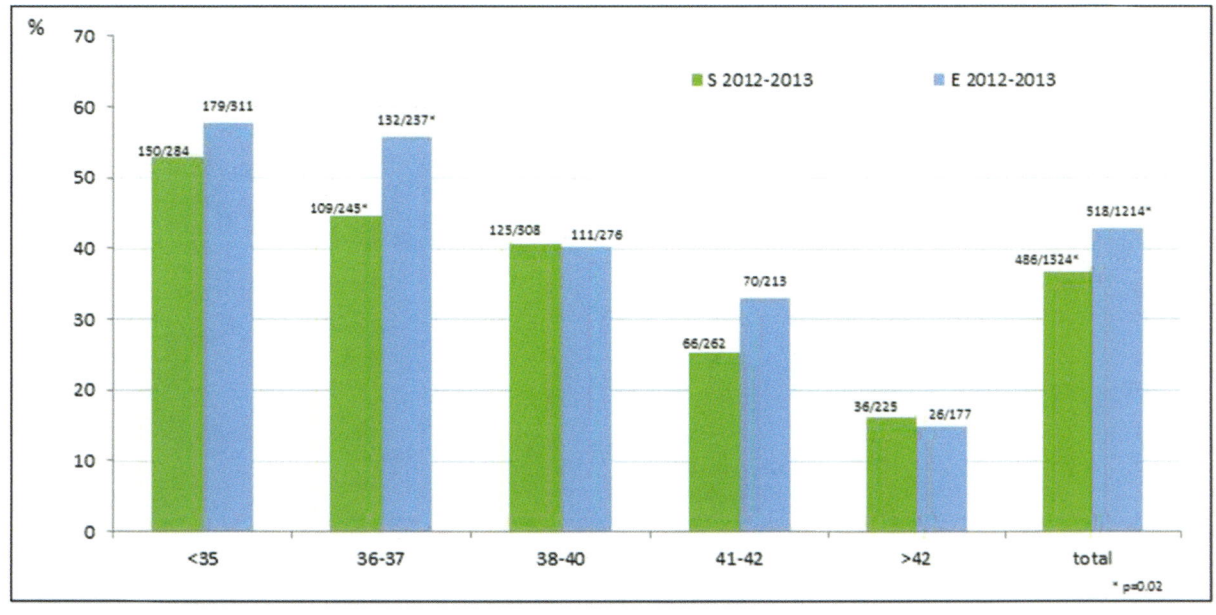

Figure 9.19 Clinical pregnancies per ET (sac seen) in day 3 ET cycles by maternal age: S 2012–13 vs. E 2012–13 groups. Chi-square test statistic.

suggests the benefits of TLM. A higher number of day 5 ET can be achieved even when the number of oocytes per recipient is fewer than ten (shared donor cycles). This is very important when using frozen oocytes and maximizing clinical outcome per DER patient.

As previously indicated, one of the secondary benefits of using the EmbryoScope is an increased number of frozen BL. We expect that by including the outcome of frozen BL cycles, the cumulative pregnancy rate per treatment cycle will also be increased.

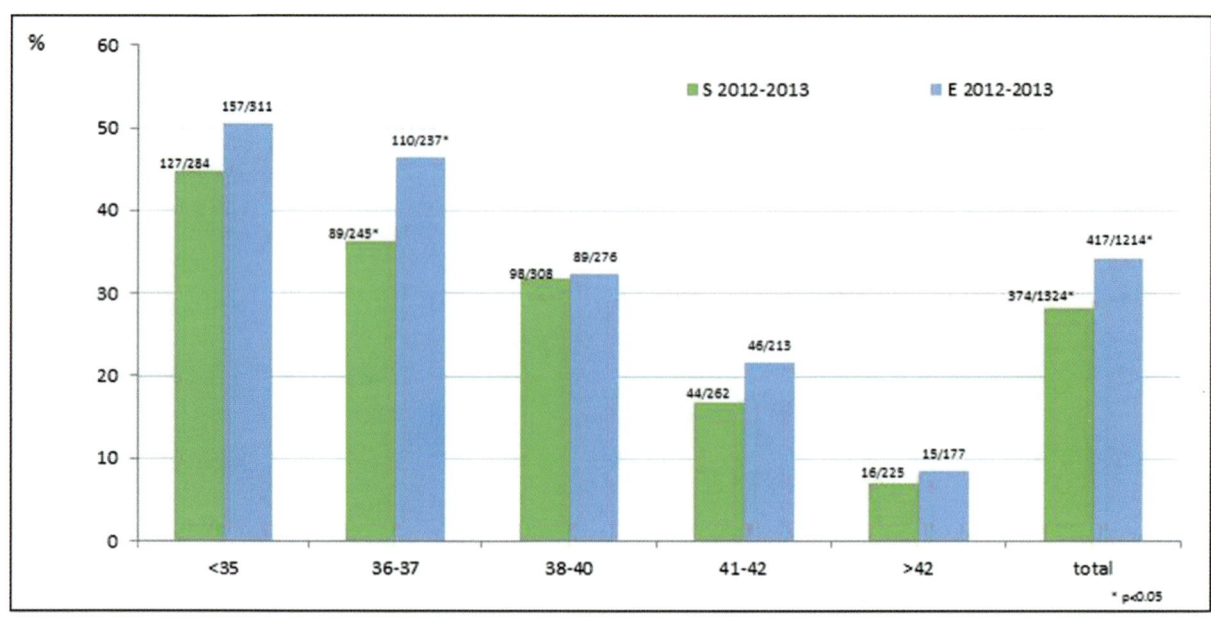

Figure 9.20 Ongoing pregnancies per ET (LB) in day 3 ET cycles by maternal age: S 2012–13 vs. E 2012–13 groups. Chi-square test statistic.

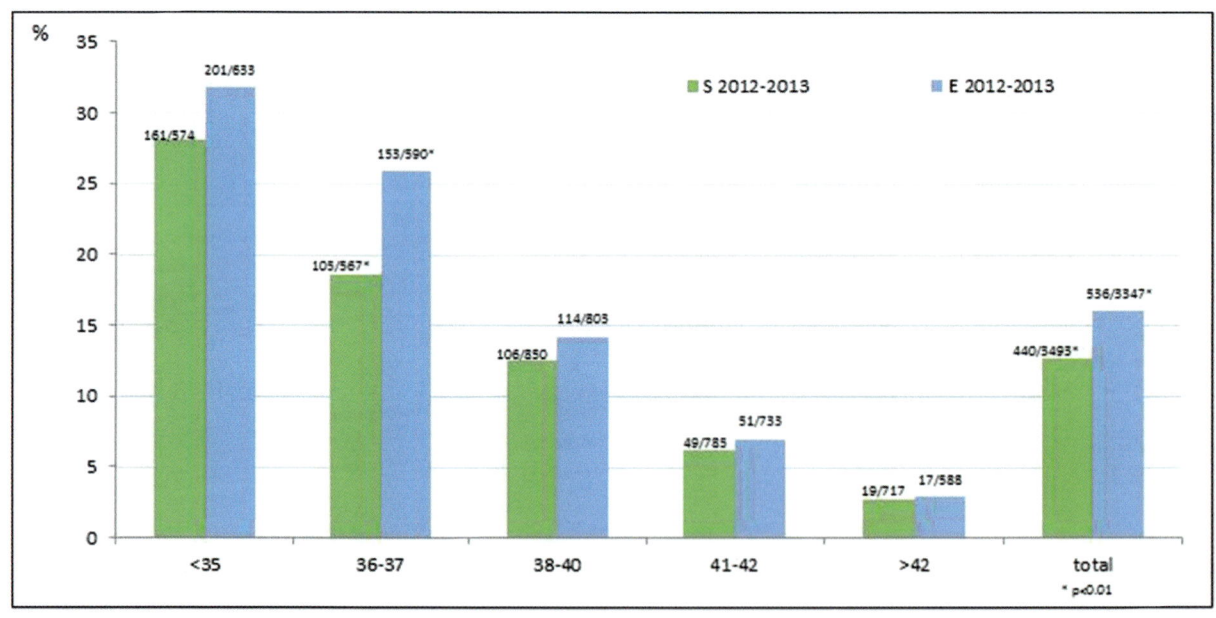

Figure 9.21 Implantation rate (number of LB/number of embryos ET) in day 3 ET cycles by maternal age: S 2012–13 vs. E 2012–13 groups. Chi-square test statistic.

9.4. Morphokinetics

The role of morphokinetics in embryo development and embryo selection will be discussed extensively in other chapters of this book. However, we would like to comment on our view of its role in clinical embryology. Notably, the current status of embryo morphokinetics and the implementation in choosing the best embryo by time-lapse has been superbly reviewed [20, 21, 22]. Morphokinetics is defined as the precise time points (t-times) in embryo cleavage

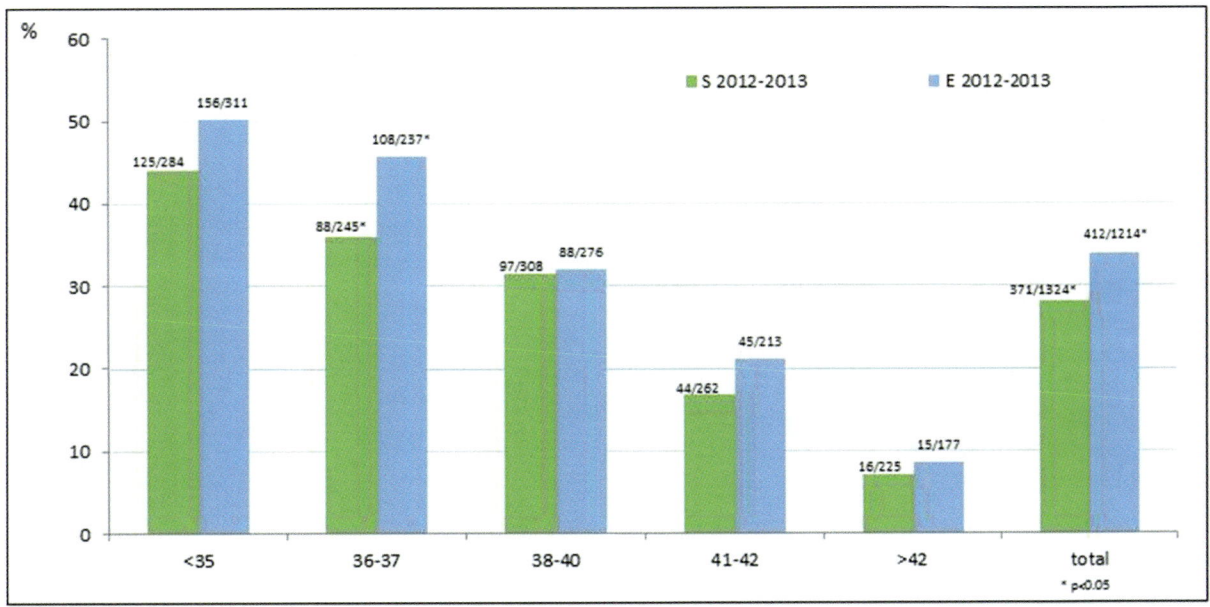

Figure 9.22 Live birth cycles (LB/ET) in day 3 ET cycles by maternal age: S 2012–13 vs. E 2012–13 groups. Chi-square test statistic.

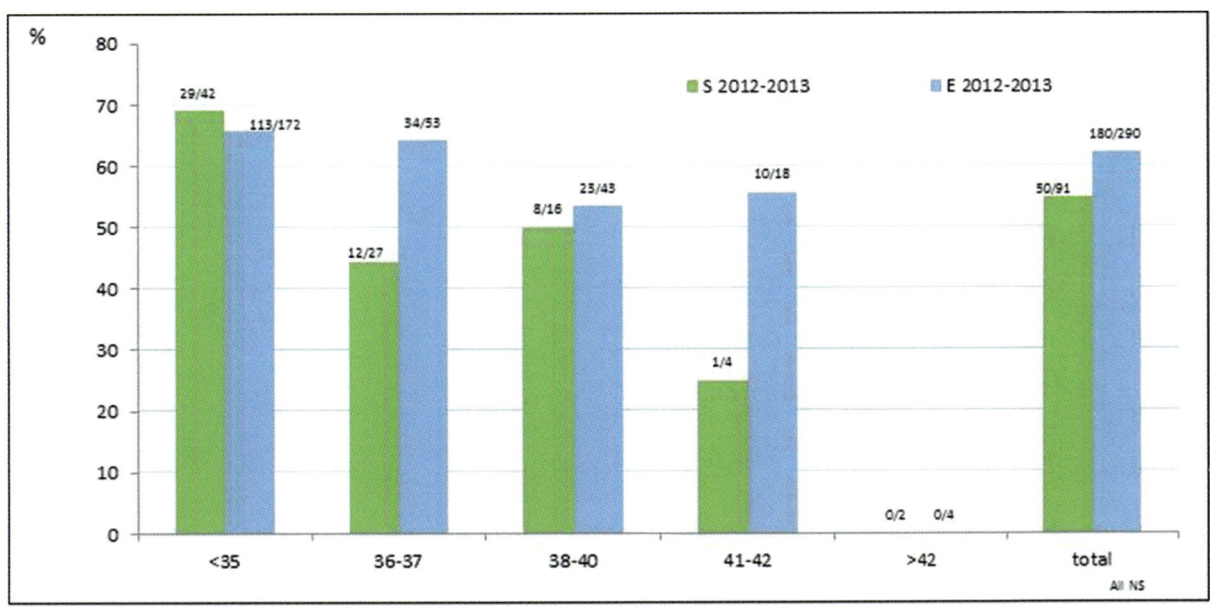

Figure 9.23 Clinical pregnancies per ET (sac seen) in day 5 ET cycles by maternal age: S 2012–13 vs. E 2012–13 groups. Chi-square test statistic.

events. It has been argued that t-times are correlated with embryo development, blastocyst development and subsequent implantation [23, 24, 25, 26, 27, 28, 29, 30]. Currently, morphokinetic parameters are evidently correlated with blastocyst development [20, 29, 30]. It is still an open question if morphokinetic parameters alone, combined within embryo selection models, can work universally for all patients and clinics. First, we have to demonstrate that TLM culture with reduced oxygen is clearly beneficial for embryo development as it appears to be in our hands. Second, we have to identify which patients, or all, will

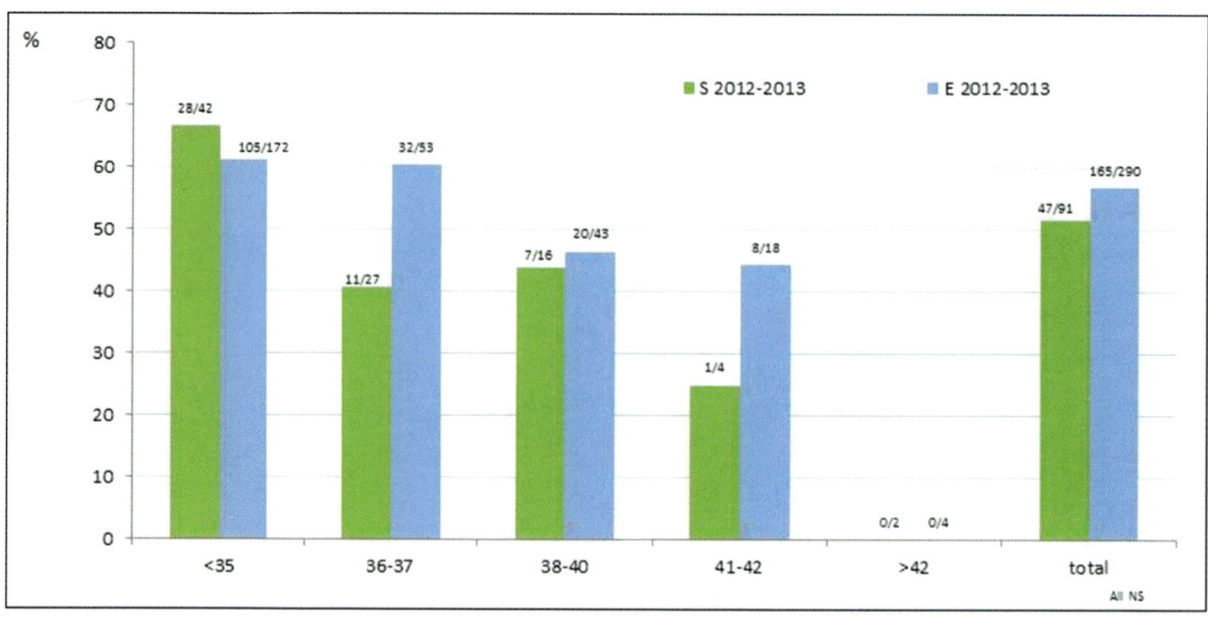

Figure 9.24 Ongoing pregnancies per ET (LB) in day 5 ET cycles by maternal age: S 2012–13 vs. E 2012–13 groups. Chi-square test statistic.

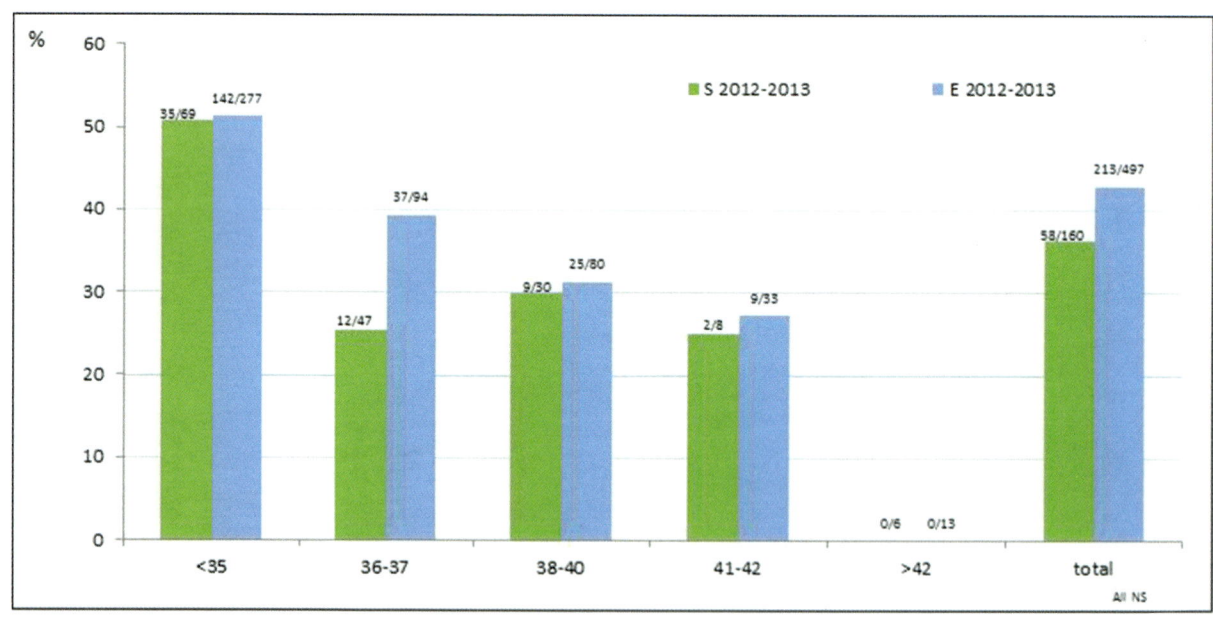

Figure 9.25 Implantation rate (number of LB/number of embryos ET) in day 5 ET cycles by maternal age: S 2012–13 vs. E 2012–13 groups. Chi-square test statistic.

benefit from TLM culture (our current study). This will require larger, preferably prospective studies [31]. Third, we need to identify normal and abnormal embryo development and apply these to embryo selection models. Fourth, embryo selection models should be developed on large data sets and prospectively evaluated.

Why do some investigators believe that embryo selection models cannot work universally? Arguably, there are different patient populations, stimulation,

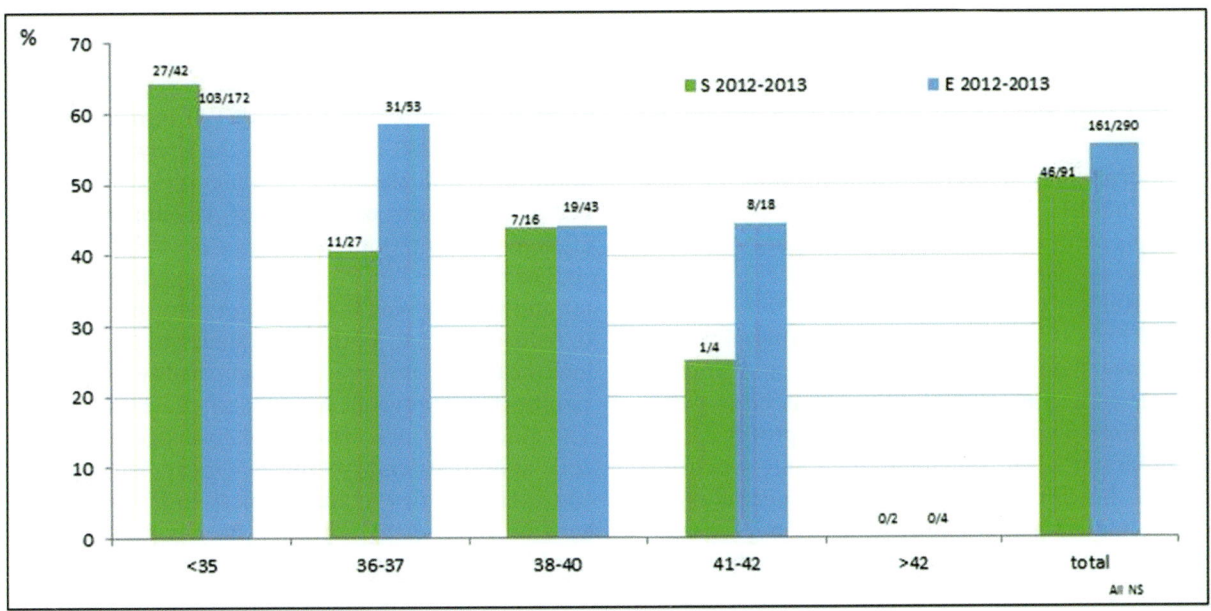

Figure 9.26 Live birth cycles (LB/ET) in day 5 ET cycles by maternal age: S 2012–13 vs. E 2012–13 groups. Chi-square test statistic.

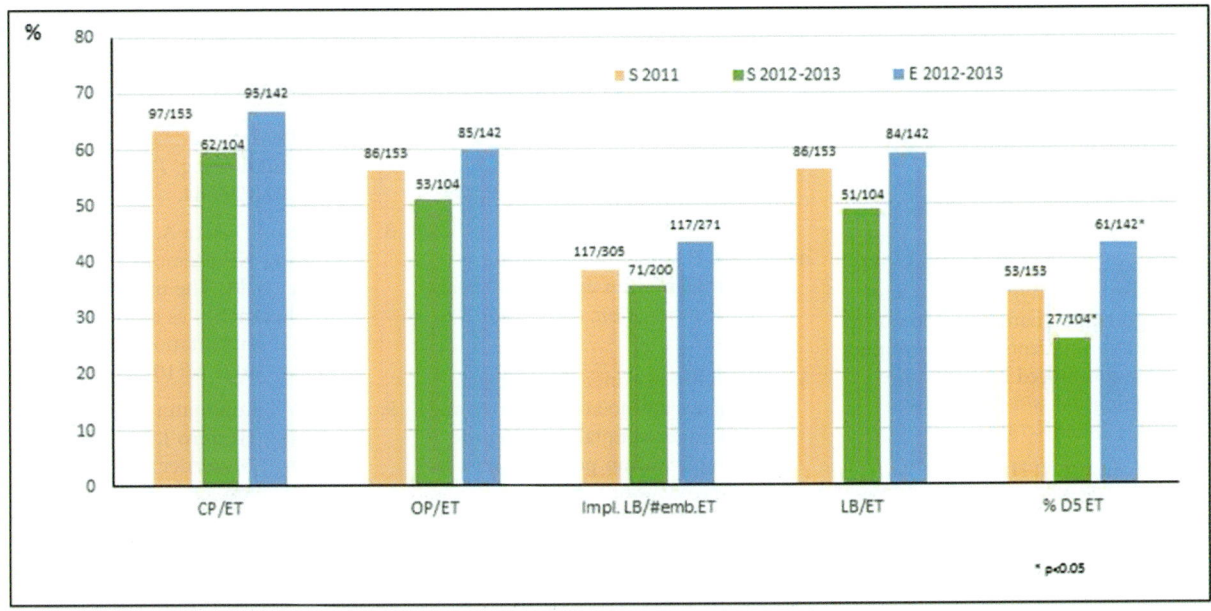

Figure 9.27 Overall clinical outcome data for clinical pregnancies (CP, sac seen), ongoing pregnancies (OP, per LB), implantation (number of LB per ET) and live birth cycles (LB/ET) between 2011 S, 2012–13 S and 2012–13 E donor egg recipients (DER) cycles. Chi-square test statistic.

culture conditions, and oxygen concentrations between centers and all these factors can influence morphokinetic parameters [32, 33, 34, 35, 36]. One of the key factors in IVF embryo culture is how to choose culture media and to identify interactions of media components as well as other subtle influences on other culture conditions (i.e., gases). Key factors in the culture media and culture environment have

161

profound effects on mammalian embryo development and morphokinetics, and can serve as a sensitive marker for human embryo culture [37]. Another possibility is the way selection models are developed. Knowledge of standard embryology and morphology should be evaluated and included in the models [38]. Statistical analyses and data modeling will allow different laboratory data sets to show similar embryo characteristics and selection criteria. Current embryo selection models should only be used as guidelines, although each laboratory should evaluate their data and apply it to their own clinical application. It is clear that a single time-point in embryo development is not as important as an outcome factor but synchronicity of the cleavages and/or developmental events and combined t-times can be used in prediction models. Embryo selection models, if usable, need to indicate by day 3 which embryos have the highest developmental and implantation potential beyond standard morphology. For example, on day 3 you will not chose a four- to five-cell embryo over an eight-cell embryo. However, models should tell us in the cohort of good embryos on day 3 which one is the best embryo and has the greatest chance to result in a viable pregnancy. In addition, it should help us to reduce the number of embryos for ET and increase SET. Our current effort is focused on comparing a large set of morphokinetic data (over 20,000 embryos) to the known implantation data with the hope of generating models for better prediction of blastocyst development and embryo implantation.

For now, the EmbryoScope is an excellent incubator and we have seen the benefit of culturing embryos in reduced oxygen. With its use, embryos grow better, look healthier, and produce high quality blastocysts resulting in an overall increase in clinical outcome. The amount of new data and knowledge of normal and abnormal embryo development is inspiring and excites us in everyday clinical embryology. It is hard to imagine that anyone who has utilized TLM would ever revert to traditional embryo incubation and evaluation techniques.

Acknowledgments

We thank Naomi Lourie for data acquisition and Abigail Nixon for editorial assistance. We thank all the CRM staff and Embryology laboratory.

References

1. Cruz M, Gadea B, Garrido N, Pedersen KS, Martínez M, Pérez-Cano I, Muñoz M, Meseguer M. Embryo quality, blastocyst and ongoing pregnancy rates in oocyte donation patients whose embryos were monitored by time-lapse imaging. *J Assist Reprod Genet.* 2011; 28: 569–73.

2. Kirkegaard K, Hindkjaer JJ, Grøndahl ML, Kesmodel US, Ingerslev HJ. A randomized clinical trial comparing embryo culture in a conventional incubator with a time-lapse incubator. *J Assist Reprod Genet.* 2012; 29(6): 565–72.

3. Park H, Bergh C, Selleskog U, Thurin-Kjellberg A, Lundin K. No benefit of culturing embryos in a closed system compared with a conventional incubator in terms of number of good quality embryos: results from an RCT. *Hum Reprod.* 2014. Epub Nov 28.

4. Leese HJ. Metabolism of the preimplantation embryo: 40 years on. *Reproduction.* 2012; 143: 417–27.

5. Wale PL, Gardner DK. Oxygen regulates amino acid turnover and carbohydrate uptake during the preimplantation period of mouse embryo development. *Biol Reprod.* 2012; 87: 1–8.

6. Bontekoe S, Mantikou E, van Wely M, Seshadri S, Repping S, Mastenbroek S. Low oxygen concentrations for embryo culture in assisted reproductive technologies. *Cochrane Database Syst Rev.* 2012; 11: 7.

7. Kirkegaard K1, Hindkjaer JJ, Ingerslev HJ. Effect of oxygen concentration on human embryo development evaluated by time-lapse monitoring. *Fertil Steril.* 2013; 99: 738–44.

8. Herrero J, Tejera A, Hilligsøe KM, Ramsing NB, Remohí J, Meseguer M. The use of morphokinetics as a predictor of embryo implantation. *Hum Reprod.* 2011; 26(10): 2658–71.

9. Rubio I, Kuhlmann R, Agerholm I, Kirk J, Herrero J, Escribá MJ, Bellver J, Meseguer M. Limited implantation success of direct-cleaved human zygotes: a time-lapse study. *Fertil Steril.* 2012; 98 (6): 1458–63.

10. Cruz M, Garrido N, Gadea B, Muñoz M, Pérez-Cano I, Meseguer M. Oocyte insemination techniques are related to alterations of embryo developmental timing in an oocyte donation model. *Reprod Biomed Online.* 2013; 27: 367–75.

11. Athayde Wirka K, Chen AA, Conaghan J, Ivani K, Gvakharia M, Behr B, Suraj V, Tan L, Shen S. Atypical embryo phenotypes identified by time-lapse microscopy: high prevalence and association with embryo development. *Fertil Steril*. 2014; 101(6): 1637–48.

12. Chavez SL, Loewke KE, Han J, Moussavi F, Colls P, Munne S, Behr B, Reijo Pera RA. Dynamic blastomere behaviour reflects human embryo ploidy by the four-cell stage. *Nat Commun*. 2012; 3: 1251.

13. Chatzimeletiou K, Morrison EE, Prapas N, Prapas Y, Handyside AH. Spindle abnormalities in normally developing and arrested human preimplantation embryos in vitro identified by confocal laser scanning microscopy. *Hum Reprod*. 2005; 20: 672–82.

14. Golubovsky MD. Postzygotic diploidization of triploids as a source of unusual cases of mosaicism, chimerism and twinning. *Hum Reprod*. 2003; 18: 236–42.

15. Veeck L, Zaninovic N. *An Atlas of Human Blastocysts*. New York: Parthenon Publishing, 2003.

16. Meseguer M, Rubio I, Cruz M, Basile N, Marcos J, Requena A. Embryo incubation and selection in a time-lapse monitoring system improves pregnancy outcome compared with a standard incubator: a retrospective cohort study. *Fertil Steril*. 2012; 98: 1481–9.

17. Herrero J, Meseguer M. Selection of high potential embryos using time-lapse imaging: the era of morphokinetics. *Fertil Steril*. 2013; 99: 1030–4.

18. Rubio I, Galán A, Larreategui Z, Ayerdi F, Bellver J, Herrero J, Meseguer M. Clinical validation of embryo culture and selection by morphokinetic analysis: a randomized, controlled trial of the EmbryoScope. *Fertil Steril*. 2014; 102: 1287–94.

19. Yang Z, Zhang J, Salem SA, Liu X, Kuang Y, Salem RD, Liu J. Selection of competent blastocysts for transfer by combining time-lapse monitoring and array CGH testing for patients undergoing preimplantation genetic screening: a prospective study with sibling oocytes. *BMC Med Genomics*. 2014; 7: 38.

20. Kirkegaard K, Ahlström A, Ingerslev HJ, Hardarson T. Choosing the best embryo by time lapse versus standard morphology. *Fertil Steril*. 2014. Epub Dec 16.

21. Chen AA, Tan L, Suraj V, Reijo Pera R, Shen S. Biomarkers identified with time-lapse imaging: discovery, validation, and practical application. *Fertil Steril*. 2013; 99(4):1035–43.

22. Aparicio B, Cruz M, Meseguer M. Is morphokinetic analysis the answer? *Reprod Biomed Online*. 2013; 27(6): 654–63.

23. Wong CC, Loewke KE, Bossert NL, Behr B, De Jonge CJ, Baer TM, Reijo Pera RA. Non-invasive imaging of human embryos before embryonic genome activation predicts development to the blastocyst stage. *Nat Biotechnol*. 2010; 28: 1115–21.

24. Meseguer M, Herrero J, Tejera A, Hilligsøe KM, Ramsing NB, Remohí J. The use of morphokinetics as a predictor of embryo implantation. *Hum Reprod*. 2011; 26: 2658–71.

25. Herrero J, Tejera A, Albert C, Vidal C, de los Santos MJ, Meseguer M. A time to look back: analysis of morphokinetic characteristics of human embryo development. *Fertil Steril*. 2013; 100(6):1602–9.

26. Basile N, Vime P, Florensa M, Aparicio Ruiz B, García Velasco JA, Remohí J, Meseguer M. The use of morphokinetics as a predictor of implantation: a multicentric study to define and validate an algorithm for embryo selection. *Hum Reprod*. 2014. Epub Dec 19.

27. Conaghan J, Chen AA, Willman SP, Ivani K, Chenette PE, Boostanfar R, Baker VL, Adamson GD, Abusief ME, Gvakharia M, Loewke KE, Shen S. Improving embryo selection using a computer-automated time-lapse image analysis test plus day 3 morphology: results from a prospective multicenter trial. *Fertil Steril*. 2013; 100(2): 412–19.

28. VerMilyea MD, Tan L, Anthony JT, Conaghan J, Ivani K, Gvakharia M, Boostanfar R, Baker VL, Suraj V, Chen AA, Mainigi M, Coutifaris C, Shen S. Computer-automated time-lapse analysis results correlate with embryo implantation and clinical pregnancy: A blinded, multi-centre study. *Reprod Biomed Online*. 2014; 29(6): 729–36.

29. Cruz M, Garrido N, Herrero J, Pérez-Cano I, Muñoz M, Meseguer M. Timing of cell division in human cleavage-stage embryos is linked with blastocyst formation and quality. *Reprod Biomed Online*. 2012; 25(4): 371–81.

30. Dal Canto M, Coticchio G, Mignini Renzini M, De Ponti E, Novara PV, Brambillasca F, Comi R, Fadini R. Cleavage kinetics analysis of human embryos predicts development to blastocyst and implantation. *Reprod Biomed Online*. 2012; 25(5): 474–80.

31. Kaser DJ, Racowsky C. Clinical outcomes following selection of human preimplantation embryos with time-lapse monitoring: a systematic review. *Hum Reprod Update*. 2014; 20: 617–31.

32. Bellver J, Mifsud A, Grau N, Privitera L, Meseguer M. Similar morphokinetic patterns in embryos derived from obese and normoweight infertile women: a

time-lapse study. *Hum Reprod.* 2013; 28(3): 794–800.

33. Leary C, Leese HJ, Sturmey RG. Human embryos from overweight and obese women display phenotypic and metabolic abnormalities. *Hum Reprod.* 2015; 30(1): 122–32.

34. Fréour T, Dessolle L, Lammers J, Lattes S, Barrière P. Comparison of embryo morphokinetics after in vitro fertilization-intracytoplasmic sperm injection in smoking and nonsmoking women. *Fertil Steril.* 2013; 99(7): 1944–50.

35. Muñoz M, Cruz M, Humaidan P, Garrido N, Pérez-Cano I, Meseguer M. The type of GnRH analogue used during controlled ovarian stimulation influences early embryo developmental kinetics: a time-lapse study. *Eur J Obstet Gynecol Reprod Biol.* 2013; 168(2): 167–72.

36. Wissing ML, Bjerge MR, Olesen AI, Hoest T, Mikkelsen AL. Impact of PCOS on early embryo cleavage kinetics. *Reprod Biomed Online.* 2014; 28(4): 508–14.

37. Morbeck DE, Krisher RL, Herrick JR, Baumann NA, Matern D, Moyer T. Composition of commercial media used for human embryo culture. *Fertil Steril.* 2014; 102: 759–66.

38. Desai N, Ploskonka S, Goodman LR, Austin C, Goldberg J, Falcone T. Analysis of embryo morphokinetics, multinucleation and cleavage anomalies using continuous time-lapse monitoring in blastocyst transfer cycles. *Reprod Biol Endocrinol.* 2014; 12: 54.

Unusual phenomena in embryonic development

Giovanni Coticchio, Mariabeatrice Dal Canto, Mario Mignini Renzini, and Rubens Fadini

10.1. Introduction

Time-lapse microscopy (TLM) is a conceptually simple photographic technique associated with live cell microscopy, by which micrographs are acquired at short time intervals and viewed sequentially at accelerated pace in order to observe phenomena and processes that otherwise would go unnoticed. In association with phase contrast optics, TLM is particularly suited to observe dynamically and non-invasively the development of the pre-implantation embryo in vitro, whose ability to establish a viable pregnancy is reflected in the achievement of specific developmental stages at precise time intervals. Although it took, unbelievably, several decades to adapt TLM to the technology of the human IVF lab, in recent years continued observation of human embryos generated in vitro has become an everyday reality [1]. Events such as embryo cleavage, fragmentation, multinucleation, compaction, blastocoel formation, and blastocyst development can be described temporally with unprecedented precision and therefore interpreted under a new light. However, the debate on the ability of TLM to improve the efficiency of assisted reproduction technology (ART) treatments remains currently unresolved [1, 2]. Instead, and perhaps more intriguingly, the contribution of this technology to a better understanding of the biology and pathology of early human development has appeared clear from the very beginning. TLM has in fact revealed an unsuspected secret life of the embryo manifested in atypical modalities of polar body II (PBII) extrusion, deregulation of pronuclear formation and breakdown, abnormal manifestations of cleavage, and other uncommon phenomena. Access to such evidence has the potential to extend our appreciation of embryo development, very much like mutant organisms offer

clues for understanding the biological rules that govern the life cycles of wild-type organisms.

Providing scope for this chapter on the editorial project "time-lapse microscopy in in-vitro fertilization," we aim to illustrate examples of a previously unrecognized phenomenology of human embryo development. Whenever possible, we will propose possible interpretations of such phenomena and discuss the potential implications. To adhere to a temporal and logical order, the reported cases are organized in four main headings, i.e., anomalies of fertilization, cleavage, compaction, and blastocyst formation.

10.2. Anomalies of fertilization

10.2.1. "Tentative" emission of polar body III

Polar body II (PBII) extrusion and chromatid segregation are highly regulated processes. In particular, the orientation of the cleavage plane of PBII emission is known to be determined by the position of the MII spindle, while the mechanics of PBII extrusion rely on a specialized actomyosin domain that forms in the oocyte cortex as an effect of a RAN GTPase signal emitted by the adjacent maternal meiotic chromatin [3–5]. PBII extrusion occurs approximately 3 hours after sperm penetration (or ICSI), but its temporal range extends over many hours (1 to more than 10). Application of TLM to ART has revealed several anomalies of PB emission occurring during fertilization of the human oocyte. Intriguingly, in rare cases extrusion of the PBII is accompanied by the apparent emission of an additional PB, for convenience referred to as polar body III (PBIII) in this chapter. However, such behaviors is only transitory and is concluded with the

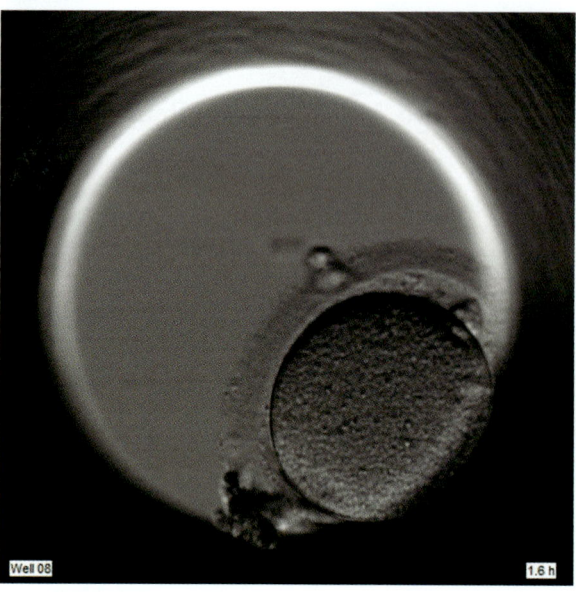

Well 08   1.6 h

Video 10.1 Transient emission of PBIII. Extrusion of the PBII is narrowly followed at the 9 o'clock position by the apparent emission of an additional PB, for convenience referred to as PBIII. Persistence of the extra PB is however transitory and concluded with resorption back into the oocyte mass.

resorption of the small mass of cytoplasm initially emerged from the cortex. In Video 10.1, the PBIII can be transiently observed in proximity of the 9 o'clock position for approximately 2 hours, from 6.8 to 8.8 hours post ICSI (p.i.), with a time frame coincident with that of PBII extrusion, which appears at the 1 o'clock position. It is difficult to establish with certainty the origin of this phenomenon. However, the sequence shown in Video 10.1 prompts some interesting reflections: (a) the PBII emerges near the PBI. Therefore, the pronucleus (PN) that appears in the immediate vicinity of the PBII at 8.4 hours p.i. is very likely to be of maternal origin (as it should be according to theory); (b) at 8.6 hours p.i., the appearance of the second PN, presumably of paternal origin, occurs immediately after and in close proximity to the site of formation of the PBIII. Together, these observations may suggest the intriguing hypothesis that the transitory phenomenon of PBIII emission is triggered by the sperm chromatin. Normally, after oocyte–sperm fusion or ICSI, the sperm chromatin undergoes a phase of decondensation required for the replacement of protamines with histones, followed by the process of pronuclear formation. However, in

a significant proportion of cases of failed fertilization, instead of organizing the formation of the male pronucleus, paternal chromosomes can elicit tubulin polymerization and give rise to a recognizable "paternal spindle" that co-exists with the maternal MII spindle in the same oocyte [6]. Although atypical for the paternal chromatin at fertilization, this phenomenon should not be entirely unexpected, because metaphase-stage chromosomes do have microtubule-organizing ability. In cases in which formation of the paternal spindle occurs eccentrically near the oocyte surface, we hypothesize that the paternal chromatin may induce in the adjacent oocyte cortex the formation of an actomyosin structure that normally forms only in the vicinity of the maternal MII spindle in preparation for the extrusion of PBII. Such a supplementary actomyosin domain may support an illegitimate attempt of extrusion of a third PB that however for some reason is not brought to completion and shortly afterwards is followed by resorption of the extruded cytoplasmic mass and re-establishment of the normal path of fertilization, as suggested by the sequence of **Video 10.1**. Downstream developmental consequences of this phenomenon are not known, although possible clinical implications cannot be ruled out. Regardless, it is interesting to appreciate the existence of previously unknown anomalies of fertilization and speculate on possible interactions between chromatin, microtubules, and oocyte cortex.

10.2.2. Late formation of a third pronucleus

Formation of one or more supernumerary pronuclei is a relatively frequent phenomenon, occurring in 4–5% of cases. In standard IVF, supernumerary pronuclei may derive from the concomitant penetration of two or more sperm. More generally, in both IVF and ICSI cases it is more likely that formation of a third pronucleus occurs as a consequence of an inability of the MII oocyte to eliminate one set of chromatids through the extrusion of the PBII. Usually, in 3PN embryos the three pronuclei start to form within a narrow time interval, if not at the same time. Video 10.2 illustrates a case of tripronuclear fertilization, but deserves particular attention for the timing of pronuclear formation. Extrusion of a relatively large PBII occurs at 3 hours p.i. Two pronuclei, presumably maternal and paternal, become visible as early as 5–6 hours p.i. in the oocyte center. Afterwards, fertilization progresses without anomalies for many hours. However, starting from

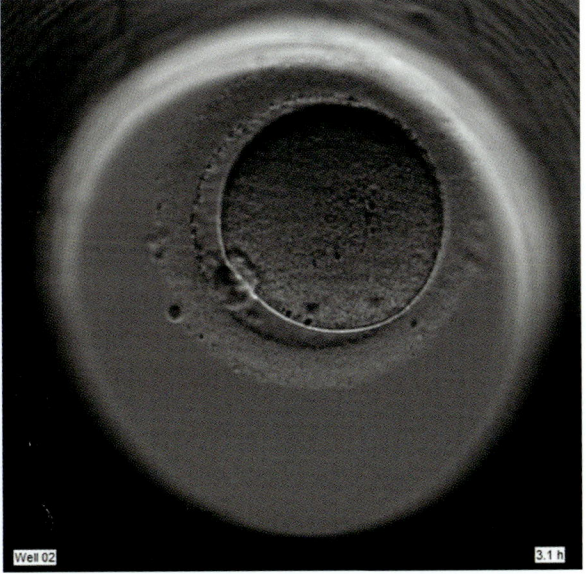

Well 02   3.1 h

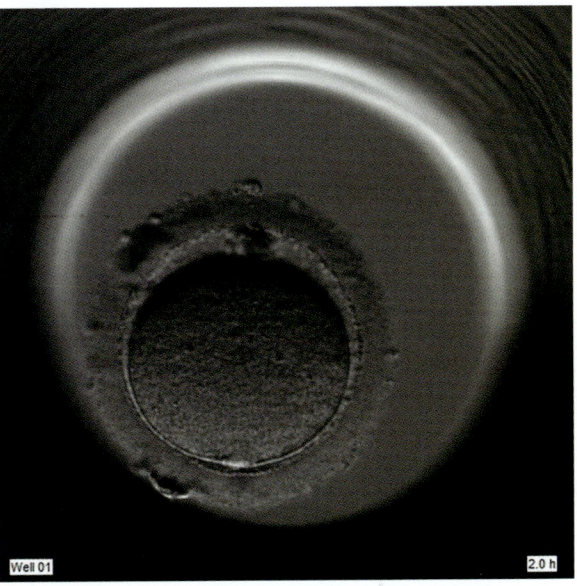

Well 01   2.0 h

Video 10.2 Late formation of a third supernumerary pronucleus. During fertilization, two pronuclei become progressively visible in the oocyte center. Unexpectedly, starting from 15 hours p.i. the PBII is taken up back into the oocyte mass and a third pronucleus forms in the same area of the oocyte cortex.

Video 10.3 PN breakdown and reappearance. Rarely, PN breakdown may be not followed by the first cleavage, but instead by reformation of two, sometimes three, pronuclei. In this example, PN breakdown occurs at 25.2 hours p.i. and two newly formed pronuclei are visible as early as 27.4 hours p.i..

15 hours p.i. the PBII is "reabsorbed" by the oocyte and a third pronucleus forms in the same area of the oocyte cortex. Over the following 2 hours the third pronucleus is repositioned towards the oocyte centre and by 18 hours p.i. it is found in close proximity to the other original two pronuclei. What is striking in this example is not the origin of the third pronucleus, clearly associable to an extra set of maternal chromatids, but rather its morphokinetics. In particular, PBII reabsorption at a late stage of fertilization is indicative of dysregulation of the forces that control the mechanical tension of the oocyte cortex. From a practical standpoint it is very important to notice that a single assessment of the oocyte pronuclear status, even when performed at approximately 16 hours p.i., may not be sufficient to rule out aberrant fertilization.

10.2.3. Pronuclear breakdown and reappearance

Formation of the male and female pronuclei is one of the most distinct phenomena of fertilization.

The former derives from the sperm chromatin, after replacement of protamine with histones. The latter forms after elimination of one set of maternal chromatids, through the extrusion of the PBII. With appropriate optics, PNs are discernible as early as 4–5 hours p.i. [7] and usually remain visible until 23–24 hours p.i. [8]. Their disappearance is followed by a phase in which the fertilized egg is engaged in the delicate process of assembly of the zygotic genome, signified by the alignment of maternal and paternal chromosomes at the equator of the mitotic spindle. Afterwards, the first embryonic cleavage occurs, within only 2 hours from PN breakdown.

Rarely and rather atypically, PN breakdown may be not followed by the first cleavage, but instead by reformation of two, sometimes three, pronuclei. This event is shown in Video 10.3, in which PN breakdown occurs at 25.2 hours p.i. and two newly formed pronuclei are visible as early as 27.4 hours p.i., in the absence of cleavage. At 43.8 hours p.i., pronuclear breakdown occurs again and this time is followed by a highly irregular cleavage by which the zygote divides into three daughter cells.

While pronuclear breakdown and reformation is a rare phenomenon, it seems to follow some rules, because it can be observed in several oocytes of the same cycle and is often associated with direct cleavage into three blastomeres. Its pathological significance is therefore evident, also considering the large delay in the start of embryo cleavage. However, the biological bases of this manifestation appear obscure. From a strictly morphokinetic standpoint, the case described in Video 10.3 is rather conventional for the stages preceding pronuclear reformation, with one exception though. In fact, it may be noticed that nucleolar precursor bodies are in very low numbers in both pronuclei. Whether this morphological trait indicates a condition of insufficient chromatin condensation at a phase shortly preceding PN breakdown is possible, but it remains speculative. Also, the speed by which pronuclei reform is suggestive that in the reported case the cell cycle does not proceed to the mitotic metaphase after the first mitotic breakdown. It is possible, therefore, that an immature chromatin, i.e., relatively decondensed as suggested by the paucity of nucleolar precursor bodies, emits a signal that causes a fall in M-phase promoting factor (MPF, the major cell cycle regulator) and a consequent reassembly of the pronuclear envelopes [9].

10.2.4. Pronuclear breakdown asynchrony

Pronuclear formation occurs approximately at the same time for both pronuclei, although preliminary observations suggested the possibility that the male pronucleus may form slightly earlier [7]. On the other hand, there is little doubt that breakdown should be strictly synchronous and swift in normal fertilization. Video 10.4 describes an infrequent case of asynchrony of pronuclear breakdown. The image sequence shows an oocyte undergoing fertilization. The process unfolds apparently with normal morphokinetics. PBII extrusion occurs at 3 hours p.i., while both pronuclei appear starting from approximately 7 hours p.i. and remain juxtaposed in a pericentric position until 23–24 hours p.i. However, something unexpected occurs in the following hours. At 26 hours p.i. one of the two pronuclei disappears and the zygote remains in a mono-pronuclear phase until 32 hours p.i. From 28 to 30 hours p.i. another atypical phenomenon occurs, consisting of a wave of distortion of the cortex running clockwise, as suggested by the displacement of the PBI.

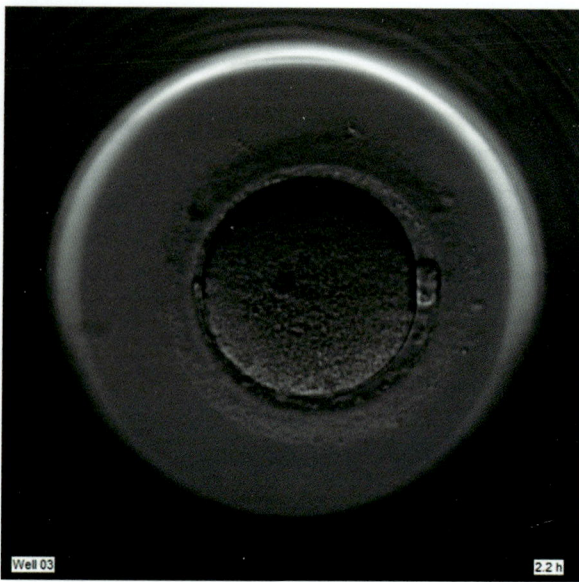

Well 03   2.2 h

Video 10.4 Asynchronous PN breakdown and direct cleavage into three blastomeres. The sequence shows multiple developmental anomalies. At 26 hours p.i. one of the two pronuclei disappears and the zygote remains in a mono-pronuclear phase until 32 hours p.i. From 28 to 30 hours p.i., a wave of distortion affects the cortex running clockwise. Finally, pronuclear breakdown asynchrony is followed by direct cleavage into three blastomeres.

Under normal conditions, the envelopes of both pronuclei break up as an effect of phosphorylation of the nuclear lamina resulting from an increased activity of MPF. Therefore, the same cell cycle signal controls breakdown of both pronuclear envelopes, which should occur synchronously. Why in the described case the two pronuclei behave so differently is unknown. It is only possible to speculate that anomalies in the chromatin of one of the two pronuclei influences the ability of the nuclear envelope to break up. In such a case, pronuclear asynchrony could represent a sign of profound developmental incompetence of at least half of the zygotic genome. Interesting in this respect is the fact that in the case described in Video 10.4 pronuclear breakdown asynchrony is followed by direct cleavage into three blastomeres, whose pathological significance is described elsewhere in this chapter. Additional morphokinetic observations not discussed in this chapter confirm that asynchrony of PN breakdown is also

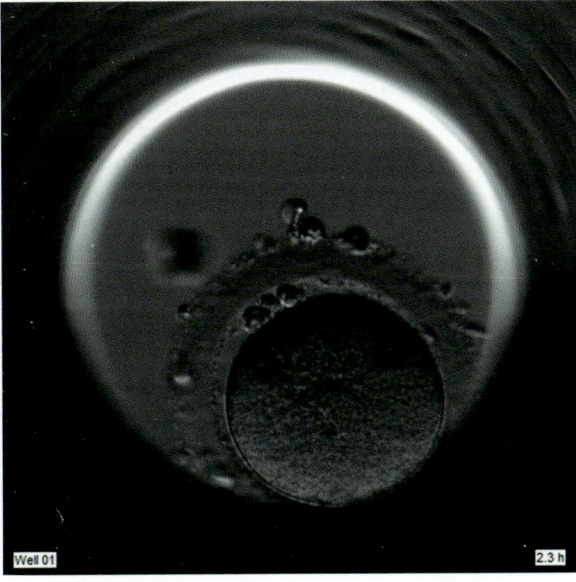

Well 01   2.3 h

Video 10.5 Cleavage in the absence of pronuclear formation. A previously microinjected oocyte undergoes a long series of contractions of small amplitude occurring approximately between 2 and 22 hours p.i. Cleavage into two blastomeres occurs by 27 hours p.i. without prior pronuclear formation.

closely associated with major cleavage anomalies (massive fragmentation and early arrest) and developmental failure.

10.2.5. First cleavage in the absence of pronuclear formation

TLM has unveiled phenomena whose existence was not suspected or even imagined only a few years ago. The occurrence of apparent cleavage into two blastomeres of a previously microinjected oocyte in the absence of pronuclear formation is typical in this respect. Video 10.5 shows a sequence in which an oocyte previously subject to ICSI does not show any morphokinetic activity for many hours, except a long series of contractions of small amplitude occurring approximately between 2 and 22 hours p.i. Rather incredibly, this oocytes cleaves into two blastomeres by 27 hours p.i. without prior pronuclear formation. It is worth noting that in the reported example pronuclei were not visible even after acquisition of the same sequence from

different focal planes (videos not shown), suggesting that the observed phenomenon is not a technical artefact. It is rather unlikely that an oocyte can skip a fundamental phase of fertilization, such as PN formation, and then cleave. Instead, it is possible that what it may be initially interpreted as cleavage of a fertilized oocyte into two blastomeres is in fact a different phenomenon. An alternative explanation may be elaborated starting from the hypothesis that in the described case the meiotic spindle was not positioned cortically but, rather, centrally. Indeed, spindles are occasionally observed in the depths of the ooplasm. In such a case, late activation (spontaneous or induced by microinjection) of the oocyte may have caused extrusion of a "giant" PBII whose size was comparable to that of the rest of the oocyte. This scenario is not unrealistic because the orientation of the cleavage plane is determined by the position of the spindle in both meiosis and mitosis. It is plausible, therefore, that completion of meiosis may have occurred according to a geometry that is typical of the first mitosis, involving a centrally positioned meiotic spindle and a cleavage plane passing from the oocyte center. Clearly, it is impossible to establish whether such a hypothesis explains what happened in reality. However, it is interesting to note that the event that divided the oocyte into two masses of comparable volume was not preceded by the extrusion of a PBII, suggesting that meiosis was not completed at the time of cleavage.

10.3. Anomalies of cleavage

10.3.1. Direct division of the fertilized egg or a blastomere into three cells

At around 25–27 hours p.i., the fertilized egg usually cleaves into two blastomeres. This initial cleavage event ideally occurs smoothly through a symmetric, well-defined, and swift division of the egg mass. However, anomalies at this stage are not infrequent: for example, the two daughter cells may differ significantly in size, cytokinesis may be accompanied by substantial formation of anucleated fragments, and the cleavage furrow may create pronounced instability of the blastomere cortex. These and other manifestations are believed to be indicative of a compromised developmental ability of the embryo [10, 11]. TLM has allowed identification of a particular

169

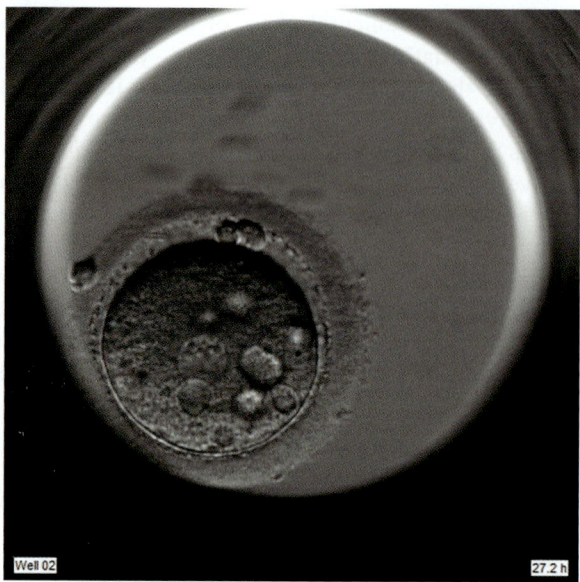

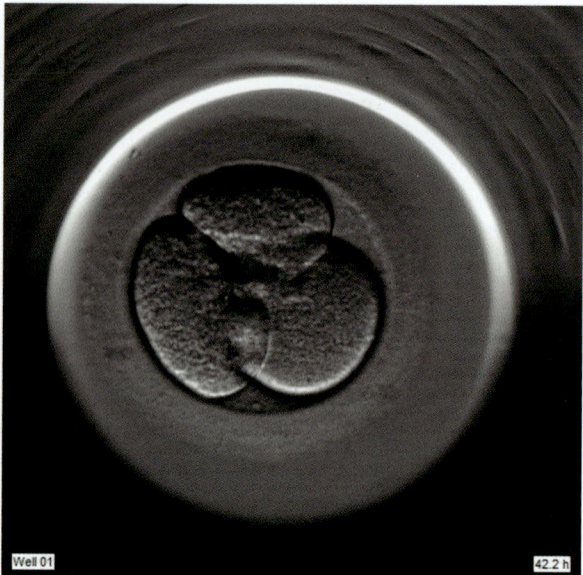

Video 10.6 Direct cleavage into five blastomeres. A fertilized egg, displaying multiple fertilization anomalies (asynchronous PN breakdown, PN displacement, and cortical instability), cleaves into five blastomeres during a single division event.

Video 10.7 Direct cleavage of a blastomere into three daughter blastomeres. Example describing a single blastomere of a three-cell stage embryo cleaving into three distinct cytoplasmic masses. It is plausible that the impact on embryo viability of direct cleavage into three cells becomes gradually smaller when occurring at progressively later stages of development.

typology of such cleavage anomalies. Video 10.4 not only shows asynchrony of PN breakdown, but also provides evidence of zygote cleavage into three blastomeres (at 35.2 hours p.i.). Video 10.6 illustrates an even more extreme situation in which a fertilized egg, after displaying multiple fertilization anomalies (asynchronous PN breakdown, PN displacement, and cortical instability), incredibly cleaves into five blastomeres. Division into three or more daughter cells requires multiple cleavage planes and, as a consequence, multipolar spindles. Documentation of tri- or multipolar spindles in human zygotes is not available (understandably, considering obvious ethical implications), but we have unpublished evidence of the presence of tripolar spindles in human oocytes, suggesting that such cytoskeletal anomalies can occur also in fertilized eggs. Direct zygote cleavage into three or more blastomeres is one of the (so far) few morphokinetic parameters that are indisputably indicative of poor embryo implantation ability. In fact, Rubio and colleagues reported that in cases in which zygote division results directly in a three-

cell embryo or progression from the two- to the three-cell stage occurs within a short time interval (< 5 hours), implantation rate falls to 1.2%, much less than that of normally cleaving embryos (20.2%) [12]. No births derived from zygotes directly cleaving into three or more cells have been reported. This is expected, because cleavage into three cells can only generate extensive aneuploidy. Therefore, the transfer of such embryos should be strongly discouraged.

Cleavage into three daughter cells can occur also at later developmental stages. Video 10.7 describes a case in which a single blastomere of a three-cell stage embryo manifests such cleavage behavior. It is plausible that the impact on embryo viability of cleavage into three cells becomes gradually smaller when occurring at progressively later stages of development, but data in this respect are presently lacking.

10.3.2. Reverse cleavage

Occasionally, the embryo cell cycle is subject to a major perturbation consisting of a inversion of the normal sequence of cleavage–DNA duplication–cleavage. Video 10.8 is representative of a phenomenon of reverse cleavage occurring at the two-cell stage. In this example, fertilization occurs in an apparently normal fashion until when the zygote cleaves into two blastomeres at 25.9 hours p.i. The embryo persists at such stage for a few hours and unexpectedly at 29.4 hours p.i. undergoes blastomere fusion. This is followed by re-formation of several pronuclei. Interestingly, at the two-cell stage nuclei are visible, but it is rather unlikely that the short persistence at such a stage (less than 4 hours) may have allowed complete duplication of DNA. Regardless, as expected considering the scale of the aberration, the embryo is fatally destined to early death, as indicated by a very asymmetric cleavage at 38 hours p.i. and subsequent developmental arrest and degeneration (not shown). Therefore, although rare and

not well described, reverse cleavage is highly likely to have a strong negative prognostic value for embryo implantation and viability.

10.4. Anomalies of compaction

10.4.1. Reverse compaction

Compaction is the process by which the blastomeres of a day 4 embryo lose their round shape, establish stronger intercellular contacts and form a mass whose cellular borders are not well defined. This creates a neat distinction between outer and inner cells and has important implications. In fact, it is believed that the former develop into trophectoderm, while the latter give rise to the inner cell mass. Ideally, compaction should occur between 80 and 85 hours p.i., and involve the entire embryo, but circumstances in which compaction occurs at later times or only partially are common and can be pinpointed even without the aid of TLM.

Video 10.9 illustrates a previously undescribed anomaly of compaction. In this example,

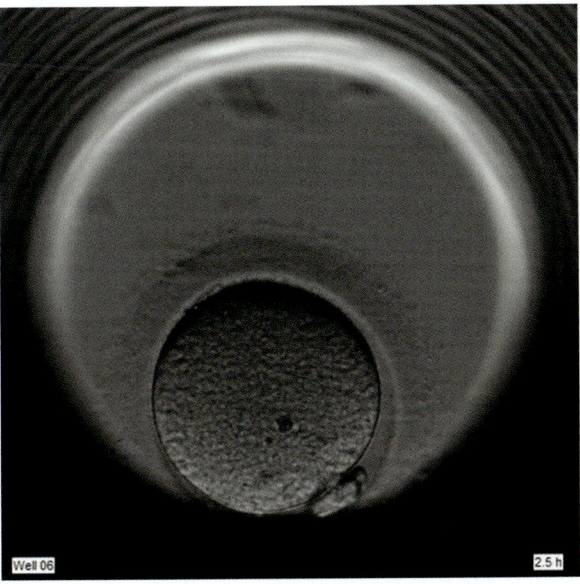

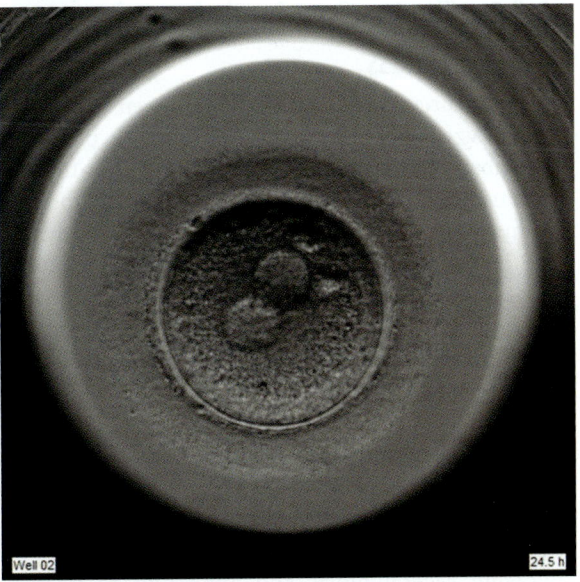

Video 10.8 Reverse cleavage. In the proposed example, fertilization occurs in an apparently normal fashion until when the zygote cleaves into two blastomeres at 25.9 hours p.i. The embryo persists at the two-cell stage for a few hours before undergoing blastomere fusion at 29.4 hours p.i.

Video 10.9 Reverse compaction. In this example, compaction starts at 84.5 hours p.i. and seems to have come to completion 4 hours later. However, at approximately 91 hours p.i. compaction is reverted and blastomeres become again rounded and well defined. This event is not followed by a possible second phase of compaction.

morphokinetics unfolds almost perfectly until approximately 75 hours p.i., with the exception of the formation of multiple nuclei at the two-cell stage. Compaction starts at 84.5 hours p.i. and seems to have come to completion 4 hours later. However, at approximately 91 hours p.i. compaction is reverted and blastomeres become again rounded and well defined. This event is not followed by a possible second phase of compaction. Nevertheless, the embryo continues to develop, forming an apparently good quality blastocyst. It is interesting to note that this blastocyst was transferred individually and gave rise to a positive hGC test. However, later examination did not confirm the establishment of a clinical pregnancy.

This phenomenon of "reverse compaction" is difficult to interpret because per se compaction is not very amenable to morphokinetic analysis. However, it suggests the concept that also this poorly observed stage of development could offer clues to predict embryo viability.

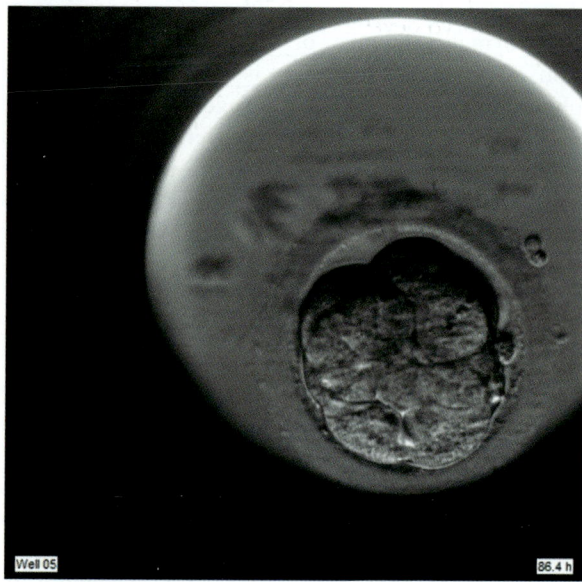

Video 10.10 Cell segregation during blastulation. Beginning from 95 hours p.i. the growing blastocyst undergoes a series of contractions, some of which are rather intense. During one of such contractions, at 112 hours p.i. a portion of the blastocyst mass is expelled at the 12 o'clock position, remaining at the same location over the following several hours and appearing as necrotic material at 119 hours p.i.

10.5. Anomalies of blastocyst formation

10.5.1. Elimination of cells from the blastocyst mass

Over the previous decades, it has been repeatedly hypothesized that during development to blastocyst the embryo has the ability to eliminate cells that are in some way abnormal and unable to contribute to the make up of the blastocyst. TLM offers evidence that indeed cells are eliminated during blastocyst development. Video 10.10 starts at 80 hours p.i., when the embryo has already undergone compaction, although not completely. Blastocoel formation starts between 90 and 91 hours p.i. Beginning from 95 hours p.i. the growing blastocyst undergoes a series of contractions, some of which are rather intense. During one of such contractions, at 112 hours p.i. a portion of the blastocyst mass is expelled at the 12 o'clock position, remaining at the same location over the following several hours and appearing as necrotic material at 119 hours p.i. Therefore, even at late stages of development in vitro, blastomere disposal is an existing phenomenon. Interestingly, between 104 and 112 hours p.i., the expelled part seems to correspond to an unusually large cell. The large size

may suggest that the expelled cell was polyploid, perhaps derived from repeated cycles of DNA duplication not separated by cytokinesis events. Clearly, the observed phenomenon may have occurred for several other reasons, but the segregation and elimination of chromosomally abnormal cells is a recurrent theme in embryo development research. Regardless of the type of anomaly, it is not known whether cell elimination during blastocyst development occurs as an effect of a cell self-destruction mechanism or recognition by the embryo of a fault in its design. It is interesting to observe that blastomere disposal occurs concomitantly with an intense activity of rhythmic contractions. Therefore, it is tempting to hypothesize that contraction represents a "stress test" by which the blastocyst assesses its own "robustness."

10.6. Conclusions

Large parts of this editorial project describe the potential, limitations, pros and cons of predicting the developmental ability of the human embryo by careful analysis of well-defined morphokinetic parameters. In this chapter, we approached the observation of embryo development by TLM from a different angle. In particular, we focused on rare and sometimes bizarre events that, as such, are of very little help for developing largely applicable algorithms able to predict embryo viability. However, these atypical phenomena (and others that were not included for brevity) have disclosed facets of human embryo development that were simply unknown until very recently, and are still largely poorly understood. Although the entire process of development to the blastocyst stage is interspersed with atypical phenomena, much of this "second life" of the human pre-implantation embryo concerns the fertilization process. This is not surprising, considering the extreme complexity and uniqueness of the interaction of the maternal and paternal complements. Through these examples, we have learned that the rules that govern cell cycle progression and cell cleavage may be broken, although the cost that comes with it often leads to developmental failure. Regardless, systematic observation and interpretation of these phenomena will offer us a more profound appreciation of fertilization and early development of the human embryo.

References

1. Kaser DJ, Racowsky C. Clinical outcomes following selection of human preimplantation embryos with time-lapse monitoring: a systematic review. *Human Reproduction Update*. 2014 Sep;20 (5):617–31.

2. Meseguer M, Herrero J, Tejera A, Hilligsøe KM, Ramsing NB, Remohí J. The use of morphokinetics as a predictor of embryo implantation. *Human Reproduction*. 2011 Oct;26 (10):2658–71.

3. Brunet S, Verlhac MH. Positioning to get out of meiosis: the asymmetry of division. *Human Reproduction Update*. 2010 Dec 13;17(1):68–75.

4. Coticchio G, Guglielmo M-C, Albertini DF, Dal Canto M, Mignini Renzini M, De Ponti E, et al. Contributions of the actin cytoskeleton to the emergence of polarity during maturation in human oocytes. *Molecular Human Reproduction*. 2014 Mar;20(3):200–7.

5. Deng M, Suraneni P, Schultz RM, Li R. The Ran GTPase mediates chromatin signaling to control cortical polarity during polar body extrusion in mouse oocytes. *Developmental Cell*. 2007 Feb;12 (2):301–8.

6. Van Blerkom J, Davis P, Alexander S. Occurrence of maternal and paternal spindles in unfertilized human oocytes: possible relationship to nucleation defects after silent fertilization. *Reproductive Biomedicine Online*. 2004 Apr;8(4):454–9.

7. Payne D, Flaherty SP, Barry MF, Matthews CD. Preliminary observations on polar body extrusion and pronuclear formation in human oocytes using time-lapse video cinematography. *Human Reproduction*. 1997 Mar;12(3):532–41.

8. Aguilar J, Motato Y, Escribá M-J, Ojeda M, Muñoz E, Meseguer M. The human first cell cycle: impact on implantation. *Reproductive Biomedicine Online*. 2014 Apr 1;28(4):475–84.

9. Hörmanseder E, Tischer T, Mayer TU. Modulation of cell cycle control during oocyte-to-embryo transitions. *The EMBO Journal*. 2013 Aug 14;32(16):2191–203.

10. Alpha Scientists in Reproductive Medicine and ESHRE Special Interest Group of Embryology, Balaban B, Brison D, Calderon G, Catt J, Conaghan J, et al. The Istanbul consensus workshop on embryo assessment: proceedings of an expert meeting. *Human Reproduction*. 2011 May 17;26 (6):1270–83.

11. Chavez SL, Loewke KE, Han J, Moussavi F, Colls P, Munné S, et al. Dynamic blastomere behaviour reflects human embryo ploidy by the four-cell stage. *Nature Communications*. 2012;3:1251.

12. Rubio I, Kuhlmann R, Agerholm I, Kirk J, Herrero J, Escribá M-J, et al. Limited implantation success of direct-cleaved human zygotes: a time-lapse study. *Fertility and Sterility*. 2012; 98(6): 1458–63.

Index